April 2012
Publication No. FHWA-HIF-12-003

Hydraulic Engineering Circular No. 18

Evaluating Scour at Bridges
Fifth Edition

U.S. Department of Transportation
Federal Highway Administration

Published by Books Express Publishing
Copyright © Books Express, 2012
ISBN 978-1-78266-121-4

Books Express publications are available from all good retail and online booksellers. For publishing proposals and direct ordering please contact us at: info@books-express.com

TABLE OF CONTENTS

LIST OF FIGURES vii

LIST OF TABLES xi

LIST OF SYMBOLS xiii

ACKNOWLEDGMENTS xxi

GLOSSARY xxiii

CHAPTER 1 INTRODUCTION 1.1

1.1 PURPOSE 1.1
1.2 BACKGROUND 1.1
1.3 COMPREHENSIVE ANALYSIS 1.2
1.4 PROCEDURAL GUIDANCE 1.4

1.4.1 Objectives of a Bridge Scour Evaluation Program 1.4
1.4.2 Bridge Scour and the NBIS 1.4

1.5 ADVANCES IN THE STATE-OF-PRACTICE FOR ESTIMATING SCOUR
     AT BRIDGES 1.5
1.6 MANUAL ORGANIZATION 1.7
1.7 DUAL SYSTEM OF UNITS 1.8

CHAPTER 2 - DESIGNING AND EVALUATING BRIDGES TO RESIST SCOUR ... 2.1

2.1 SCOUR DESIGN PHILOSOPHY AND CONCEPTS FOR NEW BRIDGES ... 2.1
2.2 GENERAL DESIGN PROCEDURE 2.2
2.3 DESIGN CONSIDERATIONS 2.6

2.3.1 General 2.7
2.3.2 Piers 2.7
2.3.3 Abutments 2.8
2.3.4 Superstructures 2.9

2.4 DETAILED PROCEDURES AND SPECIFIC DESIGN APPROACH 2.11

2.4.1 Step 1: Determine Scour Analysis Variables 2.11
2.4.2 Step 2: Determine the Magnitude of Long-Term Degradation or Aggradation ... 2.12
2.4.3 Step 3: Compute the Magnitude of Contraction Scour 2.12
2.4.4 Step 4: Compute the Magnitude of Local Scour at Piers 2.12
2.4.5 Step 5: Determine the Foundation Elevation for Abutments 2.13
2.4.6 Step 6: Plot the Total Scour Depths and Evaluate the Design 2.13

2.5 SCOUR EVALUATION PHILOSOPHY AND CONCEPTS FOR EXISTING
     BRIDGES 2.14

2.5.1 Overview 2.14
2.5.2 Countermeasures for Scour Critical Bridges 2.15
2.5.3 Bridges with Unknown Foundations 2.16

CHAPTER 3 - BASIC CONCEPTS AND DEFINITIONS OF SCOUR 3.1

3.1 GENERAL 3.1
3.2 TOTAL SCOUR 3.2

3.2.1 Aggradation and Degradation 3.2
3.2.2 Contraction Scour 3.2
3.2.3 Local Scour 3.2
3.2.4 Other Types of Scour 3.2
3.2.5 Lateral Stream Migration 3.2

3.3 LONG-TERM STREAMBED ELEVATION CHANGES
  (AGGRADATION AND DEGRADATION) 3.3
3.4 CLEAR-WATER AND LIVE-BED SCOUR 3.3
3.5 CONTRACTION SCOUR 3.5

3.5.1 Basic Conditions for Contraction Scour 3.5

3.6 LOCAL SCOUR 3.5

3.6.1 Scour at Bridge Piers and Abutments 3.5
3.6.2 Bridge Pier Flow Field 3.8
3.6.3 Bridge Abutment Flow Field 3.12

3.7 LATERAL SHIFTING OF A STREAM 3.15

CHAPTER 4 - SOILS, ROCK, AND GEOTECHNICAL CONSIDERATIONS 4.1

4.1 GENERAL 4.1
4.2 SCOUR PROCESSES 4.2

4.2.1 Cohesionless Soils 4.2
4.2.2 Cohesive Soils 4.2
4.2.3 Rock 4.3

4.3 ERODIBILITY 4.6

4.3.1 Velocity 4.7
4.3.2 Shear Stress 4.8
4.3.3 Stream Power 4.10
4.3.4 Erosion Rates 4.11
4.3.5 Devices to Measure Erodibility 4.11

4.4 SOIL PROPERTIES 4.14

4.4.1 Particle Size 4.14
4.4.2 Plasticity and the Atterberg Limits 4.16
4.4.3 Density and Compaction 4.17
4.4.4 Shear Strength 4.20
4.4.5 Hydraulic Conductivity 4.23

| 4.5 | CLASSIFICATION OF SOILS | 4.24 |

4.5.1 Unified Soil Classification System ... 4.24
4.5.2 AASHTO Classification System ... 4.25

| 4.6 | ROCK PROPERTIES | 4.27 |

4.6.1 Igneous, Sedimentary, and Metamorphic Rocks ... 4.27
4.6.2 Rock Mass Descriptions and Characteristics ... 4.27

| 4.7 | CLASSIFICATION OF ROCK | 4.35 |

4.7.1 Rock Mass Rating System ... 4.36
4.7.2 Erodibility Index Method ... 4.36

| 4.8 | SUMMARY | 4.41 |

CHAPTER 5 - LONG-TERM AGGRADATION AND DEGRADATION ... 5.1

5.1 INTRODUCTION ... 5.1
5.2 LONG-TERM BED ELEVATION CHANGES ... 5.1
5.3 ESTIMATING LONG-TERM AGGRADATION AND DEGRADATION ... 5.2

5.3.1 Bridge Inspection Records ... 5.2
5.3.2 Gaging Station Records ... 5.2
5.3.3 Geology and Stream Geomorphology ... 5.4
5.3.4 Computer Models ... 5.4
5.3.5 Aggradation, Degradation, and Total Scour ... 5.5
5.3.6 Inspection, Maintenance, and Countermeasures ... 5.5

CHAPTER 6 - CONTRACTION SCOUR ... 6.1

6.1 INTRODUCTION ... 6.1
6.2 CONTRACTION SCOUR ... 6.1

6.2.1 Contraction Scour Conditions ... 6.1
6.2.2 Contraction Scour Cases ... 6.2

6.3 LIVE-BED CONTRACTION SCOUR ... 6.10
6.4 CLEAR-WATER CONTRACTION SCOUR ... 6.12
6.5 CONTRACTION SCOUR WITH BACKWATER ... 6.12
6.6 CONTRACTION SCOUR EXAMPLE PROBLEMS ... 6.13

6.6.1 Example Problem 1 - Live-Bed Contraction Scour ... 6.13
6.6.2 Example Problem 2 - Alternate Method ... 6.14
6.6.3 Example Problem 3 - Relief Bridge Contraction Scour ... 6.14
6.6.4 Comprehensive Example ... 6.15

6.7 CONTRACTION SCOUR IN COHESIVE MATERIALS ... 6.15

6.7.1 Ultimate Scour ... 6.16
6.7.2 Time Rate of Scour ... 6.17

6.8 CONTRACTION SCOUR IN ERODIBLE ROCK ... 6.19

6.9 SCOUR AT OPEN-BOTTOM CULVERTS 6.19

6.9.1 Laboratory Investigations of Scour at Open-Bottom Culverts..............................6.19
6.9.2 Clear-Water Scour Equation for Open-Bottom Culverts 6.22
6.9.3 Example Problem 6.23

6.10 PRESSURE FLOW SCOUR (VERTICAL CONTRACTION SCOUR) 6.24

6.10.1 Estimating Pressure Flow Scour 6.24
6.10.2 Pressure Flow Scour Example Problems...6.26

CHAPTER 7 - PIER SCOUR 7.1

7.1 GENERAL 7.1
7.2 HEC-18 PIER SCOUR EQUATION 7.2
7.3 FLORIDA DOT PIER SCOUR METHODOLOGY..7.5
7.4 PIER SCOUR AT WIDE PIERS 7.10
7.5 SCOUR FOR COMPLEX PIER FOUNDATIONS 7.11

7.5.1 Introduction..7.11
7.5.2 Superposition of Scour Components Method of Analysis...................................7.12
7.5.3 Determination of the Pier Stem Scour Depth Component 7.13
7.5.4 Determination of the Pile Cap (Footing) Scour Depth Component 7.13
7.5.5 Determination of the Pile Group Scour Depth Component 7.17
7.5.6 Determination of Total Scour Depth for the Complex Pier..................................7.22

7.6 MULTIPLE COLUMNS SKEWED TO THE FLOW 7.22
7.7 SCOUR FROM DEBRIS ON PIERS 7.23

7.7.1 Debris Accumulation on Piers 7.23
7.7.2 Debris Size and Shape..7.24
7.7.3 Effective Pier Width with Debris 7.25

7.8 TOPWIDTH OF SCOUR HOLES 7.26
7.9 PHYSICAL MODEL STUDIES 7.27
7.10 PIER SCOUR EXAMPLE PROBLEMS 7.27

7.10.1 Example Problem 1 - Scour at a Simple Solid Pier 7.27
7.10.2 Example Problem 2 - Angle of Attack 7.28
7.10.3 Example Problem 3 - Scour at Complex Piers (Solid Pier on an Exposed Footing)7.28
7.10.4 Example Problem 4 - Scour at a Complex Pier with Pile Cap in the Flow 7.30
7.10.5 Example Problem 5 - Scour at Multiple Columns 7.33
7.10.6 Example Problem 6 - Florida DOT Pier Scour Methodology 7.34
7.10.7 Example Problem 7 - Pier Scour with Debris 7.35
7.10.8 Comprehensive Example 7.37

7.11 PIER SCOUR IN COARSE BED MATERIALS 7.37
7.12 PIER SCOUR IN COHESIVE MATERIALS 7.38
7.13 PIER SCOUR IN ERODIBLE ROCK 7.40

7.13.1 Quarrying and Plucking 7.40
7.13.2 Example Problem - Erodibility Index Method for Rock 7.42
7.13.3 Abrasion 7.43
7.13.4 Example Problem - Long-Term Abrasion of Rock 7.47

CHAPTER 8 - EVALUATING LOCAL SCOUR AT ABUTMENTS 8.1

8.1   GENERAL 8.1
8.2   ABUTMENT SCOUR EQUATIONS 8.2

8.2.1 Overview 8.2
8.2.2 Abutment Scour Parameter Determination 8.4

8.3   ABUTMENT SITE CONDITIONS 8.5
8.4   ABUTMENT SKEW 8.5
8.5   ABUTMENT SHAPE 8.5
8.6   ESTIMATING SCOUR AT ABUTMENTS 8.7

8.6.1 Froehlich's Abutment Scour Equation 8.7
8.6.2 HIRE Abutment Scour Equation 8.7
8.6.3 NCHRP 24-20 Abutment Scour Approach 8.8

8.7   ABUTMENT SCOUR EXAMPLE PROBLEMS 8.20

8.7.1 Example Problem 1 - Froehlich Equation 8.20
8.7.2 Example Problem 2 - HIRE Equation 8.22
8.7.3 Example Problem 3 - NCHRP Live-Bed Scour 8.23
8.7.4 Example Problem 4 - NCHRP Clear-Water Scour (Particle Size) 8.23
8.7.5 Example Problem 5 - NCHRP Clear-Water Scour (Shear Stress) 8.24
8.7.6 Comprehensive Example 8.24

CHAPTER 9 - SCOUR ANALYSIS FOR TIDAL WATERWAYS 9.1

9.1   INTRODUCTION 9.1
9.2   OVERVIEW OF TIDAL PROCESS 9.2

9.2.1 Glossary 9.2
9.2.2 Definition of Tidal and Coastal Processes 9.4
9.2.3 Aggradation, Degradation, and Scour in Tidal Waterways 9.7

9.3   LEVEL 1 ANALYSIS 9.10

9.3.1 Tidally Affected River Crossings 9.10
9.3.2 Tidal Inlets, Bays, and Estuaries 9.11

9.4   LEVEL 2 ANALYSIS 9.12

9.4.1 Introduction 9.12
9.4.2 Evaluation of Hydraulic Characteristics 9.13
9.4.3 Design Storm and Storm Tide 9.13
9.4.4 Scour Evaluation Concepts 9.14

9.5   TIME DEPENDENT CHARACTERISTICS OF TIDAL SCOUR 9.15
9.6   LEVEL 3 ANALYSIS 9.17
9.7   TIDAL HYDROLOGY, HYDRAULICS, AND SCOUR AT BRIDGES 9.17
9.8   HIGHWAYS IN THE COASTAL ENVIRONMENT 9.19

CHAPTER 10 - SCOUR EVALUATION, INSPECTION, AND PLAN OF ACTION 10.1

10.1   INTRODUCTION 10.1
10.2   OFFICE REVIEW 10.2
10.3   BRIDGE INSPECTION 10.3

10.3.1 Safety Considerations 10.3
10.3.2 FHWA Recording and Coding Guide 10.3
10.3.3 General Site Considerations 10.4
10.3.4 Assessing the Substructure Condition 10.4
10.3.5 Assessing the Condition of Countermeasures 10.5
10.3.6 Assessing Scour Potential at Bridges 10.6
10.3.7 Underwater Inspections 10.9
10.3.8 Notification Procedures 10.10

10.4   MONITORING BRIDGES FOR SCOUR 10.10

10.4.1 General 10.10
10.4.2 Portable Monitoring Devices 10.11
10.4.3 Fixed Monitoring Devices 10.12
10.4.4 Selection and Maintenance of Monitoring Devices 10.12

10.5   CASE HISTORIES OF BRIDGE INSPECTION PROBLEMS 10.13

10.5.1 Introduction 10.13
10.5.2 Schoharie Creek Bridge Failure 10.13
10.5.3 Hatchie River Bridge Failure 10.15
10.5.4 Arroyo Pasajero Bridge Failure 10.16
10.5.5 Conclusions 10.17

10.6   PLAN OF ACTION 10.17

10.6.1 Background 10.17
10.6.2 Developing a Plan of Action 10.18
10.6.3 Maintaining a Plan of Action 10.19

CHAPTER 11 - LITERATURE CITED 11.1

**APPENDIX A - METRIC SYSTEM, CONVERSION FACTORS, AND WATER
                        PROPERTIES** A.1
APPENDIX B - EXTREME EVENTS B.1
APPENDIX C - CONTRACTION SCOUR AND CRITICAL VELOCITY EQUATIONS C.1
APPENDIX D - COMPREHENSIVE SCOUR PROBLEM D.1
APPENDIX E - UNKNOWN FOUNDATIONS E.1

LIST OF FIGURES

Figure 1.1. Flow chart for scour and stream stability analysis and evaluation 1.3

Figure 3.1. Pier scour depth in a sand-bed stream as a function of time 3.4

Figure 3.2. Simple schematic representation of scour at a cylindrical pier 3.6

Figure 3.3. The main flow features forming the flow field at a narrow pier of circular cylindrical form 3.9

Figure 3.4. Variation of flow field with reducing approach flow depth; narrow to transitional pier of constant pier width 3.11

Figure 3.5. Main features of the flow field at a wide pier (y/a < 0.2) 3.12

Figure 3.6. Variation of soil and sediment types at a bridge crossing 3.13

Figure 3.7. Flow structure including macro-turbulence generated by flow around abutments in a narrow main channel 3.14

Figure 3.8. Flow structure including macro-turbulence generated by floodplain/main channel flow interaction, flow separation around abutment, and wake region on the floodplain of a compound channel 3.14

Figure 3.9. Interaction of flow features causing scour and erodibility of boundary 3.15

Figure 4.1. Photographs of scour in soil and rock 4.1

Figure 4.2. Method for accumulating the effects of scour resulting from multiple floods .. 4.4

Figure 4.3. Conceptual stream power model for geomorphically effective floods 4.6

Figure 4.4. Average cumulative erosion related to integrated stream power and abrasion number 4.7

Figure 4.5. Typical velocity distribution in an open channel 4.8

Figure 4.6. Critical shear stress vs. particle grain size 4.9

Figure 4.7. Erosion rate vs. velocity for a wide range of geomaterials 4.11

Figure 4.8. Schematic diagram of piston-type erosion rate device 4.12

Figure 4.9. Schematic diagram of rotating-type erosion rate device 4.13

Figure 4.10. Schematic diagram of jet-type erosion rate device 4.13

Figure 4.11. Typical sieves and hydrometers used for grain size analyses 4.14

Figure 4.12.	Typical grain size curves for two different soils	4.15
Figure 4.13.	Atterberg Limit tests for Liquid Limit and Plastic Limit	4.17
Figure 4.14.	Standard laboratory compaction test	4.19
Figure 4.15.	Moisture-density curves for different compactive efforts	4.19
Figure 4.16.	Permeameters: (a) Constant-head, (b) Falling-head	4.23
Figure 4.17.	Plasticity chart for the Unified Soil Classification System for fine-grained soils	4.25
Figure 4.18.	Plasticity chart for the AASHTO classification system for fine-grained soils	4.27
Figure 5.1.	Specific gage data for Cache Creek, California	5.3
Figure 6.1.	Case 1a: Abutments project into channel	6.3
Figure 6.2.	Case 1b: Abutments at edge of channel	6.4
Figure 6.3.	Case 1c: Abutments set back from channel	6.5
Figure 6.4.	Case 2a: River narrows	6.6
Figure 6.5.	Case 2b: Bridge abutments and/or piers constrict flow	6.7
Figure 6.6.	Case 3: Relief bridge over floodplain	6.8
Figure 6.7.	Case 4: Relief bridge over secondary stream	6.8
Figure 6.8.	Fall velocity of sand-sized particles with specific gravity of 2.65 in metric units	6.11
Figure 6.9.	Critical shear stress versus particle size	6.15
Figure 6.10.	Example of critical shear and erosion rate from a material test	6.16
Figure 6.11.	Generalized relationships for scour in cohesive materials	6.17
Figure 6.12.	Illustration of time-dependent scour calculations	6.18
Figure 6.13.	Open-bottom culvert on Whitehall Road over Euclid Creek in Cuyahoga County, OH	6.20
Figure 6.14.	Flow concentration and separation zone	6.20
Figure 6.15.	Rectangular model with vertical face	6.21
Figure 6.16.	Rectangular model with wing walls	6.21

Figure 6.17.	Arched model with wing walls	6.22
Figure 6.18.	Vertical contraction and definition for geometric parameters	6.24
Figure 7.1.	Comparison of scour equations for variable depth ratios (y/a)	7.2
Figure 7.2.	Definition sketch for pier scour	7.3
Figure 7.3.	Common pier shapes	7.4
Figure 7.4.	Scour for FDOT methodology	7.10
Figure 7.5.	Definition sketch for scour components for a complex pier	7.12
Figure 7.6.	Suspended pier scour ratio	7.14
Figure 7.7.	Pile cap (footing) equivalent width	7.14
Figure 7.8.	Definition sketch for velocity and depth on exposed footing	7.17
Figure 7.9.	Projected width of piles for the special case of aligned flow	7.19
Figure 7.10.	Projected width of piles for the general case of skewed flow	7.19
Figure 7.11.	Pile spacing factor	7.20
Figure 7.12.	Adjustment factor for number of aligned rows of piles	7.21
Figure 7.13.	Pile group height adjustment factor	7.21
Figure 7.14.	Multiple columns skewed to the flow	7.23
Figure 7.15.	Woody debris at a bridge pier	7.24
Figure 7.16.	Idealized dimensions of rectangular debris accumulations	7.25
Figure 7.17.	Idealized dimensions of triangular debris accumulations	7.25
Figure 7.18.	Topwidth of scour hole	7.27
Figure 7.19.	Conceptual model of quarrying and plucking at a bridge pier	7.40
Figure 7.20.	Example data from modified slake durability test	7.44
Figure 7.21.	Pier scour in rock as a function of stream power and Geotechnical Scour Number	7.45
Figure 7.22.	Transforming a mean daily flow series to mean daily effective stream power	7.46
Figure 7.23.	Cumulative stream power for the water year in Figure 7.22	7.47

Figure 7.24.	Mean daily flow, Sacramento River near Redding, CA 1938 - 2009	7.47
Figure 7.25.	Cumulative daily stream power, Sacramento River near Redding, CA 1938 - 2009	7.48
Figure 8.1.	Schematic representation of abutment scour in a compound channel	8.1
Figure 8.2.	Scour of bridge abutment and approach embankment	8.2
Figure 8.3.	Comparison of (a) laboratory flow characteristics to (b) field flow conditions	8.3
Figure 8.4.	Determination of length of embankment blocking live flow for abutment scour estimation	8.5
Figure 8.5.	Orientation of embankment angle, θ, to the flow	8.6
Figure 8.6.	Abutment shape	8.6
Figure 8.7.	Abutment scour conditions	8.9
Figure 8.8.	Conceptual geotechnical failures resulting from abutment scour	8.10
Figure 8.9.	Scour amplification factor for spill-through abutments and live-bed conditions	8.11
Figure 8.10.	Scour amplification factor for wingwall abutments and live-bed conditions	8.12
Figure 8.11.	Scour amplification factor for spill-through abutments and clear-water conditions	8.14
Figure 8.12.	Scour amplification factor for wingwall abutments and clear-water conditions	8.15
Figure 8.13.	Velocity and streamlines at a bridge constriction	8.16
Figure 8.14.	Velocity for SBR<5	8.17
Figure 8.15.	Velocity for SBR>5	8.18
Figure 8.16.	Velocity for SBR>5 and SBR<5	8.19
Figure 9.1.	Types of tidal waterway crossings	9.5
Figure 9.2.	Principal tidal terms	9.6
Figure 9.3.	Sediment transport in tidal inlets	9.9
Figure 9.4.	Time development of clear-water scour	9.16
Figure 9.5.	Initial clear-water scour development	9.16
Figure 9.6.	Contraction scour development with sediment supply	9.17
Figure 10.1.	Photograph of riprap at pier 2, October 1956	10.14
Figure 10.2.	Photograph of riprap at pier 2, August 1977	10.14

LIST OF TABLES

Table 1.1.	Commonly Used Engineering Terms in English and SI Units	1.8
Table 2.1.	Hydraulic Design, Scour Design, and Scour Design Check Flood Frequencies	2.1
Table 2.2.	Bridge Scour Evaluation Program Status - 2011	2.15
Table 2.3.	Hydraulic Design, Scour Design, and Scour Countermeasure Design Flood Frequencies	2.15
Table 4.1.	Factors Influencing the Erodibility of Cohesive Soils	4.3
Table 4.2.	Parameters Influencing the Rate of Scour in Rock	4.5
Table 4.3.	Gradation Based on C_u and C_c Parameters	4.16
Table 4.4.	Soil Density	4.20
Table 4.5.	Soil Strength	4.22
Table 4.6.	Typical Void Ratio, Porosity, and Hydraulic Conductivity of Geomaterials	4.24
Table 4.7.	Unified Soil Classification System	4.26
Table 4.8.	AASHTO Soil Classification System	4.28
Table 4.9.	Rock Groups and Types	4.29
Table 4.10.	Terms Used to Describe Grain Size	4.30
Table 4.11.	Terms Used to Describe Grain Shape	4.30
Table 4.12.	Terms Used to Describe Stratum Thickness	4.31
Table 4.13.	Terms Used to Describe Rock Weathering and Alteration	4.31
Table 4.14.	Terms Used to Describe the Strength of Rock	4.32
Table 4.15.	Terms Used to Describe Rock Hardness	4.32
Table 4.16.	Terms to Describe Discontinuities	4.33
Table 4.17.	Terms to Classify Discontinuities Based on Aperture Size	4.34
Table 4.18.	Recommended Allowable Bearing Pressure for Footings on Rock	4.35
Table 4.19.	Geomechanics Classification of Rock Masses	4.37
Table 4.20.	Geomechanics Rating Adjustment for Joint Orientations	4.38

Table 4.21.	Geomechanics Rock Mass Classes Determined From Total Ratings	4.38
Table 4.22.	Values of the Rock Mass Strength Parameter M_s	4.39
Table 4.23.	Rock Joint Set Number J_n	4.39
Table 4.24.	Joint Roughness Number J_r	4.40
Table 4.25.	Joint Alteration Number J_a	4.40
Table 4.26.	Relative Orientation Parameter J_s	4.42
Table 7.1.	Correction Factor, K_1, for Pier Nose Shape	7.4
Table 7.2.	Correction Factor, K_2, for Angle of Attack, θ, of the Flow	7.4
Table 7.3.	Increase in Equilibrium Pier Scour Depths, K_3, for Bed Condition	7.5
Table 7.4.	Hyperbolic Tangent of X	7.9
Table 7.5.	Calculation Results for Example Problem	7.43
Table 8.1.	Abutment Shape Coefficients	8.6
Table 10.1.	Tips for Inspecting Riprap	10.6
Table 10.2.	Assessing the Scour Potential at Bridges	10.7

LIST OF SYMBOLS

a	=	Pier width, ft (m)
a^*	=	Effective pier width, ft (m)
a^*_d	=	Effective width of pier when debris is present, ft (m)
a_{proj}	=	Sum of non-overlapping projected widths of piles in pile group, ft (m)
A	=	Maximum amplitude of elevation of the tide or storm surge, ft (m)
A_e	=	Flow area of approach cross section obstructed by the embankment, ft² (m²)
A_c	=	Cross-sectional area of the waterway at mean tide elevation--half between high and low tide, ft² (m²)
	=	Net cross-sectional area in the inlet at the crossing, at mean water surface elevation, ft² (m²)
b	=	Pier width perpendicular to flow direction, ft (m)
C_d	=	Coefficient of drag
D	=	Diameter of the bed material, ft (m)
	=	Diameter of smallest nontransportable particle in the bed material, m (ft)
D_m	=	Effective mean diameter of bed material in the bridge, mm or m
	=	$1.25\ D_{50}$
D_r	=	Relative density of soil
D_{50}	=	Median diameter of the bed material, diameter which 50% of the sizes are smaller, mm or m
D_{84}	=	Diameter of the bed material of which 84% are smaller, mm or m
D_{90}	=	Diameter of the bed material of which 90% are smaller, mm or m
e	=	Void ratio of soil
f	=	Distance between front edge of pile cap or footing and pier, ft (m)
F	=	Impact imparted by debris, lb (N)
F_d	=	Drag force per unit length of bridge, lb/ft (N/m)
Fr	=	Froude Number $[V/(gy)^{½}]$
	=	Froude Number of approach flow upstream of the abutment

Fr	=	Froude Number based on the velocity and depth adjacent to and upstream of the abutment
Fr_1	=	Froude Number directly upstream of a pier
g	=	Acceleration of gravity, ft/s² (m/s²)
h_o	=	Height of pile cap above bed at beginning of computation, ft (m)
h_1	=	Height of the pier stem above the bed before scour, ft (m)
h_2	=	Height of pile cap after pier stem scour component has been computed, ft (m)
h_3	=	Height of pile group after the pier stem and pile cap scour components have been computed, ft (m)
h_{1-2}	=	Head loss between sections 1 and 2, ft (m)
h_b	=	Bridge opening height, ft (m)
h_c	=	Average depth of flow in the waterway at mean water elevation, ft (m)
h_u	=	Upstream channel flow depth for vertical contraction scour
h_{ue}	=	Effective upstream channel flow depth for live-bed conditions and bridge overtopping, ft (m)
H	=	Height (i.e., height of a dune), ft (m)
	=	Height (thickness) of debris, ft (m)
	=	Depth of submergence, ft (m)
	=	Densimetric Froude Number
H_b	=	Distance from the low chord of the bridge to the average elevation of the stream bed before scour, ft (m)
	=	Hydraulic gradient of soil, ft/ft (m/m)
J_s	=	Relative orientation parameter
k_s	=	Grain roughness of bed, ft (m)
K	=	Various coefficients in equations as described below
	=	Conveyance in Manning equation $\frac{(AR^{2/3})}{n}$, ft³/s (m³/s)
	=	Bottom width of the scour hole as a fraction of scour depth, ft (m)
	=	Erodibility index of rock

K_b	=	Bend coefficient (dimensionless)
	=	Block size parameter
K_d	=	Shear strength parameter
K_s	=	Shields coefficient
K_w	=	Correction factor for pier width
K_1	=	Correction factor for pier nose shape
	=	Coefficient for abutment shape
K_2	=	Correction factor for angle of attack of flow (pier)
	=	Correction factor for angle of attack of flow (abutment)
K_3	=	Correction factor for increase in equilibrium pier scour depth for bed condition
k_1 & k_2	=	Exponents determined in Laursen live-bed contraction equation, depends on the mode of bed material transport
K_s	=	Dimensionless Shields parameter
k_s	=	Grain roughness of the bed, ft (m)
k_u	=	Units conversion factor
$K_{h\,pg}$	=	Pile group height factor
K_m	=	Coefficient for number of aligned rows in pile group
K_{sp}	=	Coefficient for pile spacing in pile group
L	=	Length of pier, ft (m)
L_c	=	Length of the waterway, ft (m)
L' or L	=	Length of abutment (embankment) projected normal to flow, ft (m)
M	=	Mass of debris, slugs (kg)
M_s	=	Intact rock mass strength parameter
n	=	Manning n
n_1	=	Manning n for upstream main channel
n_2	=	Manning n for contracted section
P	=	Instantaneous stream power, ft-lb/s per ft^2 (KW/m^2)

Q	=	Discharge through the bridge or on the overbank at the bridge, ft³/s (m³/s)
Q_{B1}	=	Discharge blocked by road embankment on one side of open-bottom culvert, ft³/s (m³/s)
Q_e	=	Flow obstructed by the abutment and approach embankment, ft³/s (m³/s)
Q_{max}	=	Maximum discharge in the tidal cycle, ft³/s (m³/s)
	=	Maximum discharge in the inlet, ft³/s (m³/s)
Q_t	=	Discharge at any time, t, in the tidal cycle, ft³/s (m³/s)
Q_{ue}	=	Effective channel discharge for live-bed conditions and bridge overtopping flow, ft³/s (m³/s)
Q_1	=	Flow in the upstream main channel transporting sediment, ft³/s (m³/s)
Q_2	=	Flow in the contracted channel, ft³/s (m³/s). Often this is equal to the total discharge unless the total flood flow is reduced by relief bridges or water overtopping the approach roadway
Q_{100}	=	Storm-event having a probability of occurrence of one every 100 years, ft³/s (m³/s)
Q_{500}	=	Storm-event having a probability of occurrence of one every 500 years, ft³/s (m³/s)
q	=	Discharge per unit width, ft³/s (m³/s)
	=	Discharge in conveyance tube, ft³/s (m³/s)
q_{2c}	=	Unit discharge in constructed bridge opening accounting for nonuniform flow, ft²/s (m²/s)
R	=	Hydraulic radius
	=	Coefficient of resistance
S	=	Spacing between columns of piles, pile center to pile center, ft (m)
	=	Stopping distance for debris mass, ft (m)
SBR	=	Set-back ratio of each abutment
S_1	=	Slope of energy grade line of main channel, ft/ft (m/m)
S_f	=	Slope of the energy grade line, ft/ft (m/m)
S_o	=	Average bed slope, ft/ft (m/m)

S_s	=	Specific gravity of bed material. For most bed material this is equal to 2.65
t	=	Time from the beginning of total cycle, min
	=	Duration of flow, hr
	=	Boundary layer thickness, ft (m)
T	=	Total time for one complete tidal cycle, min
	=	Tidal period between successive high or low tides, s
	=	Thickness of pile cap or footing, ft (m)
V	=	Average velocity, ft/s (m/s)
	=	Characteristic average velocity in the contracted section for estimating a median stone diameter, D_{50}, ft/s (m/s)
V_{max}	=	Q_{max}/A', or maximum velocity in the inlet, ft/s (m/s)
V_1	=	Average velocity at upstream main channel, ft/s (m/s)
	=	Mean velocity of flow directly upstream of the pier, ft/s (m/s)
	=	Approach velocity used at the beginning of computations, ft/sec (m/sec)
V_2	=	Average velocity in the contracted section, ft/s (m/s)
	=	Adjusted velocity for pile cap computations, ft/sec (m/sec)
V_3	=	Adjusted velocity for pile group computations, ft/sec (m/sec)
V_c	=	Critical velocity, m/s (ft/s), above which the bed material of size D, D_{50}, etc. and smaller will be transported
V_{c50}	=	Critical velocity for D_{50} bed material size, ft/s (m/s)
V_{c90}	=	Critical velocity for D_{90} bed material size, ft/s (m/s)
V_e	=	Q_e/A_e, ft/s (m/s)
V_f	=	Average velocity of flow zone below the top of the footing, ft/s (m/s)
V_i	=	Approach velocity when particles at a pier begin to move, ft/s (m/s)
V_{ip}	=	Velocity of the live-bed peak scour, ft/s (m/s)
V_{max}	=	Maximum average velocity in the cross section at Q_{max}, ft/s (m/s)
V_R	=	Velocity ratio

V_*	=	Shear velocity in the upstream section, ft/s (m/s)
	=	$(\tau_o/\rho) = (gy_1S_1)^{1/2}$
VOL	=	Volume of water in the tidal prism between high and low tide levels, ft³ (m³)
W	=	Bottom width of the bridge less pier widths, or overbank width (set back distance less pier widths, ft (m)
	=	Topwidth of the scour hole from each side of the pier of footing, ft (m)
	=	Width of debris perpendicular to the flow direction, ft (m)
W_1	=	Bottom width of the upstream main channel, ft (m)
W_2	=	Bottom width of the main channel in the contracted section less pier widths, ft (m)
W_c	=	Width of open-bottom culvert, ft (m)
y	=	Depth of flow, ft (m)
	=	Depth of flow in the contracted bridge opening for estimating a median stone diameter, D_{50}, ft (m)
	=	Amplitude or elevation of the tide above mean water level, ft (m), at time t
y_1	=	Approach flow depth at the beginning of computations, ft (m)
	=	Average depth in the upstream main channel or on the floodplain prior to contraction scour, ft (m)
	=	Depth of flow directly upstream of the pier, ft (m)
	=	Depth of flow at the abutment, on the overbank or in the main channel for abutment scour, ft (m)
y_2	=	Adjusted flow depth for pile cap computations ft (m)
	=	Average depth in the contracted section (bridge opening) or on the overbank at the bridge, ft (m)
	=	Average depth under lower cord, ft (m)
y_3	=	Adjusted flow depth for pile group computations, ft (m)
y_a	=	Average depth of flow on the floodplain, ft (m)
y_f	=	Distance from the bed to the top of the footing, ft (m)
y_o	=	Existing depth of flow, ft (m)

y_o	=	Existing depth of flow in the contracted bridge section before scour
y_{ps}	=	Depth of pier scour, ft (m)
y_s	=	Average contraction scour depth, ft (m)
	=	Local scour depth, ft (m)
	=	Depth of vertical contraction scour relative to mean bed elevation, ft (m)
	=	Total scour depth, ft (m)
y_{sc}	=	Depth of contraction scour, ft (m)
$y_{s\ pier}$	=	Scour component for the pier stem in the flow, ft (m)
$y_{s\ pc}$	=	Scour component for the pier cap or footing in the flow, ft (m)
$y_{s\ pg}$	=	Scour component for the piles exposed to the flow, ft (m)
Z	=	Vertical offset to datum, ft (m)
$\dot{z}_i$	=	Initial rate of scour, ft/hr (m/hr)
α_A	=	Amplification factor for abutment scour
σ	=	Sediment gradation coefficient (D_{84}/D_{50})
Ω	=	Cumulative stream power, ft-lb/day per sec per ft^2 (KW-hr/m^2)
τ	=	Design shear stress, lbs/ft^2 (Pa or N/m^2)
τ_2, τ_o	=	Average bed shear stress at the contracted section, Pa or lbs/ft^2 (N/m²)
τ_c	=	Critical bed shear stress at incipient motion, lbs/ft^2 (N/m²)
γ, γ_w	=	Specific weight of water, lbs/ft^3 (N/m^3)
ρ_w	=	Density of water, slugs/ft^3 (kg/m^3)
ρ_s	=	Density of sediment, slugs/ft^3 (kg/m^3)
θ	=	Angle of repose of the bed material (ranges from about 30° to 44°)
	=	Skew angle of flow with respect to pier
	=	Skew angle of abutment (embankment) with respect to flow
	=	Angle, in degrees, subdividing the tidal cycle
ω	=	Fall velocity of the bed material of a given size, ft/s (m/s)
ΔE	=	Energy loss per unit distance, ft/ft (m/m)
ΔH	=	Maximum difference in water surface elevation between the bay and ocean side of the inlet or channel, ft (m)

(page intentionally left blank)

ACKNOWLEDGMENTS

This manual is a major revision of the fourth edition of HEC-18 which was published in 2001. The writers wish to acknowledge the long-term contributions of Dr. E.V. Richardson and Mr. S.R. Davis, co-authors of all previous editions of HEC-18. Dr. Richardson and Mr. Davis provided technical assistance to FHWA at the inception of the national bridge scour program and continued their significant contributions over the last 20 years as the scour program evolved and additional technical guidance became available.

The writers also wish to recognize the National Cooperative Highway Research Program (NCHRP) and its researchers and the Federal Highway Administration and its researchers for the many projects completed that contributed to this document. We also recognize State Departments of Transportation for the research they have sponsored and completed as well as research completed by other federal agencies, universities, and private researchers.

All of this work will enable improved bridge design practice and bridge maintenance and inspection procedures resulting in greater safety for the users of the nation's bridges.

DISCLAIMER

Mention of a manufacturer, registered or trade name does not constitute a guarantee or warranty of the product by the U.S. Department of Transportation or the Federal Highway Administration and does not imply their approval and/or endorsement to the exclusion of other products and/or manufacturers that may also be suitable.

GLOSSARY

abrasion: Removal of streambank material due to entrained sediment, ice, or debris rubbing against the bank.

aggradation: General and progressive buildup of the longitudinal profile of a channel bed due to sediment deposition.

alluvial channel: Channel wholly in alluvium; no bedrock is exposed in channel at low flow or likely to be exposed by erosion.

alluvial fan: A fan-shaped deposit of material at the place where a stream issues from a narrow valley of high slope onto a plain or broad valley of low slope. An alluvial cone is made up of the finer materials suspended in flow while a debris cone is a mixture of all sizes and kinds of materials.

alluvial stream: A stream which has formed its channel in cohesive or noncohesive materials that have been and can be transported by the stream.

alluvium: Unconsolidated material deposited by a stream in a channel, floodplain, alluvial fan, or delta.

alternating bars: Elongated deposits found alternately near the right and left banks of a channel.

anabranch: Individual channel of an anabranched stream.

anabranched stream: A stream whose flow is divided at normal and lower stages by large islands or, more rarely, by large bars; individual islands or bars are wider than about three times water width; channels are more widely and distinctly separated than in a braided stream.

anastomosing stream: An anabranched stream.

angle of repose: The maximum angle (as measured from the horizontal) at which gravel or sand particles can stand.

annual flood: The maximum flow in one year (may be daily or instantaneous).

apron: Protective material placed on a streambed to resist scour.

apron, launching: An apron designed to settle and protect the side slopes of a scour hole after settlement.

armor (armoring): Surfacing of channel bed, banks, or embankment slope to resist erosion and scour. (a) natural process whereby an erosion- resistant layer of relatively large particles is formed on a streambed due to the removal of finer particles by streamflow; (b) placement of a covering to resist erosion.

GLOSSARY (continued)

average velocity: Velocity at a given cross section determined by dividing discharge by cross sectional area.

avulsion: A sudden change in the channel course that usually occurs when a stream breaks through its banks; usually associated with a flood or a catastrophic event.

backfill: The material used to refill a ditch or other excavation, or the process of doing so.

backwater: The increase in water surface elevation relative to the elevation occurring under natural channel and floodplain conditions. It is induced by a bridge or other structure that obstructs or constricts the free flow of water in a channel.

backwater area: The low-lying lands adjacent to a stream that may become flooded due to backwater.

bank: The sides of a channel between which the flow is normally confined.

bank, left (right): The side of a channel as viewed in a downstream direction.

bankfull discharge: Discharge that, on the average, fills a channel to the point of overflowing.

bank protection: Engineering works for the purpose of protecting streambanks from erosion.

bank revetment: Erosion-resistant materials placed directly on a streambank to protect the bank from erosion.

bar: Elongated deposit of alluvium within a channel, not permanently vegetated.

base floodplain: Floodplain associated with the flood with a 100-year recurrence interval.

bay: Body of water connected to the ocean with an inlet.

bed: Bottom of a channel bounded by banks.

bed form: A recognizable relief feature on the bed of a channel, such as a ripple, dune, plane bed, antidune, or bar. Bed forms are a consequence of the interaction between hydraulic forces (boundary shear stress) and the bed sediment.

GLOSSARY (continued)

bed layer: A flow layer, several grain diameters thick (usually two) immediately above the bed.

bed load: Sediment that is transported in a stream by rolling, sliding, or skipping along the bed or very close to it; considered to be within the bed layer (contact load).

bed load discharge (or bed load): The quantity of bed load passing a cross section of a stream in a unit of time.

bed material: Material found in and on the bed of a stream (May be transported as bed load or in suspension).

bedrock: The solid rock exposed at the surface of the earth or overlain by soils and unconsolidated material.

bed sediment discharge: The part of the total sediment discharge that is composed of grain sizes found in the bed and is equal to the transport capability of the flow.

bed material: Material found in and on the bed of a stream (May be transported as bed load or in suspension).

bedrock: The solid rock exposed at the surface of the earth or overlain by soils and unconsolidated material.

bed sediment discharge: The part of the total sediment discharge that is composed of grain sizes found in the bed and is equal to the transport capability of the flow.

bed shear (tractive force): The force per unit area exerted by a fluid flowing past a stationary boundary.

bed slope: The inclination of the channel bottom.

blanket: Material covering all or a portion of a streambank to prevent erosion.

boulder: A rock fragment whose diameter is greater than 250 mm.

braid: A subordinate channel of a braided stream.

braided stream: A stream whose flow is divided at normal stage by small mid-channel bars or small islands; the individual width of bars and islands is less than about three times water width; a braided stream has the aspect of a single large channel within which are subordinate channels.

GLOSSARY (continued)

bridge opening:
: The cross-sectional area beneath a bridge that is available for conveyance of water.

bridge substructural:
: Structural elements supporting a bridge in contact with the stream or channel bed, including bridge abutments, piers, and footings.

bridge waterway:
: The area of a bridge opening available for flow, as measured below a specified stage and normal to the principal direction of flow.

bulk density:
: Density of the water sediment mixture (mass per unit volume), including both water and sediment.

bulkhead:
: A vertical, or near vertical, wall that supports a bank or an embankment; also may serve to protect against erosion.

bulking:
: Increasing the water discharge to account for high concentrations of sediment in the flow.

catchment:
: See drainage basin.

causeway:
: Rock or earth embankment carrying a roadway across water.

caving:
: The collapse of a bank caused by undermining due to the action of flowing water.

cellular-block:
: Interconnected concrete blocks with regular cavities placed mattress: directly on a streambank or filter to resist erosion. The cavities can permit bank drainage and the growth of vegetation where synthetic filter fabric is not used between the bank and mattress.

channel:
: The bed and banks that confine the surface flow of a stream.

channelization:
: Straightening or deepening of a natural channel by artificial cutoffs, grading, flow-control measures, or diversion of flow into an engineered channel.

channel diversion:
: The removal of flows by natural or artificial means from a natural length of channel.

channel pattern:
: The aspect of a stream channel in plan view, with particular reference to the degree of sinuosity, braiding, and anabranching.

GLOSSARY (continued)

channel process:
: Behavior of a channel with respect to shifting, erosion and sedimentation.

check dam:
: A low dam or weir across a channel used to control stage or degradation.

choking (of flow):
: Excessive constriction of flow which may cause severe backwater effect.

clay (mineral):
: A particle whose diameter is in the range of 0.00024 to 0.004 mm.

clay plug:
: A cutoff meander bend filled with fine grained cohesive sediments.

clear-water scour:
: Scour at a pier or abutment (or contraction scour) when there is no movement of the bed material upstream of the bridge crossing at the flow causing bridge scour.

cobble:
: A fragment of rock whose diameter is in the range of 64 to 250 mm.

confluence:
: The junction of two or more streams.

constriction:
: A natural or artificial control section, such as a bridge crossing, channel reach or dam, with limited flow capacity in which the upstream water surface elevation is related to discharge.

contact load:
: Sediment particles that roll or slide along in almost continuous contact with the streambed (bed load).

contraction:
: The effect of channel or bridge constriction on flow streamlines.

contraction scour:
: Contraction scour, in a natural channel or at a bridge crossing, involves the removal of material from the bed and banks across all or most of the channel width. This component of scour results from a contraction of the flow area at the bridge which causes an increase in velocity and shear stress on the bed at the bridge. The contraction can be caused by the bridge or from a natural narrowing of the stream channel.

countermeasure:
: A measure intended to prevent, delay or reduce the severity of hydraulic problems.

GLOSSARY (continued)

critical shear stress:	The minimum amount of shear stress required to initiate soil particle motion.
crossing:	The relatively short and shallow reach of a stream between bends; also crossover or riffle.
cross section:	A section normal to the trend of a channel or flow.
current:	Water flowing through a channel.
current meter:	An instrument used to measure flow velocity.
cut bank:	The concave wall of a meandering stream.
cutoff:	(a) A direct channel, either natural or artificial, connecting two points on a stream, thereby shortening the original length of the channel and increasing its slope; (b) A natural or artificial channel which develops across the neck of a meander loop (neck cutoff) or across a point bar (chute cutoff).
cutoff wall:	A wall, usually of sheet piling or concrete, that extends down to scour-resistant material or below the expected scour depth.
daily discharge:	Discharge averaged over one day (24 hours).
debris:	Floating or submerged material, such as logs, vegetation, or trash, transported by a stream.
degradation (bed):	A general and progressive (long-term) lowering of the channel bed due to erosion, over a relatively long channel length.
deep water (for waves):	Water of such a depth that surface waves are little affected by bottom conditions; customarily, water deeper than half the wavelength.
densimetric Froude Number:	Froude number where length scale is particle size.
depth of scour:	The vertical distance a streambed is lowered by scour below a reference elevation.
design flow (design flood):	The discharge that is selected as the basis for the design or evaluation of a hydraulic structure including a hydraulic design flood, scour design flood, and scour design check flood.

GLOSSARY (continued)

dike:
: An impermeable linear structure for the control or containment of overbank flow. A dike-trending parallel with a streambank differs from a levee in that it extends for a much shorter distance along the bank, and it may be surrounded by water during floods.

dike (groin, spur, jetty):
: A structure extending from a bank into a channel that is designed to: (a) reduce the stream velocity as the current passes through the dike, thus encouraging sediment deposition along the bank (permeable dike); or (b) deflect erosive current away from the streambank (impermeable dike).

discharge:
: Volume of water passing through a channel during a given time.

dominant discharge:
: (a) The discharge of water which is of sufficient magnitude and frequency to have a dominating effect in determining the characteristics and size of the stream course, channel, and bed; (b) That discharge which determines the principal dimensions and characteristics of a natural channel. The dominant formative discharge depends on the maximum and mean discharge, duration of flow, and flood frequency. For hydraulic geometry relationships, it is taken to be the bankfull discharge which has a return period of approximately 1.5 years in many natural channels.

drainage basin:
: An area confined by drainage divides, often having only one outlet for discharge (catchment, watershed).

drift:
: Alternative term for vegetative "debris."

eddy current:
: A vortex-type motion of a fluid flowing contrary to the main current, such as the circular water movement that occurs when the main flow becomes separated from the bank.

entrenched stream:
: Stream cut into bedrock or consolidated deposits.

ephemeral stream:
: A stream or reach of stream that does not flow for parts of the year. As used here, the term includes intermittent streams with flow less than perennial.

equilibrium scour:
: Scour depth in sand-bed stream with dune bed about which live bed pier scour level fluctuates due to variability in bed material transport in the approach flow.

GLOSSARY (continued)

erosion:
: Displacement of soil particles due to water or wind action.

erosion control matting:
: Fibrous matting (e.g., jute, paper, etc.) placed or sprayed on a stream-bank for the purpose of resisting erosion or providing temporary stabilization until vegetation is established.

fall velocity:
: The velocity at which a sediment particle falls through a column of still water.

fill slope:
: Side or end slope of an earth-fill embankment. Where a fill-slope forms the streamward face of a spill-through abutment, it is regarded as part of the abutment.

filter:
: Layer of fabric (geotextile) or granular material (sand, gravel, or graded rock) placed between bank revetment (or bed protection) and soil for the following purposes: (1) to prevent the soil from moving through the revetment by piping, extrusion, or erosion; (2) to prevent the revetment from sinking into the soil; and (3) to permit natural seepage from the streambank, thus preventing the buildup of excessive hydrostatic pressure.

filter fabric (cloth):
: Geosynthetic fabric that serves the same purpose as a granular filter blanket.

fine sediment load:
: That part of the total sediment load that is composed of particle sizes finer than those represented in the bed (wash load). Normally, the fine-sediment load is finer than 0.062 mm for sand-bed channels. Silts, clays and sand could be considered wash load in coarse gravel and cobble-bed channels.

flanking:
: Erosion around the landward end of a stream stabilization countermeasure.

flashy stream:
: Stream characterized by rapidly rising and falling stages, as indicated by a sharply peaked hydrograph. Typically associated with mountain streams or highly disturbed urbanized catchments. Most flashy streams are ephemeral, but some are perennial.

flood-frequency curve:
: A graph indicating the probability that the annual flood discharge will exceed a given magnitude, or the recurrence interval corresponding to a given magnitude.

floodplain:
: A nearly flat, alluvial lowland bordering a stream, that is subject to frequent inundation by floods.

GLOSSARY (continued)

flow-control structure: A structure either within or outside a channel that acts as a countermeasure by controlling the direction, depth, or velocity of flowing water.

flow hazard: Flow characteristics (discharge, stage, velocity, or duration) that are associated with a hydraulic problem or that can reasonably be considered of sufficient magnitude to cause a hydraulic problem or to test the effectiveness of a countermeasure.

flow slide: Saturated soil materials which behave more like a liquid than a solid. A flow slide on a channel bank can result in a bank failure.

fluvial geomorphology: The science dealing with the morphology (form) and dynamics of streams and rivers.

fluvial system: The natural river system consisting of (1) the drainage basin, watershed, or sediment source area, (2) tributary and mainstem river channels or sediment transfer zone, and (3) alluvial fans, valley fills and deltas, or the sediment deposition zone.

freeboard: The vertical distance above a design stage that is allowed for waves, surges, drift, and other contingencies.

fresh water: Water that is not salty as compared to sea water which generally has a salinity of 35 000 parts per million.

Froude Number: A dimensionless number that represents the ratio of inertial to gravitational forces in open channel flow.

geomorphology/morphology: That science that deals with the form of the Earth, the general configuration of its surface, and the changes that take place due to erosion and deposition.

grade-control structure (sill, check dam): Structure placed bank to bank across a stream channel (usually with its central axis perpendicular to flow) for the purpose of controlling bed slope and preventing scour or headcutting.

graded stream: A geomorphic term used for streams that have apparently achieved a state of equilibrium between the rate of sediment transport and the rate of sediment supply throughout long reaches.

gravel: A rock fragment whose diameter ranges from 2 to 64 mm.

grout: A fluid mixture of cement and water or of cement, sand, and water used to fill joints and voids.

guide bank: A dike extending upstream from the approach embankment at either or both sides of the bridge opening to direct the flow through the opening. Some guide banks extend downstream from the bridge.

GLOSSARY (continued)

hardpoint: A streambank protection structure whereby "soft" or erodible materials are removed from a bank and replaced by stone or compacted clay. Some hard points protrude a short distance into the channel to direct erosive currents away from the bank. Hard points also occur naturally along streambanks as passing currents remove erodible materials leaving nonerodible materials exposed.

headcutting: Channel degradation associated with abrupt changes in the bed elevation (headcut) that generally migrates in an upstream direction.

helical flow: Three-dimensional movement of water particles along a spiral path in the general direction of flow. These secondary-type currents are of most significance as flow passes through a bend; their net effect is to remove soil particles from the cut bank and deposit this material on a point bar.

hydraulics: The applied science concerned with the behavior and flow of liquids, especially in pipes, channels, structures, and the ground.

hydraulic model: A small-scale physical or mathematical representation of a flow situation.

hydraulic problem: An effect of streamflow, tidal flow, or wave action such that the integrity of the highway facility is destroyed, damaged, or endangered.

hydraulic radius: The cross-sectional area of a stream divided by its wetted perimeter.

hydraulic structures: The facilities used to impound, accommodate, convey or control the flow of water, such as dams, weirs, intakes, culverts, channels, and bridges.

hydrograph: The graph of stage or discharge against time.

hydrology: The science concerned with the occurrence, distribution, and circulation of water on the earth.

imbricated: In reference to stream bed sediment particles, having an overlapping or shingled pattern.

icing: Masses or sheets of ice formed on the frozen surface of a river or floodplain. When shoals in the river are frozen to the bottom or otherwise dammed, water under hydrostatic pressure is forced to the surface where it freezes.

GLOSSARY (continued)

incised reach: A stretch of stream with an incised channel that only rarely overflows its banks.

incised stream: A stream which has deepened its channel through the bed of the valley floor, so that the floodplain is a terrace.

invert: The lowest point in the channel cross section or at flow control devices such as weirs, culverts, or dams.

ineffective flow: An area of flow where water is not being conveyed in a downstream direction (e.g., ponding above or below an embankment).

island: A permanently vegetated area, emergent at normal stage, that divides the flow of a stream. Islands originate by establishment of vegetation on a bar, by channel avulsion, or at the junction of minor tributary with a larger stream.

jetty: (a) An obstruction built of piles, rock, or other material extending from a bank into a stream, so placed as to induce bank building, or to protect against erosion; (b) A similar obstruction to influence stream, lake, or tidal currents, or to protect a harbor (also spur).

lateral erosion: Erosion in which the removal of material is extended horizontally as contrasted with degradation and scour in a vertical direction.

launching: Release of undercut material (stone riprap, rubble, slag, etc.) downslope or into a scoured area.

levee: An embankment, generally landward of top bank, that confines flow during high-water periods, thus preventing overflow into lowlands.

live flow: Area of flow where water is actively conveyed in a downstream direction (e.g., channel flow and unobstructed floodplain flow.

live-bed scour: Scour at a pier or abutment (or contraction scour) when the bed material in the channel upstream of the bridge is moving at the flow causing bridge scour.

load (or sediment load): Amount of sediment being moved by a stream.

local scour: Removal of material from around piers, abutments, spurs, and embankments caused by an acceleration of flow and resulting vortices induced by obstructions to the flow.

GLOSSARY (continued)

longitudinal profile:
: The profile of a stream or channel drawn along the length of its centerline. In drawing the profile, elevations of the water surface or the thalweg are plotted against distance as measured from the mouth or from an arbitrary initial point.

lower bank:
: That portion of a streambank having an elevation less than the mean water level of the stream.

mathematical model:
: A numerical representation of a flow situation using mathematical equations (also computer model).

mattress:
: A blanket or revetment of materials interwoven or otherwise lashed together and placed to cover an area subject to scour.

meander or full meander:
: A meander in a river consists of two consecutive loops, one flowing clockwise and the other counter-clockwise.

meander amplitude:
: The distance between points of maximum curvature of successive meanders of opposite phase in a direction normal to the general course of the meander belt, measured between center lines of channels.

meander belt:
: The distance between lines drawn tangent to the extreme limits of successive fully developed meanders.

meander length:
: The distance along a stream between corresponding points of successive meanders.

meander loop:
: An individual loop of a meandering or sinuous stream lying between inflection points with adjoining loops.

meander ratio:
: The ratio of meander width to meander length.

meander radius of curvature:
: The radius of a circle inscribed on the centerline of a meander loop.

meander scrolls:
: Low, concentric ridges and swales on a floodplain, marking the successive positions of former meander loops.

meander width:
: The amplitude of a fully developed meander measured from midstream to midstream.

GLOSSARY (continued)

meandering stream:
: A stream having a sinuosity greater than some arbitrary value. The term also implies a moderate degree of pattern symmetry, imparted by regularity of size and repetition of meander loops. The channel generally exhibits a characteristic process of bank erosion and point bar deposition associated with systematically shifting meanders.

median diameter:
: The particle diameter of the 50th percentile point on a size distribution curve such that half of the particles (by weight, number, or volume) are larger and half are smaller (D_{50}.)

mid-channel bar:
: A bar lacking permanent vegetal cover that divides the flow in a channel at normal stage.

middle bank:
: The portion of a streambank having an elevation approximately the same as that of the mean water level of the stream.

migration:
: Change in position of a channel by lateral erosion of one bank and simultaneous accretion of the opposite bank.

mud:
: A soft, saturated mixture mainly of silt and clay.

natural levee:
: A low ridge that slopes gently away from the channel banks that is formed along streambanks during floods by deposition.

nominal diameter:
: Equivalent spherical diameter of a hypothetical sphere of the same volume as a given sediment particle.

nonalluvial channel:
: A channel whose boundary is in bedrock or non-erodible material.

normal stage:
: The water stage prevailing during the greater part of the year.

open-bottom culvert:
: Bridge/culvert structures with natural channel materials as the bottom.

overbank flow:
: Water movement that overtops the bank either due to stream stage or to overland surface water runoff.

oxbow:
: The abandoned former meander loop that remains after a stream cuts a new, shorter channel across the narrow neck of a meander. Often bow-shaped or horseshoe-shaped.

pavement:
: Streambank surface covering, usually impermeable, designed to serve as protection against erosion. Common pavements used on streambanks are concrete, compacted asphalt, and soil-cement.

GLOSSARY (continued)

paving: Covering of stones on a channel bed or bank (used with reference to natural covering).

peaked stone dike: Riprap placed parallel to the toe of a streambank (at the natural angle of repose of the stone) to prevent erosion of the toe and induce sediment deposition behind the dike.

perennial stream: A stream or reach of a stream that flows continuously for all or most of the year.

phreatic line: The upper boundary of the seepage water surface landward of a streambank.

pile: An elongated member, usually made of timber, concrete, or steel, that serves as a structural component of a river-training structure or bridge.

piping: Removal of soil material through subsurface flow of seepage water that develops channels or "pipes" within the soil bank.

point bar: An alluvial deposit of sand or gravel lacking permanent vegetal cover occurring in a channel at the inside of a meander loop, usually somewhat downstream from the apex of the loop.

poised stream: A stream which, as a whole, maintains its slope, depths, and channel dimensions without any noticeable raising or lowering of its bed (stable stream). Such condition may be temporary from a geological point of view, but for practical engineering purposes, the stream may be considered stable.

pressure flow/scour: See vertical contraction scour.

probable maximum flood: A very rare flood discharge value computed by hydro-meteorological methods, usually in connection with major hydraulic structures.

rapid drawdown: Lowering the water against a bank more quickly than the bank can drain without becoming unstable.

reach: A segment of stream length that is arbitrarily bounded for purposes of study.

recurrence interval: The reciprocal of the annual probability of exceedance of a hydrologic event (also return period, exceedance interval).

GLOSSARY (continued)

regime:
: The condition of a stream or its channel with regard to stability. A stream is in regime if its channel has reached an equilibrium form as a result of its flow characteristics. Also, the general pattern of variation around a mean condition, as in flow regime, tidal regime, channel regime, sediment regime, etc. (used also to mean a set of physical characteristics of a river).

regime change:
: A change in channel characteristics resulting from such things as changes in imposed flows, sediment loads, or slope.

regime channel:
: Alluvial channel that has attained, more or less, a state of equilibrium with respect to erosion and deposition.

regime formula:
: A formula relating stable alluvial channel dimensions or slope to discharge and sediment characteristics.

relief bridge:
: An opening in an embankment on a floodplain to permit passage of overbank flow.

revetment:
: Rigid or flexible armor placed to inhibit scour and lateral erosion.

riffle:
: A natural, shallow flow area extending across a streambed in which the surface of flowing water is broken by waves or ripples. Typically, riffles alternate with pools along the length of a stream channel.

riparian:
: Pertaining to anything connected with or adjacent to the banks of a stream (corridor, vegetation, zone, etc.).

riprap:
: Layer or facing of rock or broken concrete dumped or placed to protect a structure or embankment from erosion; also the rock or broken concrete suitable for such use. Riprap has also been applied to almost all kinds of armor, including wire-enclosed riprap, grouted riprap, sacked concrete, and concrete slabs.

river training:
: Engineering works with or without the construction of embankment, built along a stream or reach of stream to direct or to lead the flow into a prescribed channel. Also, any structure configuration constructed in a stream or placed on, adjacent to, or in the vicinity of a streambank that is intended to deflect currents, induce sediment deposition, induce scour, or in some other way alter the flow and sediment regimes of the stream.

GLOSSARY (continued)

rock:	Indurated geomaterial that requires drilling, wedging, blasting, or other methods of applying force for excavation.
roughness coefficient:	Numerical measure of the frictional resistance to flow in a channel, as in the Manning or Chezy's formulas.
rubble:	Rough, irregular fragments of materials of random size used to retard erosion. The fragments may consist of broken concrete slabs, masonry, or other suitable refuse.
runoff:	That part of precipitation which appears in surface streams of either perennial or intermittent form.
saltation load:	Sediment bounced along the streambed by energy and turbulence of flow, and by other moving particles.
sand:	A rock fragment whose diameter is in the range of 0.062 to 2.0 mm.
scour:	Erosion of streambed or bank material due to flowing water; often considered as being localized (see local scour, contraction scour, total scour).
scour prism:	Total volume of stream bed material removed by scour in the bridge reach for design flood conditions.
sediment or fluvial sediment:	Fragmental material transported, suspended, or deposited by water.
sediment concentration:	Weight or volume of sediment relative to the quantity of transporting (or suspending) fluid.
sediment discharge:	The quantity of sediment that is carried past any cross section of a stream in a unit of time. Discharge may be limited to certain sizes of sediment or to a specific part of the cross section.
sediment load:	Amount of sediment being moved by a stream.
sediment yield:	The total sediment outflow from a watershed or a drainage area at a point of reference and in a specified time period. This outflow is equal to the sediment discharge from the drainage area.
seepage:	The slow movement of water through small cracks and pores of the bank material.

GLOSSARY (continued)

shallow water (for waves): Water of such a depth that waves are noticeably affected by bottom conditions; customarily, water shallower than half the wavelength.

shear stress: See unit shear force.

shoal: A relatively shallow submerged bank or bar in a body of water.

silt: A particle whose diameter is in the range of 0.004 to 0.062 mm.

sinuosity: The ratio between the thalweg length and the valley length of a stream.

slope (of channel or stream): Fall per unit length along the channel centerline or thalweg.

slope protection: Any measure such as riprap, paving, vegetation, revetment, brush or other material intended to protect a slope from erosion, slipping or caving, or to withstand external hydraulic pressure.

sloughing: Sliding or collapse of overlying material; same ultimate effect as caving, but usually occurs when a bank or an underlying stratum is saturated.

slope-area method: A method of estimating unmeasured flood discharges in a uniform channel reach using observed high-water levels.

slump: A sudden slip or collapse of a bank, generally in the vertical direction and confined to a short distance, probably due to the substratum being washed out or having become unable to bear the weight above it.

soil: Any unconsolidated geomaterial composed of discrete particles with gases and liquids in between.

sorting: Progressive reduction of size (or weight) of particles of the sediment load carried down a stream.

spill-through abutment: A bridge abutment having a fill slope on the streamward side. The term originally referred to the "spill-through" of fill at an open abutment but is now applied to any abutment having such a slope.

spread footing: A pier or abutment footing that transfers load directly to the earth.

GLOSSARY (continued)

stability:
A condition of a channel when, though it may change slightly at different times of the year as the result of varying conditions of flow and sediment charge, there is no appreciable change from year to year; that is, accretion balances erosion over the years.

stable channel:
A condition that exists when a stream has a bed slope and cross section which allows its channel to transport the water and sediment delivered from the upstream watershed without aggradation, degradation, or bank erosion (a graded stream).

stage:
Water-surface elevation of a stream with respect to a reference elevation.

still-water elevation:
Flood height to which water rises as a result of barometric pressure changes occurring during a storm event.

stone riprap:
Natural cobbles, boulders, or rock dumped or placed as protection against erosion.

stream:
A body of water that may range in size from a large river to a small rill flowing in a channel. By extension, the term is sometimes applied to a natural channel or drainage course formed by flowing water whether it is occupied by water or not.

streambank erosion:
Removal of soil particles or a mass of particles from a bank surface due primarily to water action. Other factors such as weathering, ice and debris abrasion, chemical reactions, and land use changes may also directly or indirectly lead to bank erosion.

streambank failure:
Sudden collapse of a bank due to an unstable condition such as removal of material at the toe of the bank by scour.

streambank protection:
Any technique used to prevent erosion or failure of a streambank.

suspended sediment discharge:
The quantity of sediment passing through a stream cross section above the bed layer in a unit of time suspended by the turbulence of flow (suspended load).

sub-bed material:
Material underlying that portion of the streambed which is subject to direct action of the flow. Also, substrate.

subcritical, supercritical flow:
Open channel flow conditions with Froude Number less than and greater than unity, respectively.

GLOSSARY (continued)

thalweg: The line extending down a channel that follows the lowest elevation of the bed.

tidal waterways: Scour at bridges over tidal waterways, i.e., in the coastal zone.

toe of bank: That portion of a stream cross section where the lower bank terminates and the channel bottom or the opposite lower bank begins.

toe protection: Loose stones laid or dumped at the toe of an embankment, groin, etc., or masonry or concrete wall built at the junction of the bank and the bed in channels or at extremities of hydraulic structures to counteract erosion.

total scour: The sum of long-term degradation, general (contraction) scour, and local scour.

total sediment load: The sum of suspended load and bed load or the sum of bed material load and wash load of a stream (total load).

tractive force: The drag or shear on a streambed or bank caused by passing water which tends to move soil particles along with the streamflow.

turbulence: Motion of fluids in which local velocities and pressures fluctuate irregularly in a random manner as opposed to laminar flow where all particles of the fluid move in distinct and separate lines.

ultimate scour: The maximum depth of scour attained for a given flow condition. May require multiple flow events and in cemented or cohesive soils may be achieved over a long time period.

uniform flow: Flow of constant cross section and velocity through a reach of channel at a given time. Both the energy slope and the water slope are equal to the bed slope under conditions of uniform flow.

unit discharge: Discharge per unit width (may be average over a cross section, or local at a point).

unit shear force (shear stress): The force or drag developed at the channel bed by flowing water. For uniform flow, this force is equal to a component of the gravity force acting in a direction parallel to the channel bed on a unit wetted area. Usually in units of stress, Pa (N/m^2) or (lb/ft^2).

GLOSSARY (continued)

unsteady flow: Flow of variable discharge and velocity through a cross section with respect to time.

upper bank: The portion of a streambank having an elevation greater than the average water level of the stream.

velocity: The time rate of flow usually expressed in m/s (ft/sec). The average velocity is the velocity at a given cross section determined by dividing discharge by cross-sectional area.

vertical abutment: An abutment, usually with wingwalls, that has no fill slope on its streamward side.

vertical contraction scour: Scour resulting from flow impinging on bridge superstructure elements (e.g., low chord).

vortex: Turbulent eddy in the flow generally caused by an obstruction such as a bridge pier or abutment (e.g., horseshoe vortex).

wandering channel: A channel exhibiting a more or less non-systematic process of channel shifting, erosion and deposition, with no definite meanders or braided pattern.

wandering thalweg: A thalweg whose position in the channel shifts during floods and typically serves as an inset channel that conveys all or most of the stream flow at normal or lower stages.

wash load: Suspended material of very small size (generally clays and colloids) originating primarily from erosion on the land slopes of the drainage area and present to a negligible degree in the bed itself.

watershed: See drainage basin.

waterway opening width (area): Width (area) of bridge opening at (below) a specified stage, measured normal to the principal direction of flow.

CHAPTER 1

INTRODUCTION

1.1 PURPOSE

The purpose of this document is to provide guidelines for the following:

1. Designing new and replacement bridges to resist scour
2. Evaluating existing bridges for vulnerability to scour
3. Inspecting bridges for scour
4. Improving the state-of-practice of estimating scour at bridges

1.2 BACKGROUND

The most common cause of bridge failures is from floods scouring bed material from around bridge foundations. Scour is the engineering term for the erosion caused by water of the soil surrounding a bridge foundation (piers and abutments). During the spring floods of 1987, 17 bridges in New York and New England were damaged or destroyed by scour. In 1985, 73 bridges were destroyed by floods in Pennsylvania, Virginia, and West Virginia. A 1973 national study for the Federal Highway Administration (FHWA) of 383 bridge failures caused by catastrophic floods showed that 25 percent involved pier damage and 75 percent involved abutment damage (FHWA 1973). A second more extensive study in 1978 indicated local scour at bridge piers to be a problem about equal to abutment scour problems (FHWA 1978). A number of case histories on the causes and consequences of scour at major bridges are presented in Transportation Research Record 950 (TRB 1984).

From available information, the 1993 flood in the upper Mississippi basin, caused 23 bridge failures for an estimated damage of $15 million. The modes of bridge failures were 14 from abutment scour, two from pier scour, three from pier and abutment scour, two from lateral bank migration, one from debris load, and one from unknown cause.

In the 1994 flooding from storm Alberto in Georgia, there were over 500 state and locally owned bridges with damage attributed to scour. Thirty-one of state-owned bridges experienced from 15 to 20 feet of contraction scour and/or long-term degradation in addition to local scour. These bridges had to be replaced. Of more than 150 bridges identified as scour damaged, the Georgia Department of Transportation (GADOT) also recommended that 73 non-federal aid bridges be repaired or replaced. Total damage to the GADOT highway system was approximately $130 million.

The American Association of State Highway and Transportation Officials (AASHTO) standard specifications for highway bridges has the following requirements to address the problem of stream stability and scour (AASHTO 1992a):

- Hydraulic studies are a necessary part of the preliminary design of a bridge and should include. . .estimated scour depths at piers and abutments of proposed structures.
- The probable depth of scour shall be determined by subsurface exploration and hydraulic studies. Refer to Article 1.3.2 and FHWA Hydraulic Engineering Circular (HEC) 18 for general guidance regarding hydraulic studies and design.
- . . .in all cases, the pile length shall be determined such that the design structural load may be safely supported entirely below the probable scour depth.

1.3 COMPREHENSIVE ANALYSIS

This manual is part of a set of HECs issued by FHWA to provide guidance for bridge scour and stream stability analyses. The three manuals in this set are:

    HEC-18   Evaluating Scour at Bridges
    HEC-20   Stream Stability at Highway Structures (FHWA 2012b)
    HEC-23   Bridge Scour and Stream Instability Countermeasures (FHWA 2009)

The Flow Chart of Figure 1.1 illustrates graphically the interrelationship between these three documents and emphasizes that they should be used as a set. A comprehensive scour analysis or stability evaluation should be based on information presented in all three documents.

While the flow chart does not attempt to present every detail of a complete stream stability and scour evaluation, it has sufficient detail to show the major elements in a complete analysis, the logical flow of a typical analysis or evaluation, and the most common decision points and feedback loops. It clearly shows how the three documents tie together, and recognizes the differences between design of a new bridge and evaluation of an existing bridge.

The HEC-20 block of the flow chart outlines initial data collection and site reconnaissance activities leading to an understanding of the problem, evaluation of river system stability and potential future response. The HEC-20 procedures include both qualitative and quantitative geomorphic and engineering analysis techniques which help establish the level of analysis necessary to solve the stream instability and scour problem for design of a new bridge, or for the evaluation of an existing bridge that may require rehabilitation or countermeasures. The "Classify Stream," "Evaluate Stability," and "Assess Response" portions of the HEC-20 block are expanded in HEC-20 into a six-step Level 1 and an eight-step Level 2 analysis procedure. In some cases, the HEC-20 analysis may be sufficient to determine that stream instability or scour problems do not exist, i.e., the bridge has a "low risk" of failure regarding scour susceptibility.

In most cases, the analysis or evaluation will progress to the HEC-18 block of the flow chart. Here more detailed hydrologic and hydraulic data are developed, with the specific approach determined by the level of complexity of the problem and waterway characteristics (e.g., tidal or riverine). The "Scour Analysis" portion of the HEC-18 block encompasses a seven-step specific design approach which includes evaluation of the components of total scour (see Chapter 2).

Since bridge scour evaluation requires multidisciplinary inputs, it is often advisable for the hydraulic engineer to involve structural and geotechnical engineers at this stage of the analysis. **Once the total scour prism is plotted, then all three disciplines must be involved in a determination of structural stability**.

For a new bridge design, if the structure is stable the design process can proceed to consideration of environmental impacts, cost, constructability, and maintainability. If the structure is unstable, revise the design and repeat the analysis. For an existing bridge, a finding of structural stability at this stage will result in a "low risk" evaluation, with no further action required. However, a Plan of Action should be developed for an unstable existing bridge (scour critical) to correct the problem as discussed in Chapter 10 and HEC-23 (FHWA 2009).

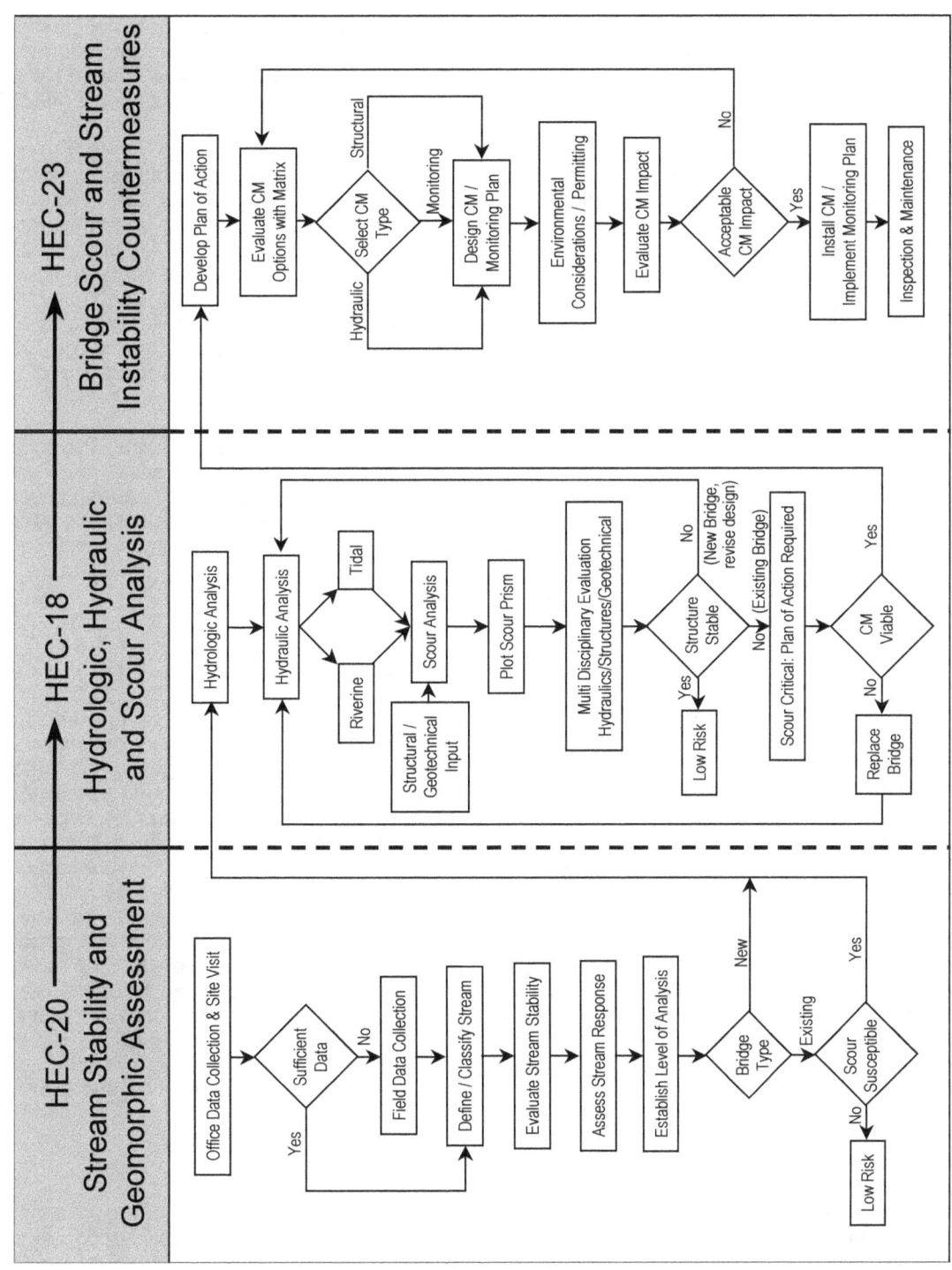

Figure 1.1. Flow chart for scour and stream stability analysis and evaluation.

The scour problem may be so serious that installing countermeasures would not provide a viable solution and a replacement or substantial bridge rehabilitation would be required. If countermeasures would correct the stream instability or scour problem at a reasonable cost and with acceptable environmental impacts, the analysis would progress to the HEC-23 block of the flow chart.

HEC-23 provides a range of resources to support bridge scour or stream instability countermeasure selection and design. A countermeasure matrix in HEC-23 presents a variety of countermeasures that have been used by State departments of transportation (DOTs) to control scour and stream instability at bridges. The matrix is organized to highlight the various groups of countermeasures and identifies distinctive characteristics of each countermeasure. The matrix identifies most countermeasures used and lists information on their functional applicability to a particular problem, their suitability to specific river environments, the general level of maintenance resources required, and which DOTs have experience with specific countermeasures. Finally, a reference source for design guidelines is noted.

HEC-23 includes specific design guidelines for the most common (and some uncommon) countermeasures used by DOTs, or references to sources of design guidance. Inherent in the design of any countermeasure is an evaluation of potential environmental impacts, permitting for countermeasure installation, and redesign, if necessary, to meet environmental requirements. As shown in the flow chart, to be effective most countermeasures will require a monitoring plan, inspection, and maintenance.

1.4 PROCEDURAL GUIDANCE

1.4.1 Objectives of a Bridge Scour Evaluation Program

The need to minimize future flood damage to the nation's bridges requires that additional attention be devoted to developing and implementing improved procedures for designing and inspecting bridges for scour (USDOT 2004). Approximately 500,000 bridges in the National Bridge Inventory are built over waterways. Statistically, we can expect hundreds of these bridges to experience floods in the magnitude of a 100-year flood or greater each year. Because it is not economically feasible to construct all bridges to resist all conceivable floods, or to install scour countermeasures at all existing bridges to ensure absolute invulnerability from scour damage, some risks of failure from future floods may have to be accepted. **However, every bridge over water, whether existing or under design, should be assessed as to its vulnerability to floods in order to determine the prudent measures to be taken.** The added cost of making a bridge less vulnerable to scour is small when compared to the total cost of a failure which can easily be two to ten times the cost of the bridge itself. Moreover, the need to ensure public safety and minimize the adverse effects resulting from bridge closures requires our best efforts to improve the state-of-practice for designing and maintaining bridge foundations to resist the effects of scour. The design of bridge waterway capacity is typically based on flood frequencies somewhat less than those recommended for scour analysis (see Chapter 2).

1.4.2 Bridge Scour and the NBIS

The Federal requirements for bridge inspection are set forth in the National Bridge Inspection Standards (NBIS). The NBIS require bridge owners to maintain a bridge inspection program that includes procedures for underwater inspection, bridge scour evaluation, and plans of action for scour critical bridges (see Chapter 10). This information may be found in the FHWA Federal Register, Title 23, Code of Federal Regulations, Highways, Part 650, Bridges, Structures, and Hydraulics, Subpart C, National Bridge Inspection Standards (23 CFR 650, Subpart C). The most recent ruling was enacted on January 13, 2005.

The primary purpose of NBIS is to identify and evaluate existing bridge deficiencies to ensure the safety of the traveling public. The NBIS sets national policy regarding bridge inspection and rating procedures, frequency of inspections, inspector qualifications, report formats, and the preparation and maintenance of a State bridge inventory. Each State or Federal agency must prepare and maintain an inventory of all bridges subject to the NBIS. Certain Structure Inventory and Appraisal (SI&A) data must be collected and retained by the State or Federal agency and reported to FHWA on an annual basis. A tabulation of this data is contained in the SI&A sheet which may be found in the FHWA's "Recording and Coding Guide for the Structure Inventory and Appraisal of the Nation's Bridges" (FHWA 1995). The National Bridge Inventory (NBI) is the aggregation of Structure Inventory and Appraisal data collected to fulfill the requirements of the NBIS.

A national scour evaluation program as an integral part of the NBIS was established in 1988 by Technical Advisory (TA) T5140.20 (USDOT 1988). This TA was published following the April 1987 collapse of New York's Schoharie Bridge due to scour. In 1991 T5140.20 was superseded by T5140.23, "Evaluating Scour at Bridges." This Technical Advisory provides more guidance on the development and implementation of procedures for evaluating bridge scour to meet the requirements of 23 CFR 650, Subpart C. Specifically, Technical Advisory T5140.23 provides guidance on:

1. Developing and implementing a scour evaluation for designing new bridges
2. Evaluating existing bridges for scour vulnerability
3. Using scour countermeasures
4. Improving the state-of-practice for estimating scour at bridges

The procedures presented in this manual serve as guidance for implementing the recommendations contained in FHWA Technical Advisory T5140.23. The recommendations have been developed to summarize the essential elements which should be addressed in developing a comprehensive scour evaluation program. A key element of the program is the identification of scour-critical bridges which will be entered into the National Bridge Inventory using the Coding Guide (FHWA 1995).

1.5 ADVANCES IN THE STATE-OF-PRACTICE FOR ESTIMATING SCOUR AT BRIDGES

The Fourth Edition of HEC-18 was published in May 2001. Since that time significant advances have been made in the state-of-practice for estimating scour at bridges. Research under the National Cooperative Highway Research Program (NCHRP) of the Transportation Research Board (TRB), research conducted and sponsored by FHWA and other Federal agencies, State DOT sponsored research, and publications resulting from academic research have all contributed to advances in the state-of-practice. The following list indicates some of the primary technical resources and research that contributed to this Fifth Edition of HEC-18:

- HDS 6, River Engineering and Sediment Transport (1st Edition)
- HEC-25, Tidal Hydraulics (1st and 2nd Editions)
- NCHRP 24-27(01) Scour Research - Piers
- NCHRP 24-27(02) Scour Research - Abutments and Contraction
- NCHRP 24-27(03) Scour Research - Geomorphology
- NCHRP 24-15 Scour in Cohesive Materials
- NCHRP 24-15(a) Scour in Cohesive Materials

- NCHRP 24-20 Predicting Scour at Abutments
- NCHRP 24-24 Selecting Hydraulic Modeling Software
- NCHRP 24-25 Risk-Based Unknown Foundations
- NCHRP 24-26 Effects of Debris on Pier Scour
- NCHRP 24-29 Pier Scour in Rock
- NCHRP 24-32 Scour at Wide and Skewed Piers
- NCHRP 24-34 Risk-Based Approach for Bridge Scour Prediction
- NCHRP Tri-Panel Meeting - Abutment Scour and HEC-18
- Florida DOT Bridge Scour Manual
- FHWA-RD-02-078 Bottomless Culvert Scour, Phase I
- FHWA-HRT-07-026 Bottomless Culvert Scour, Phase II
- FHWA-HRT-12-034 Submerged-Flow Bridge Scour Under Clear-Water Conditions
- FHWA-HRT-12-022 Pier Scour in Clear-Water Conditions with Nonuniform Materials
- FHWA-NHI-08-106, "Stream Instability, Bridge Scour, and Countermeasures: A Field Guide for Bridge Inspectors"
- On-Line NHI Modules 135085 (Plan of Action), 135086 (Stream Stability), and 135087 (Bridge Scour)

The NCHRP 24-27(01), (02), and (03) studies were intended to provide recommendations to AASHTO and FHWA on advances in technology that should be considered for revisions to technical guidance documents, including HEC-18 and HEC-20. Each of these projects included an in-depth literature search of advances in technology in the last 10 years and each made recommendations on technology that is ready and validated for current practice. The results of these studies influenced the form and content of the next editions of HEC-18 and HEC-20.

In June 2008, TRB organized a Joint Workshop on Abutment Scour: Present knowledge and future needs. This workshop included the Principal Investigators and Panels from NCHRP 24-15, 24-20, and 24-27 with the primary objective of discussing the state of knowledge on bridge abutment scour. A secondary objective was to make recommendations on potential changes to HEC-18, not just in the abutment scour area, but across the board. The results were published in NCHRP Research Results Digest 334 (TRB 2009).

Based on updated policy guidance and the studies listed above, this edition of HEC-18 contains:

- Expanded discussion on the policy and regulatory basis for the FHWA Scour Program, including risk-based approaches for evaluations, and developing Plans of Action (POAs) for scour critical bridges design philosophies, and technical approaches
- Expanded discussion on countermeasure design philosophy (new vs. existing bridges)
- New chapter on soils, rock and geotechnical considerations related to scour
- New section on contraction scour in cohesive materials
- Updated abutment scour section
- Alternative abutment design approaches
- Alternative procedures for estimating pier scour

- New guidance on pier scour with debris loading
- New approach for pier scour in coarse material
- New section on pier scour in cohesive materials
- New section on pier scour in erodible rock
- New guidance on scour at wide and skewed piers
- Revised guidance for vertical contraction scour (pressure flow) conditions
- Guidance for predicting scour at bottomless culverts
- Deletion of the "General Scour" term
- Revised discussion on scour at tidal bridges to reflect material now covered in HEC-25 (1st and 2nd Editions)

1.6 MANUAL ORGANIZATION

The procedures presented in this document contain the state-of-knowledge and practice for dealing with scour at highway bridges.

- Chapter 1 gives the background of the scour problem, a flowchart for a comprehensive analysis using HEC-18, HEC-20, and HEC-23, procedural guidance, and sources for advances in the state-of-practice.
- Chapter 2 gives updated policy guidance and recommendations for designing bridges to resist scour.
- Basic concepts and definitions for scour are presented in Chapter 3.
- Geotechnical considerations relevant to estimating bridge scour are summarized in Chapter 4.
- Methods for estimating long-term aggradation and degradation are given in Chapter 5.
- Chapter 6 provides procedures and equations for determining contraction scour.
- Chapter 7 provides equations for calculating and evaluating local scour depths at piers, including scour from debris and scour in cohesive materials and erodible rock.
- Chapter 8 discusses local scour at abutments, equations for predicting scour depths at abutments, and alternative abutment design approaches.
- Chapter 9 provides an introduction to tidal processes and scour analysis methods for bridges over tidal waterways.
- Chapter 10 explains how the National Scour Evaluation program relates to the National Bridge Inspection Standards (NBIS) and presents guidelines for inspecting bridges for scour. The need for and details of a Plan of Action for scour critical bridges are also addressed.
- Chapter 11 lists references cited.
- Appendix D provides a comprehensive example of scour analysis for a river crossing.

1.7 DUAL SYSTEM OF UNITS

This edition of HEC-18 uses dual units (English and SI metric). The "English" system of units as used throughout this manual refers to U.S. Customary units. **In Appendix A, the metric (SI) unit of measurement is explained. The conversion factors, physical properties of water in the SI and English systems of units, sediment particle size grade scale, and some common equivalent hydraulic units are also given.** This edition uses for the unit of length the foot (ft) or meter (m); of mass the slug or kilogram (kg); of weight/force the pound (lb) or newton (N); of pressure the lb/ft^2 or Pascal (Pa, N/m^2); and of temperature degrees Fahrenheit (°F) or Centigrade (°C). The unit of time is the same in English as in SI system (seconds, s). Sediment particle size is given in millimeters (mm), but in calculations the decimal equivalent of millimeters in meters is used (1 mm = 0.001 m) or for the English system feet (ft). The value of some hydraulic engineering terms used in the text in English units and their equivalent SI units are given in Table 1.1.

Table 1.1. Commonly Used Engineering Terms in English and SI Units.		
Term	English Units	SI Units
Length	3.28 ft	1 m
Volume	35.31 ft^3	1 m^3
Discharge	35.31 ft^3/s	1 m^3/s
Acceleration of Gravity	32.2 ft/s^2	9.81 m/s^2
Unit Weight of Water	62.4 lb/ft^3	9800 N/m^3
Density of Water	1.94 slugs/ft^3	1000 kg/m^3
Density of Quartz	5.14 slugs/ft^3	2647 kg/m^3
Specific Gravity of Quartz	2.65	2.65
Specific Gravity of Water	1	1
Temperature	°F	°C = 5/9 (°F - 32)

CHAPTER 2

DESIGNING AND EVALUATING BRIDGES TO RESIST SCOUR

2.1 SCOUR DESIGN PHILOSOPHY AND CONCEPTS FOR NEW BRIDGES

Bridge foundations for **new** bridges should be designed to withstand the effects of scour caused by hydraulic conditions from floods larger than the design flood. In 2010, the U.S. Congress recommended that FHWA apply risk-based and data-driven approaches to infrastructure initiatives and other FHWA bridge program goals. This included the FHWA Scour Program. Risk-based approaches factor in the importance of the structure and are defined by the need to provide safe and reliable waterway crossings and consider the economic consequences of failure. For example, principles of economic analysis and experience with actual flood damage indicate that it is almost always cost-effective to provide a foundation that will not fail, even from very large events. However, for smaller bridges designed for lower frequency floods that have lower consequences of failure, it may not be necessary or cost effective to design the bridge foundation to withstand the effects of extraordinarily large floods. Prior to the use of these risk-based approaches, all bridges would have been designed for scour using the Q_{100} flood magnitude and then checked with the Q_{500} flood magnitude. Table 2.1 presents recommended **minimum** scour design flood frequencies and scour design check flood frequencies based on hydraulic design flood frequencies.

Table 2.1. Hydraulic Design, Scour Design, and Scour Design Check Flood Frequencies.		
Hydraulic Design Flood Frequency, Q_D	Scour Design Flood Frequency, Q_S	Scour Design Check Flood Frequency, Q_C
Q_{10}	Q_{25}	Q_{50}
Q_{25}	Q_{50}	Q_{100}
Q_{50}	Q_{100}	Q_{200}
Q_{100}	Q_{200}	Q_{500}

The Hydraulic Design Flood Frequencies outlined in Table 2.1 assume an inherent level of risk. There is a direct association between the level of risk that is assumed to be acceptable at a structure as defined by an agency's standards and the frequency of the floods they are designed to accommodate.

The Scour Design Flood Frequencies presented in Table 2.1 are larger than the Hydraulic Design Flood Frequencies because there is a reasonably high liklihood that the hydraulic design flood will be exceeded during the service life of the bridge. For example, using Table B.1 (Appendix B) on "Probability of Flood Exceedance of Various Flood Levels" it can be seen that during a 50-year design life there is a 39.5 percent chance that a bridge designed to pass the Q_{100} flood will experience that flood or one that is larger. Similarly, there is a 63.6 percent chance that a bridge that is designed to pass the Q_{50} flood will experience that or a larger flood during a 50-year design life. Using the larger values for the Scour Design Flood Frequency for the 200-year flood and a 50-year design life reduces the exceedance value to 22.2 percent. This is considered to be an acceptable level of risk reduction. In other words, a bridge **must** be designed to a higher level for scour than for the hydraulic design because if the hydraulic design flood is exceeded then a greater amount of scour will occur which could lead to bridge failure. Also, designing for a higher level of scour than the hydraulic design flood ensures a level of redundancy after the hydraulic design event occurs.

The Scour Design Check Flood Frequencies are larger than the Scour Design Flood Frequencies using the same logic and for the same reasons as outlined above.

If there is a flood event greater than the Hydraulic Design Flood but less than the Scour Design Flood that causes greater stresses on the bridge, e.g., overtopping flood, it should be used as the Scour Design Flood. For this condition there would not be a Scour Design Check Flood since the overtopping flood is the one that causes the greatest stress on the bridge. Similarly, if there is a flood event greater than the Scour Design Flood but less than the Scour Design Check Flood that causes greater stresses on the bridge, it should be used as the Scour Design Check Flood. Balancing the risk of failure from hydraulic and scour events against providing safe, reliable, and economic waterway crossings requires careful evaluation of the hydraulic, structural, and geotechnical aspects of bridge foundation design.

Guidance in this chapter is based on the following concepts:

1. The foundation should be designed by an interdisciplinary team of engineers with expertise in hydraulic, geotechnical, and structural design.

2. Hydraulic studies of bridge sites are a necessary part of a bridge design. These studies should address both the sizing of the bridge waterway opening to minimize adverse impacts to upstream and downstream landowners and are required such that the foundations can be designed to be safe from scour. The scope of the hydraulic analysis should be commensurate with the complexity of the situation, the importance of the highway, and consequences of failure.

3. Consideration must be given to the limitations and gaps in existing knowledge when using currently available formulas for estimating scour. **The interdisciplinary team needs to apply engineering judgment in comparing results obtained from scour computations with available hydrologic and hydraulic data and conditions at the site to achieve a reasonable and prudent design.** Such data should include:

    a. Performance of existing structures during past floods,
    b. Effects of regulation and control of flood discharges,
    c. Hydrologic characteristics and flood history of the stream and similar streams, and
    d. Whether the bridge has redundant structural elements.

4. It must be recognized that occasional damage to highway approaches from rare floods can be repaired quickly to restore traffic service. On the other hand, a bridge which collapses or suffers major structural damage from scour can create safety hazards to motorists as well as significant social impacts and economic losses over a longer period of time. Aside from the costs to the DOTs of replacing or repairing the bridge and constructing and maintaining detours, there can be significant costs to communities or entire regions due to additional detour travel time, inconvenience, and lost business opportunities. Therefore, a higher hydraulic standard is warranted for the design of bridge foundations to resist scour than is usually required for sizing of the bridge waterway. These concepts are reflected in the following general design procedure.

2.2 GENERAL DESIGN PROCEDURE

The general design procedure for scour outlined in the following steps is recommended for determining bridge type, size, and location (TS&L) of substructure units:

Step 1. Select the flood event(s) that are expected to produce appropriately severe scour conditions. Balancing risk of failure against safety, reliability, and economic requirements suggests that scour should be evaluated for an event larger than the hydraulic design flood. For example, if a bridge is designed for a hydraulic capacity of Q_{25}, then the Scour Design Flood Frequency would be for a Q_{50} flood and the Scour Design Check Flood Frequency would be the Q_{100} flood. In all cases, if there is an overtopping event that causes greater hydraulic stresses to the bridge than the hydraulic design event then that flood should be used for computing scour and designing the foundations. Overtopping refers to flow over the approach embankment(s), the bridge itself, or both. See Appendix B for a discussion of extreme event combinations and design flood exceedance probabilities.

Step 2. Develop hydraulic parameters necessary to estimate scour for the flood flows in Step 1. This is typically done by the application of a one- or two-dimensional hydraulic model. Care must be taken to evaluate the full range of hydraulic conditions that could impact the flow conditions at and near the bridge being designed. These conditions could include the effects of downstream tail water, confluences with other streams, etc.. For one-dimensional hydraulic analysis the U.S. Army Corps of Engineers (USACE) Hydrologic Engineering Center's River Analysis System (HEC-RAS) is recommended for this task (USACE 2010a). For bridges with complex flow characteristics such as flow on embankments skewed to the flood flows, multiple floodplain openings, wide flood plains, highly contracted flows, etc., it is recommended that the FHWA's FST2DH (FHWA 2003b) two-dimensional hydraulic analysis model be used.

Step 3. Using the six-step Specific Design Approach for Scour in Section 2.4, estimate total scour for the hydraulic conditions identified from Steps 1 and 2 above. The resulting scour computed from the selected flood event should be considered in the design of a foundation. For this condition, minimum geotechnical safety factors commonly accepted by FHWA, AASHTO, and DOTs should be applied. For example, for a pile designed to have its bearing capacity through friction, a commonly applied factor of safety ranges from two to three.

Step 4. Plot the total scour depths obtained in Step 3 on a cross section of the stream channel and floodplain at the bridge site.

Step 5. Evaluate the results obtained in Steps 3 and 4 for reasonableness. Based on the judgment of a multi-disciplinary team comprised of hydraulic, geotechnical, and structural engineers, the reasonableness of the results **must** be evaluated. There are many factors that could affect the magnitude of the overall scour estimate. They could include storm duration, erodibility of channel materials, flow conditions, ice and debris, and many others. In order to assure the most reliable estimates of scour, one must also have an understanding of the theory and development of the procedures used to determine scour. **Based on the factors mentioned above, the scour depth(s) adopted for use in design may differ from the computed value(s).**

Step 6. Evaluate the proposed bridge size, configuration, and foundation elements on the basis of the scour analysis performed in Steps 3 through 5. Modify the design as necessary based on the following discussion.

　　1. Develop an understanding of the overall flood flow pattern at the bridge site for

the design conditions. Also, develop an understanding of the dynamic channel and floodplain characteristics for the reach of stream that contains the bridge. Use the understanding of these factors to identify those bridge elements most vulnerable to flood flows, channel change, and resulting scour.

2. To the extent possible, modify components of the bridge length, location, configuration, and sub-structure elements to minimize scour. The following factors can lead to reduced scour depths.

   a. Increase the bridge length. Increasing the bridge length generally reduces depths of flow and velocities through the bridge opening which reduces the magnitude of scour.

   b. Locate new or replacement bridges such as they experience as little scour as possible. This means the bridge would cross the flood plain as perpendicular as possible to the flood flows and would be located in the flood plain where the conveyance is highest.

   c. Provide substructure elements that are not as susceptible to scour as others. For example, piers that are aligned with the flow do not experience as much scour as piers that are not aligned with the flow. Also, certain pier configurations are not as susceptible to scour as others. Round nosed piers are not as susceptible to scour as square nosed piers and circular piers are not as susceptible to scour caused from non-aligned flow as are solid-wall piers.

   d. Design and install guide banks to reduce scour at the abutments. Guide banks help align flow with the abutments and minimize the adverse flow conditions at the abutments that contribute to scour. Guide banks also provide an additional benefit in that they make the bridge opening more hydraulically efficient.

Step 7. Perform the bridge foundation analysis on the basis that all streambed material in the scour prism above the total scour line (Step 4) has been removed and is not available for bearing or lateral support. All foundations should be designed in accordance with the AASHTO Standard Specifications for Highway Bridges (AASHTO 1992a). In the case of a pile foundation, the piling should be designed for additional lateral restraint and column action because of the increase in unsupported pile length during and after scour. In areas where the local scour is confined to the proximity of the footing, the lateral ground stresses on the pile length which remains embedded may be significantly reduced from the pre-local scour conditions.

1. <u>Spread Footings on Soil - Piers</u>

   a. At piers, ensure that the top of the footing is referenced to the thalweg of the channel and is below the sum of the long-term degradation and contraction scour, and considers the potential for lateral channel migration.

   b. Place the bottom of the footing below the total scour line from Step 4

c. In some cases, the top of the footing can act as a local scour arrester. **Additional analysis must be undertaken before reducing the amount of local scour at a pier based on the assumption that the top of the footing can act as a local scour arrester.**

2. Spread Footings on Soil – Abutments

   a. At abutments, ensure that the top of the footing is referenced to the thalweg of the channel or appropriate overbank elevation (if the abutment is significantly setback from the channel and it can be assured that the channel will not migrate) and is below the sum of the long-term degradation and contraction scour. **This approach requires the use of a designed countermeasure to prevent scour from developing at the base of the abutment.** Procedures for designing abutment scour countermeasures can be found in HEC-23, Bridge Scour and Stream Instability Countermeasures Experience, Selection, and Design Guidance Third Edition, Volumes 1 and 2 (FHWA 2009).

   Abutment foundations can also be designed without the use of a designed countermeasure but will need to be placed at an elevation where the top of the spread footing corresponds to the computed abutment scour depth plus long term degradation referenced to the thalweg of the channel or appropriate overbank elevation (if the abutment is significantly setback from the channel and it can be assured that the channel will not migrate). In practice this is rarely done due to the depth and foundation expense that are necessary to meet the above criteria.

3. Spread Footings On Rock Highly Resistant To Scour

   Place the bottom of the footing directly on the cleaned rock surface for massive rock formations (such as granite) that are highly resistant to scour. Small embedments (keying) should be avoided since blasting or chiseling to achieve keying frequently damages the sub-footing rock structure and makes it more susceptible to scour. If footings on smooth massive rock surfaces require lateral constraint, steel dowels should be drilled and grouted into the rock below the footing level.

4. Spread Footings On Erodible Rock

   a. Weathered or other potentially erodible rock formations need to be carefully assessed for scour. An engineering geologist familiar with the area geology should be consulted to determine if rock or soil or other criteria should be used to calculate the support for the spread footing foundation. The decision should be based on an analysis of intact rock cores, including rock quality designations and local geology, as well as hydraulic data and anticipated structure life. An important consideration may be the existence of a high quality rock formation below a thin weathered zone.

   b. For deep deposits of weathered rock, the potential scour depth should be estimated (Steps 4 and 5) and the footing base placed below that depth. Excavation into weathered rock should be made with care. If blasting is required, light, closely spaced charges should be used to minimize overbreak beneath the footing level. Loose rock pieces should be removed and the zone filled with clean concrete. In any event, the final footing should be poured in contact with the sides of the excavation for the full designed footing thickness to minimize water intrusion below footing level. Guidance on scourability of rock formations is given in FHWA memorandum "Scourability of Rock Formations" dated July 19, 1991 and the results of NCHRP Project 24-29 (NCHRP 2011e).

5. Spread Footings Placed On Tremie Seals And Supported On Soil

   a. Ensure that the top of the footing is below the sum of the long-term degradation, contraction scour, and considers the potential for lateral channel migration.

   b. Place the bottom of the footing below the total scour line as determined in Step 4.

6. For Deep Foundations (Drilled Shaft And Driven Piling) With Footings Or Caps

   Placing the top of the footing or pile cap below the streambed at a depth equal to the estimated long-term degradation and contraction scour depth will minimize obstruction to flood flows and resulting local scour. Even lower footing elevations may be desirable for pile supported footings when the piles could be damaged by erosion and corrosion from exposure to riverine flows or tidal currents. For more discussion on pile and drilled shaft foundations, see the manuals on Design and Construction of Driven Pile Foundations and Drilled Shafts (FHWA 2005) and (FHWA 2010).

7. Stub Abutments on Piling

   Stub abutments positioned in the embankment should be founded on piling driven to a depth that assumes an unsupported pile length above an elevation that is referenced to the thalweg of the channel and long-term degradation. The potential for lateral channel migration must also be considered. If an abutment is set back far enough from the main channel that it can be determined with a reasonably high degree of certainty that it will not migrate to the abutment, then only the flow characteristics and scour components in the overbank should be used to determine the abutment foundation piling elevations.

Step 8. Repeat the procedure in Steps 2 through 6 above and calculate the scour for a Scour Design Check Flood (See Table 2.1). The foundation design determined under Step 7 should be reevaluated for the Scour Design Check Flood and, if necessary, make design modifications when required.

   a. Check to make sure that the bottom of spread footings on soil or weathered rock is below the total scour depth for the Scour Design Check Flood.

   b. **All foundations should have a minimum factor of safety of 1.0 (ultimate load) under the Scour Design Check Flood conditions.** Note that in actual practice, the calculations for Step 8 would be performed concurrently with Steps 1 through 7 for efficiency of operation.

2.3 DESIGN CONSIDERATIONS

The information presented in this section presents practices and ideas that eliminate or minimize scour at bridges. There are factors that can be considered for the superstructure, piers, and abutments that can minimize or eliminate the scour that could occur during floods.

2.3.1 General

1. Raise the bridge superstructure elevation above the general elevation of the approach roadways wherever practicable. This provides for overtopping of approach embankments and relief from the hydraulic forces acting at the bridge. This is particularly important for streams carrying large amounts of debris which could clog the waterway at the bridge. It is recommended that the elevation of the lower cord of the bridge be increased a minimum of 3 feet (.9 m) above the normal freeboard for the Hydraulic Design Frequency Flood for streams that carry a large amount of debris.

2. Superstructures should be securely anchored to the substructure if the potential to trap air in or under the superstructure exists that can cause the bridge elements to be buoyant. The superstructure elements should be vented to allow trapped air to escape and help avoid buoyancy. Superstructures should also be anchored if debris and ice forces are probable. Further, the superstructure should be shallow and open to minimize resistance to the flow where overtopping is likely.

3. Continuous span bridges withstand forces due to scour and resultant foundation movement better than simple span bridges. Continuous spans provide alternate load paths (redundancy) for unbalanced forces caused by settlement and/or rotation of the foundations. This type of structural design is recommended for bridges where there is a significant scour potential.

4. Local scour holes at piers and abutments may overlap one another in some instances. If local scour holes do overlap, the scour is indeterminate and may possibly be deeper than independent estimates at one or the other. The topwidth of a local scour hole on each side of the pier ranges from 1.0 to 2.8 times the depth of local scour. A top width value of 2.0 times the depth of local scour on each side of a pier is suggested for practical applications.

5. For pile and drilled shaft supported substructures subjected to scour, a reevaluation of the foundation design may require a change in the pile or shaft length, number, cross-sectional dimension and type based on the loading and performance requirements and site-specific conditions.

6. At some bridge sites, hydraulics and traffic conditions may necessitate consideration of a bridge that will be partially or even totally inundated during high flows. This consideration results in submerged-flow vertical contraction scour through the bridge waterway. Section 6.10 of Chapter 6 has a discussion on submerged-flow vertical contraction scour for these cases.

2.3.2 Piers

1. Since the thalweg of channels migrate within a bridge opening, all piers in the main channel should be designed to the same elevation. Pier foundations on floodplains should be designed to the same elevation as pier foundations in the stream channel unless it can be determined with a reasonable degree of certainty over the life of the bridge that the overbanks are stable and the main channel will not migrate toward the overbank areas.

2. Align piers with the direction of flood flows. Assess the hydraulic advantages of circular piers, particularly where flood patterns are complex and change with flood stage.

3. Streamline piers to decrease scour and minimize potential for buildup of ice and debris. Use ice and debris deflectors where appropriate.

4. Evaluate the hazards of ice and debris buildup when considering use of multiple pile bents in stream channels. Where ice and debris buildup is a problem, consider the bent a solid pier for purposes of estimating scour. Consider use of other pier types where clogging of the waterway area could be a major problem.

5. Scour analyses of piers near abutments need to consider the potential of larger velocities and higher skew angles from the flow coming around the abutment.

2.3.3 Abutments

1. The methods used to estimate the magnitude of abutment scour were developed in a laboratory under ideal conditions and for the most part lack field verification. Because conditions in the field are different from those in the laboratory, these methods can over predict the magnitude of scour that may be expected to develop. Recognizing this, it is recommended that one of several approaches for accommodating abutment scour be used to assure that abutments or the fill material placed around them does not fail.

    a. The first and most widely used method relies on the use of a **designed scour countermeasure** to keep scour from developing at the base of the abutment or adjacent embankments. This method provides an advantage in that a reasonable and cost effective approach for determining abutment foundation depth is used, but relies on a properly designed and inspected scour countermeasure. Procedures for designing and configuring scour countermeasures can be found in HEC-23, Bridge Scour and Stream Instability Countermeasures Experience, Selection, and Design Guidance Third Edition, Volumes 1 and 2 (FHWA 2009).

    b. The second method assumes all embankment fill material has washed away and that the abutment essentially behaves as a pier. This method provides an advantage in that the failed embankment can be more easily repaired than a failed abutment, but provides a disadvantage due to the adverse flow conditions in the floodplain and channel near the abutment. Consequently, treating the abutment as a pier and estimating scour accordingly could lead to deep foundation depths. Information on computing scour for pier foundations is found in Chapter 7 of this manual.

    c. The third method relies on using procedures specifically developed for estimating abutment scour. If these methods are used it is imperative that the hydraulic variables used by the empirical methods be accurately and realistically determined. Information on computing abutment scour is presented in Chapter 8 of this manual.

Engineering judgment must be used to determine which approach provides the most reasonable and cost effective results and if the abutment foundation should be protected with a designed countermeasure, treated as a pier, or be designed using one of the empirical methods presented in Chapter 8 of this manual.

1. Relief bridges, guide banks, and river training works should be used, where needed, to minimize the effects of adverse flow conditions at abutments.

2. Where ice build-up is likely to be a problem, set the toe of spill-through slopes or vertical abutments back from the edge of the channel bank to facilitate the passage of ice.

3. Wherever possible, use sloping spill-through abutments. Scour at spill-through abutments is about 50 percent of the scour that occurs at vertical wall abutments.

4. Use riprap or other bank protection methods on the upstream side of an abutment or approach embankment to protect against accelerating flows and on the downstream side of an abutment and approach embankment to protect them from erosion by flow expansions and wake vortices. A guide bank (upstream and/or downstream) is also a useful countermeasure for moving the scour location from the toe of the abutment to the toe of the guide bank.

2.3.4 Superstructures

The design of the superstructure can have a significant impact on scour at the foundation. Hydraulic forces that should be considered in the design of a bridge superstructure include buoyancy, drag, and impact from ice and floating debris. The configuration of the superstructure should be influenced by the highway profile, the probability of submergence, expected issues with ice and debris, and flow velocities, as well as the usual economic, structural and geometric considerations. Superstructures over waterways should provide structural redundancy by using continuous spans rather than simple spans.

<u>Buoyancy</u>. The weight of a submerged or partially submerged bridge superstructure is the weight of the superstructure less the weight of the volume of water displaced. The volume of water displaced may be much greater than the volume of the superstructure components if air is trapped between girders. Also, solid parapet rails and curbs on the bridge deck can increase the volume of water displaced and increase buoyant forces. The volume of air trapped under the superstructure can be reduced by providing vents through the deck between structural members. Superstructures should be anchored to piers to counter buoyant forces and to resist drag forces. Continuous span designs are also less susceptible to failure from buoyancy than simple span designs.

<u>Drag Forces</u>. Drag forces on a submerged or partially submerged superstructure can be calculated by Equation 2.1:

$$F_d = C_d \rho H \frac{V^2}{2} \tag{2.1}$$

where:

F_d = Drag force per unit of length of bridge, lb/ft (N/m)
C_d = Coefficient of drag (2.0 to 2.2)
ρ = Density of water, 1.94 slugs/ft^3 (1000 kg/m^3)
H = Depth of submergence, ft (m)
V = Velocity of flow, ft/s (m/s)

<u>Floating Debris and Ice</u>. Where bridges are damaged by debris and ice, it usually is due to accumulations or impacts against bridge components. Waterways may be partially or totally blocked by ice and debris, creating hydraulic conditions that cause or increase scour at pier foundations and bridge abutments, structural damage from impact and uplift, and overtopping of roadways and bridges. Floating debris is a common hydraulic problem at highway stream crossings nationwide. Debris hazards occur more frequently in unstable streams where bank erosion is active and in streams with mild to moderate slopes, as contrasted with headwater streams. Debris hazards are often associated with large floods,

and most debris is derived locally along the streambanks upstream from the bridge. After being mobilized, debris typically moves as individual logs which tend to concentrate in the thalweg of the stream. Hydraulic Engineering Circular 20 provides guidance for evaluating the abundance of debris upstream of a bridge crossing and then to implement mitigation measures, such as removal and or containment, to minimize potential problems during a major flood (HEC-20, FHWA 2012b).

Ice Forces. Superstructures may be subjected to impact forces from floating ice, static pressure from thermal movements, ice jams, or uplift from adhered ice in water of fluctuating levels. The latter is usually associated with relatively large bodies of water. Superstructures in these locations should normally be high enough to be unaffected. Research is needed to define the static and dynamic loads that can be expected from ice under various conditions of ice strength and streamflow.

In addition to forces imposed on bridge superstructures by ice loads, ice jams at bridges can cause exaggerated backwater and a sluicing action under the ice. There are numerous examples of foundation scour from this orifice flow under ice as well as superstructure damage and failure from ice forces. Accumulations of ice or drift may substantially increase local pier and abutment scour, especially if they are allowed to extend downward near the channel bed. Ice also has serious effects on bank stability. For example, ice may form in bank stabilization materials, and large quantities of rock and other material embedded in the ice may be floated downstream and dumped randomly when the ice breaks up. Banks are subjected to piping forces during the drawdown of water surface elevation after the breakup.

Debris Forces. Information regarding methods for computing forces imposed on bridge superstructures by floating debris is also lacking despite the fact that debris causes or contributes to many failures. Floating debris may consist of logs, trees, house trailers, automobiles, storage tanks, lumber, houses, and many other items representative of floodplain usage. This complicates the task of computing impact forces since the mass and the resistance to crushing of the debris contribute to the impact force.

A general equation for computing impact forces is:

$$F = M\,dv/dt = \frac{MV^2}{2S} \qquad (2.2)$$

where:

F = Impact imparted by the debris, lb (N)
M = Mass of the debris, slugs (kg)
S = Stopping distance, ft (m)
V = Velocity of the floating debris prior to impact, ft/s (m/s)

In addition to impact forces, a buildup of debris increases the effective depth of the superstructure and the drag coefficient may also be increased. Perhaps the most hazardous result of debris buildup is partial or total clogging of the waterway. This can result in a sluicing action of flow under the debris which can result in scour and foundation failure or a shift in the channel location from under the bridge.

2.4 DETAILED PROCEDURES AND SPECIFIC DESIGN APPROACH

2.4.1 Step 1: Determine Scour Analysis Variables

1. Determine the magnitude of the discharges for the floods in Steps 1 and 8 of the General Design Procedure in Section 2.2, including the overtopping flood when applicable. Most states have procedures for determining the floods necessary to design for scour, but if specific guidance is needed contact the U.S. Geological Survey Water Resources District Office for assistance or clarification. Experience has shown that the incipient overtopping discharge often puts the most stress on a bridge. However, special conditions (angle of attack, submerged-flow, decrease in velocity or discharge resulting from high flows overtopping approaches or going through relief bridges, ice jams, etc.) may cause a more severe condition for scour with a flow smaller than the overtopping or design flood.

2. Determine if there are current or possible future factors that will produce a combination of high discharge and/or low tailwater control. Assess whether or not bedrock or other controls (old diversion structures, erosion control checks, other bridges, etc.) are in place and might be lowered or removed. Determine whether or not there dams or locks downstream that would seasonally control the tailwater elevation. Assess whether or not there are there dams upstream or downstream that could control the elevation of the water surface at the bridge. Select the lowest reasonable downstream water-surface elevation and the largest discharge to estimate the greatest scour potential. Assess the distribution of the velocity and discharge for the design, scour design, and scour design check flows through the bridge opening. Also, consider the contraction and expansion of the flow in the bridge waterway, as well as present conditions and anticipated future changes in the river.

3. Determine the water-surface profiles for the discharges determined in Step 1. For routine situations use a one-dimensional model such as HEC-RAS and for complex flow situations use a two-dimensional hydraulic model such as FST2DH to evaluate flood conditions and determine hydraulic parameters. In some instances, the designer may wish to use sediment transport relationships and/or a sediment transport model to most accurately determine bed elevation changes. Hydraulic studies by the USACE, USGS, the Federal Emergency Management Agency (FEMA), etc. are potentially useful sources of hydraulic data to calibrate, verify, and evaluate results from the hydraulic models that are used. The engineer should anticipate future conditions at the bridge, in the upstream watershed, and at downstream water-surface elevation controls as outlined in HEC-20 (FHWA 2012b). From computer analysis and from other hydraulic studies, determine input variables such as the discharge, velocity and depth needed for the scour calculations.

4. Collect and summarize the following information as appropriate (see HEC-20 for a step-wise analysis procedures):

    a. Boring logs to define geologic substrata at the bridge site,

    b. Bed material size, gradation, and distribution in the bridge reach,

    c. Existing stream and floodplain cross section through the reach,

    d. Stream planform,

e. Watershed characteristics,

f. Scour history from other bridges in the area,

g. Slope of energy grade line upstream and downstream of the bridge,

h. History of flooding,

i. Location of bridge site with respect to other bridges in the area, confluence with tributaries close to the site, bed rock controls, man-made controls (dams, old check structures, river training works, etc.), and confluence with another stream downstream,

j. Character of the stream (perennial, flashy, intermittent, gradual peaks, etc.),

k. Geomorphology of the site (floodplain stream; crossing of a delta, youthful, mature or old age stream; crossing of an alluvial fan; meandering, straight or braided stream; etc.).

l. Erosion history of the stream,

m. Development history (consider present and future conditions) of the stream and watershed, collect maps, ground photographs, aerial photographs; interview local residents; check for water resource projects planned or contemplated,

n. Sand and gravel mining from the streambed or floodplain up- and downstream from site,

o. Other unanticipated factors not included in the above discussion that could affect the bridge, and

p. Make a qualitative evaluation of the site with an estimate of the potential for stream movement and its effect on the bridge.

2.4.2 Step 2: Determine the Magnitude of Long-Term Degradation or Aggradation

Using the information collected in Step 1, above, determine the magnitude of long-term degradation at the bridge. Use historic records, observational data, or other empirical methods to determine the potential for long-term degradation and then factor that value into the total scour depth. If the analysis concludes that there will be long-term aggradation, it should be noted in the records but should **not** be included in the total scour depth outlined in Step 6 below.

2.4.3 Step 3: Compute the Magnitude of Contraction Scour

Using the information collected in Step 1, above, compute the magnitude of the contraction scour using the equations and procedures in Chapter 6 of this manual.

2.4.4 Step 4: Compute the Magnitude of Local Scour at Piers

Using the information collected in Step 1, above, compute the magnitude of local pier scour using the equations and procedures in Chapter 7 of this manual.

2.4.5 Step 5: Determine the Foundation Elevation for Abutments

Using the information collected in Step 1, above, compute the magnitude of abutment scour (as appropriate) using the information and procedures in Chapter 8 of this manual.

2.4.6 Step 6: Plot the Total Scour Depths and Evaluate the Design

Plot the Total Scour Depths. On the cross section of the stream channel or other general floodplain at the bridge crossing, plot the estimate of long-term bed elevation change, contraction scour, and local scour at the piers and abutments (as appropriate). Use a distorted scale so that the scour determinations will be easy to evaluate. Make a sketch of any planform changes (lateral stream channel movement due to meander migration, etc.) that might be reasonably expected to occur.

1. Long-term elevation changes may be either aggradation or degradation. However, only degradation is considered in the total scour assessment.
2. Contraction scour is plotted from and below the long-term degradation line.
3. Pier scour is plotted from and below the contraction scour and long-term degradation lines.
4. Abutment scour (as appropriate) is plotted from and below the long-term degradation line.
5. Plot the depth of scour and scour hole width at each pier and/or abutment. Use 2.0 times the depth of local scour, y_s, to estimate scour hole width on each side of the pier and/or abutment.

Evaluate the Total Scour Depths.

1. Evaluate whether the computed scour depths are reasonable and consistent with the interdisciplinary team's previous experience, and engineering judgment. If not, carefully review the calculations and design assumptions in order to modify the depths. These possible modifications must reflect sound engineering judgment.

2. Evaluate whether the local scour holes from the piers or abutments overlap between spans. If so, local scour depths can be larger because the scour holes overlap. For new or replacement bridges, the length of the bridge opening should be reevaluated and the opening increased or the number of piers decreased as necessary to avoid overlapping scour holes.

3. Evaluate the impact from factors such as lateral movement of the stream, stream flow hydrograph, velocity and discharge distribution, movement of the thalweg, shifting of the flow direction, channel changes, type of stream, or other items.

4. Evaluate whether the calculated scour depths appear reasonable for the conditions in the field, relative to the laboratory conditions under which the equations were developed. The first thing to be done in evaluating the reasonableness of the scour results is to confirm that the results of the hydrologic and hydraulic analysis are reasonable and accurate. All of the methods used to compute scour rely on accurate input to the procedures.

    If the calculated scour depths appear too deep, consider an iterative approach for computing scour by recalculating the hydraulic variables after long-term degradation and/or contraction scour are accounted for. This may provide a more realistic scour evaluation and decrease the total scour depth.

5. Evaluate cost, safety, etc. Also, account for additional opening requirements and factors that may complicate the scour computations due to ice and/or debris effects.

6. In the design of bridge foundations, the foundation design cannot rely on material above the total scour line to provide load capacity.

Reevaluate the Bridge Design. Reevaluate the bridge design on the basis of the foregoing scour computations and evaluation. Revise the design as necessary. This evaluation should consider the following questions:

1. Is the waterway area large enough (e.g., are contraction, pier, and abutment scour amounts too large)?

2. Are the piers too close to each other or to the abutments (i.e., do the scour holes overlap)? Estimate the topwidth of a scour hole on each side of a pier at 2.0 times the depth of scour. If scour holes overlap, local scour can be deeper.

3. Is there a need for relief structures? If so, how large should they be?

4. Are bridge piers and abutments properly aligned with the flow and located properly in regard to the stream channel and floodplain?

5. Is the bridge crossing of the stream and floodplain in a desirable location? If the location presents problems:

    a. Can it be changed?
    b. Can river training works, guide banks, abutment setback from the channel, or relief bridges serve to provide for an acceptable flow pattern at the bridge?

6. Is the hydraulic study adequate to provide the necessary information for foundation design?

    a. Are flow patterns complex and should a two-dimensional, water-surface profile model be used for analysis?
    b. Is the foundation design safe and cost-effective?
    c. Is a physical model study needed/warranted?

2.5 SCOUR EVALUATION PHILOSOPHY AND CONCEPTS FOR EXISTING BRIDGES

2.5.1 Overview

There are many bridges in the United States that were not designed to withstand the effects of scour at the foundations. These bridges may or may not be in jeopardy of failing if they experience a flood. In addition, there are significant numbers of bridges that are located on a reach of stream where the channel geometry has changed or the channel has migrated away from the location where it was when the bridge was designed and constructed. There are also significant numbers of bridges in existence whose foundation types are unknown, which makes it nearly impossible to assess their vulnerability to scour unless further actions are taken.

The National Bridge Inspection Standards (NBIS) regulation, 23 CFR 650.313, requires that bridge owners identify bridges that are scour critical (coded 0, 1, 2, 3, or U in Item 113) and to prepare a Plan of Action (POA) to monitor and correct known and potential deficiencies. Bridge owners have been working on completing evaluations to determine which bridges over waterways are vulnerable to scour.

The information presented in Table 2.2 shows the status of the Bridge Scour Evaluation Program as of 2011 reporting year.

Table 2.2. Bridge Scour Evaluation Program Status - 2011.						
Total Number of Bridges	Factor	Interstate Bridges	NHS Bridges	Non NHS Bridges	Total	Percent of Total
493,473	Needing Evaluation	80	136	3,701	3,917	0.8%
493,473	Foundation Unknown	55	703	40,067	40,825	8.3%
493,473	Scour Critical	937	1,936	20,181	23,034	4.7%

A scour critical bridge is a bridge that is predicted to fail from a certain magnitude flood either from analysis or observation. Once a bridge has been determined to be scour critical, FHWA policy requires that a plan of action (POA) be developed for that bridge that initiates the implementation of corrective measures and/or monitoring. Measures that can make a bridge no longer scour critical include, bridge replacement, and design and installation of bridge scour countermeasures. **Monitoring, however, is not a long-term solution and does not make a scour critical bridge a non-scour critical bridge. In addition, it does not change the National Bridge Inspection Standards (NBIS) Item 113 Coding Guide rating from a scour critical rating to a non scour critical rating.** More information on Plans of Action for scour critical bridges can be found in Chapter 10.

2.5.2 Countermeasures for Scour Critical Bridges

As shown in Table 2.2, there are more than 23,000 scour critical bridges in the United States. Since these bridges are already in place, options for structural or physical modifications such as replacement or foundation strengthening are limited and expensive. Unless these bridges are programmed for replacement, they are ultimately going to require the design and installation of a scour countermeasure in accordance with Hydraulic Engineering Circular 23 (FHWA 2009). Table 2.3 presents recommended minimum scour countermeasure design flood frequencies based on hydraulic design and scour design flood frequencies.

Table 2.3. Hydraulic Design, Scour Design, and Scour Countermeasure Design Flood Frequencies.		
Hydraulic Design Flood Frequency (Q_D)	Scour Design Flood Frequency (Q_S)	Scour Countermeasure Design Flood Frequency (Q_{CM})
Q_{10}	Q_{25}	Q_{50}
Q_{25}	Q_{50}	Q_{100}
Q_{50}	Q_{100}	Q_{200}
Q_{100}	Q_{200}	Q_{500}

Using the same logic as outlined in Section 2.1 and presented in Section B.4 (Appendix B) on Design Flood Exceedance Probability, the Scour Countermeasure Design Flood Frequencies used for the design of bridge scour countermeasures recognizes that these designed countermeasures must be stable at floods larger than those associated with the Scour Design Flood Frequency.

If there is a flood event smaller than the Scour Countermeasure Design Flood that causes greater stresses on the bridge, e.g., overtopping flood, then it should be used as the Scour Countermeasure Design Flood.

2.5.3 Bridges With Unknown Foundations

Bridges with Item 113 coded U (unknown foundation) represent a unique subset of bridges that were originally exempted from being evaluated for scour vulnerability due to the lack of a process and guidance that would have allowed owners to determine the necessary foundation characteristics. Table 2.2 shows that there are more than 40,000 bridges with unknown foundations in the United States as of the 2011 reporting year. Since the foundation type of these bridges is unknown it is not possible to assess them for their scour vulnerability using conventional procedures. Therefore, each of these bridges must have a Plan of Action developed for it. FHWA guidance recommends that a risk-based approach be used to prioritize and determine the level of effort that should be devoted to developing POAs and determining corrective actions for these bridges. More information on bridges with unknown foundations can be found in Appendix E.

CHAPTER 3

BASIC CONCEPTS AND DEFINITIONS OF SCOUR

3.1 GENERAL

Scour is the result of the erosive action of flowing water, excavating and carrying away material from the bed and banks of streams and from around the piers and abutments of bridges. Different materials scour at different rates. Loose granular soils are rapidly eroded by flowing water, while cohesive or cemented soils are more scour-resistant. **However, ultimate scour in cohesive or cemented soils can be as deep as scour in sand-bed streams.** Under constant flow conditions, scour will reach maximum depth in sand- and gravel-bed material in hours; cohesive bed material in days; glacial till, sandstones, and shale in months; limestone in years, and dense granite in centuries. Under flow conditions typical of actual bridge crossings, several floods may be needed to attain maximum scour.

Determining the magnitude of scour is complicated by the cyclic nature of some scour processes. Scour can be deepest near the peak of a flood, but hardly visible as floodwaters recede and scour holes refill with sediment.

Designers and inspectors need to carefully study site-specific subsurface information in evaluating scour potential at bridges, giving particular attention to foundations on rock. Massive rock formations with few discontinuities are highly resistant to scour during the lifetime of a typical bridge.

All of the equations for estimating contraction and local scour are based on laboratory experiments with limited field verification. However, contraction and local scour depths at piers as deep as computed by these equations have been observed in the field. The equations recommended in this document are considered to be the most applicable for estimating scour depths.

A factor in scour at highway crossings and encroachments is whether it is **clear-water** or **live-bed** scour. Clear-water scour occurs where there is no transport of bed material upstream of the crossing or encroachment or the material being transported from the upstream reach is transported through the downstream reach at less than the capacity of the flow. Live-bed scour occurs where there is transport of bed material from the upstream reach into the crossing or encroachment. This subject is discussed further in Section 3.4.

This document presents procedures, equations, and methods to analyze scour in both riverine and coastal areas. In riverine environments, scour results from flow in one direction (downstream). In coastal areas, highways that cross waterways and/or encroach longitudinally on them are subject to tidal fluctuation and scour may result from flow in two directions. In waterways influenced by tidal fluctuations, flow velocities do not necessarily decrease as scour occurs and the waterway area increases. In tidal waterways as waterway area increases, the discharge may increase. This is in sharp contrast to riverine waterways where the principle of flow continuity and a constant discharge requires that velocity be inversely proportional to the waterway area. **However, the methods and equations for determining stream instability, scour and associated countermeasures can be applied to both riverine and coastal streams.** The difficulty in tidal streams is in determining the hydraulic parameters (such as discharge, velocity, and depth) that are to be used in the scour equations. Tidal scour is discussed in Chapter 9.

3.2 TOTAL SCOUR

Total scour at a highway crossing considers three primary components:

1. **Long-term degradation** of the river bed
2. **Contraction scour** at the bridge
3. **Local scour** at the piers or abutments

These three scour components are added to obtain the total scour at a pier or abutment. This assumes that each component occurs independent of the other. Considering the components additive adds some conservatism to the design. In addition, there are **other types of scour** that occur in specific situations as well as **lateral migration** of the stream that must be assessed when evaluating total scour at bridge piers and abutments.

3.2.1 Aggradation and Degradation

Aggradation and degradation are long-term streambed elevation changes due to natural or man-induced causes which can affect the reach of the river on which the bridge is located. Aggradation involves the deposition of material eroded from the channel or watershed upstream of the bridge, but is not considered a component of total scour. Degradation involves the lowering or scouring of the streambed over relatively long reaches due to a deficit in sediment supply from upstream and contributes to total scour.

3.2.2 Contraction Scour

Contraction scour is a lowering of the streambed across the stream or waterway bed at the bridge. This lowering may be uniform across the bed or non-uniform, that is, the depth of scour may be deeper in some parts of the cross section. Contraction scour results from contraction (or constriction) of the flow, which results in removal of material from the bed across all or most of the channel width. Contraction scour is different from long-term degradation in that contraction scour occurs in the vicinity of the constriction or bridge, may be cyclic, and/or related to the passing of a flood.

3.2.3 Local Scour

Local scour involves removal of material from around piers, abutments, spurs, and embankments. It is caused by an acceleration of flow and resulting vortices induced by obstructions to the flow.

3.2.4 Other Types of Scour

Other scour conditions such as flow around a bend where the scour may be concentrated near the outside of the bend, scour resulting from stream planform characteristics, scour at confluences, or a variable downstream control can also influence the total scour in a bridge reach.

3.2.5 Lateral Stream Migration

In addition to the types of scour mentioned above, naturally occurring lateral migration of the main channel of a stream within a floodplain may affect the stability of piers in a floodplain, erode abutments or the approach roadway, or change the total scour by changing the flow angle of attack at piers and abutments. Factors that affect lateral stream movement also affect the stability of a bridge foundation. These factors are the geomorphology of the stream, location of the crossing on the stream, flood characteristics, and the characteristics of the bed and bank materials (see HEC-20, and HDS 6) (FHWA 2012b, FHWA 2001).

The following sections provide a more detailed discussion of the various components of and contributions to total scour.

3.3 LONG-TERM STREAMBED ELEVATION CHANGES (AGGRADATION AND DEGRADATION)

Long-term bed elevation changes may be the natural trend of the stream or the result of some modification to the stream or watershed. The streambed may be aggrading, degrading, or in dynamic equilibrium above, below, or in the vicinity of the bridge crossing. Long-term aggradation and degradation do not include the cutting and filling of the streambed in the vicinity of the bridge that might occur during a runoff event (contraction and local scour). A long-term trend may change during the life of the bridge. These long-term changes are the result of modifications to the stream or watershed. Such changes may be the result of natural processes or human activities. The engineer must assess the present state of the stream and watershed and then evaluate potential future changes in the river system. From this assessment, the long-term streambed changes must be estimated. Methods to estimate long-term streambed elevation changes are discussed in Chapter 5. Since long-term streambed elevation changes are essentially a channel instability problem, quantitative techniques for estimating aggradation and degradation can be found in HEC-20 (FHWA 2012b). Detailed discussion of sediment transport and sediment continuity concepts can be found in HDS 6 (FHWA 2001).

3.4 CLEAR-WATER AND LIVE-BED SCOUR

There are two conditions for contraction and local scour: **clear-water** and **live-bed** scour. Clear-water scour occurs when there is no movement of the bed material in the flow upstream of the crossing or the bed material being transported in the upstream reach is transported in suspension through the scour hole at the pier or abutment at less than the capacity of the flow. At the pier or abutment the acceleration of the flow and vortices created by these obstructions cause the bed material around them to move. Live-bed scour occurs when there is transport of bed material from the upstream reach into the crossing. Live-bed local scour is cyclic in nature; that is, the scour hole that develops during the rising stage of a flood refills during the falling stage.

Typical clear-water scour situations include (1) coarse-bed material streams, (2) flat gradient streams during low flow, (3) local deposits of larger bed materials that are larger than the biggest fraction being transported by the flow (rock riprap is a special case of this situation), (4) armored streambeds where the only locations that tractive forces are adequate to penetrate the armor layer are at piers and/or abutments, and (5) vegetated channels or floodplain overbank areas.

During a flood event, bridges over streams with coarse-bed material are often subjected to clear-water scour at low discharges, live-bed scour at the higher discharges and then clear-water scour at the lower discharges on the falling stages. Clear-water scour reaches its maximum over a longer period of time than live-bed scour (Figure 3.1). This is because clear-water scour occurs mainly in coarse-bed material streams. In fact, local clear-water scour may not reach a maximum until after several floods. For example, maximum local clear-water pier scour is about 10 percent greater than the equilibrium local live-bed pier scour.

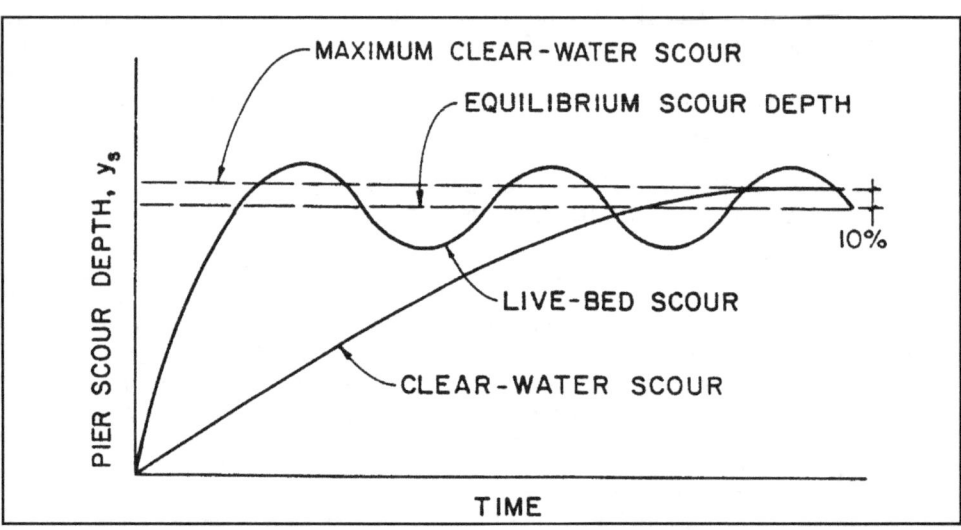

Figure 3.1. Pier scour depth in a sand-bed stream as a function of time.

Critical velocity equations with the reference particle size (D) equal to D_{50} are used to determine the velocity associated with the initiation of motion. They are used as an indicator for clear-water or live-bed scour conditions. If the mean velocity (V) in the upstream reach is equal to or less than the critical velocity (V_c) of the median diameter (D_{50}) of the bed material, then contraction and local scour will be clear-water scour. Also, if the ratio of the shear velocity of the flow to the fall velocity of the D_{50} of the bed material ($V*/\omega$) is greater than 2, contraction and local scour may be clear-water. If the mean velocity is greater than the critical velocity of the median bed material size, live-bed scour will occur. An equation to determine the critical velocity for a given flow depth and size of bed material is derived in Appendix C and given in Chapter 6.

This technique can be applied to any unvegetated channel or overbank area to determine whether scour is clear-water or live-bed. This procedure should be used with caution for assessing whether or not scour in the overbank will be clear-water or live-bed. For most cases, the presence of vegetation on the overbank will effectively bind and protect the overbank from erosive velocities. Also, in the overbank, generally the velocities are small and the bed material so fine (i.e., cohesive) that most overbank areas will experience clear-water scour.

For example, live-bed pier scour in sand-bed streams with a dune bed configuration fluctuates about the equilibrium scour depth as shown in Figure 3.1. This is due to the variability of the bed material sediment transport in the approach flow when the bed configuration of the stream is dunes. In this case (dune bed configuration in the channel upstream and through the bridge), maximum depth of pier scour is about 30 percent larger than equilibrium depth of scour. However, with the exception of crossings over large rivers (i.e., the Mississippi, Columbia, etc.), the bed configuration in sand-bed streams will plane out during flood flows due to the increase in velocity and shear stress. For general practice, the maximum depth of pier scour is approximately 10 percent greater than equilibrium scour.

For a discussion of bedforms in alluvial channel flow, see Chapter 3 of HDS 6 (FHWA 2001). Equations for estimating local scour at piers or abutments are given in Chapters 7 and 8 of this document. These equations were developed from laboratory experiments and limited field data for both clear-water and live-bed scour.

3.5 CONTRACTION SCOUR

3.5.1 Basic Conditions for Contraction Scour

Contraction scour occurs when the flow area of a stream at flood stage is reduced, either by a natural contraction (or constriction) of the stream channel or by a bridge. It also occurs when overbank flow is forced back to the channel by roadway embankments at the approaches to a bridge. From continuity, a decrease in flow area results in an increase in average velocity and bed shear stress through the contraction. Hence, there is an increase in erosive forces in the contraction and more bed material is removed from the contracted reach than is transported into the reach. This increase in transport of bed material from the reach lowers the natural bed elevation. As the bed elevation is lowered, the flow area increases and, in the riverine situation, the velocity and shear stress decrease until relative equilibrium is reached; i.e., the quantity of bed material that is transported into the reach is equal to that removed from the reach, or the bed shear stress is decreased to a value such that no sediment is transported out of the reach. Contraction scour, in a natural channel or at a bridge crossing, involves removal of material from the bed across all or most of the channel width. Methods to estimate live-bed and clear-water contraction scour are presented in Chapter 6.

In coastal waterways which are affected by tides, as the cross-sectional area increases the discharge from the ocean may increase and thus the velocity and shear stress may not decrease. Consequently, relative equilibrium may not be reached. Thus, at tidal inlets contraction scour may result in a continual lowering of the bed.

Live-bed contraction scour is typically cyclic; for example, the bed scours during the rising stage of a runoff event and fills on the falling stage. The cyclic nature of contraction scour causes difficulties in determining contraction scour depths after a flood. The contraction of flow at a bridge can be caused by either a natural decrease in flow area of the stream channel or by abutments projecting into the channel and/or piers blocking a portion of the flow area. Contraction can also be caused by the approaches to a bridge cutting off floodplain flow. This can cause clear-water scour on a setback portion of a bridge section or a relief bridge because the floodplain flow does not normally transport significant concentrations of bed material sediments. This clear-water picks up additional sediment from the bed upon reaching the bridge opening. In addition, local scour at abutments may well be greater due to the clear-water floodplain flow returning to the main channel at the end of the abutment.

Other factors that can cause contraction scour are (1) natural stream constrictions, (2) long highway approaches to the bridge over the floodplain, (3) ice formations or jams, (4) natural berms along the banks due to sediment deposits, (5) debris, (6) vegetative growth in the channel or floodplain, and (7) pressure flow.

3.6 LOCAL SCOUR

3.6.1 Scour at Bridge Piers and Abutments

The basic mechanism causing local scour at piers or abutments is the formation of vortices at their base. Figure 3.2 presents a simple schematic of the flow field at a narrow cylindrical pier. The horseshoe vortex at a bridge pier results from the pileup of water on the upstream surface of the obstruction and subsequent acceleration of the flow around the nose of the pier. The action of the vortex removes bed material from around the base of the pier. The transport rate of sediment away from the base region is greater than the transport rate into the region, and, consequently, a scour hole develops. As the depth of scour increases, the strength of the horseshoe vortex is reduced, thereby reducing the transport rate from the base region. Eventually, for live-bed local scour, equilibrium is reestablished between bed material inflow and outflow and scouring ceases. For clear-water scour, scouring ceases when the shear stress caused by the horseshoe vortex equals the critical shear stress of the sediment particles at the bottom of the scour hole.

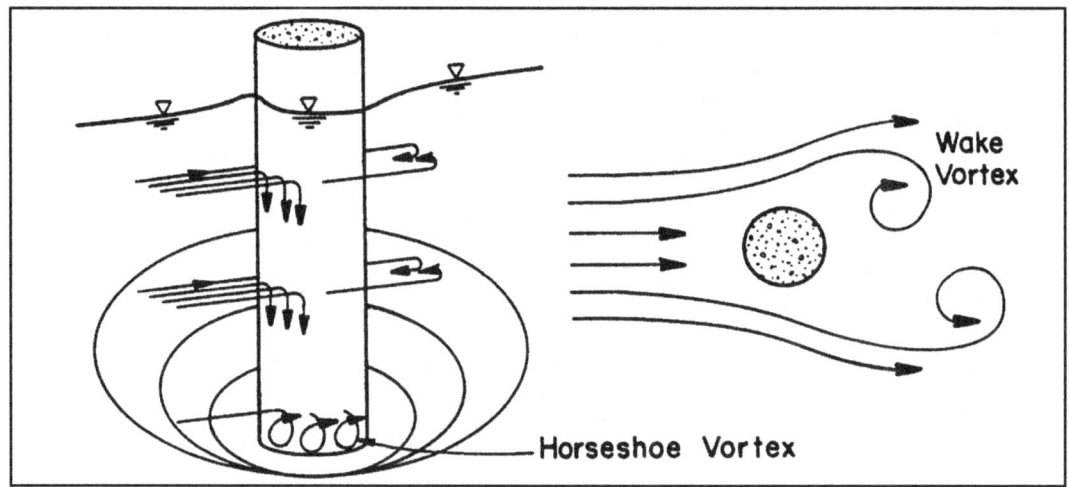

Figure 3.2. Simple schematic representation of scour at a cylindrical pier.

In addition to the horseshoe vortex around the base of a pier, there are vertical vortices downstream of the pier called the wake vortex (Figure 3.2). Both the horseshoe and wake vortices remove material from the pier base region. However, the intensity of wake vortices diminishes rapidly as the distance downstream of the pier increases. Therefore, immediately downstream of a long pier there is often deposition of material.

Factors which affect the magnitude of local scour depth at piers and abutments are (1) velocity of the approach flow, (2) depth of flow, (3) width of the pier, (4) discharge intercepted by the abutment and returned to the main channel at the abutment (in laboratory flumes this discharge is a function of projected length of an abutment into the flow), (5) length of the pier if skewed to flow, (6) size and gradation of bed material, (7) angle of attack of the approach flow to a pier or abutment, (8) shape of a pier or abutment, (9) bed configuration, and (10) ice formation or jams and debris.

1. Flow velocity affects local scour depth. The greater the velocity, the deeper the scour. There is a high probability that scour is affected by whether the flow is subcritical or supercritical. However, most research and data are for subcritical flow (i.e., flow with a Froude Number less than 1.0, Fr < 1).

2. Flow depth also has an influence on the depth of local scour. An increase in flow depth can increase scour depth by as much as a factor of 2 or greater for piers. With abutments, the increase is somewhat less depending on the shape of the abutment.

3. Pier width has a direct influence on depth of local scour. As pier width increases, there is an increase in scour depth. There is a limit to the increase in scour depth as width increases. Very wide piers (see Chapter 7) do not have scour depths as deep as predicted by existing equations.

4. In laboratory flume studies, an increase in the projected length of an abutment (or embankment) into the flow increased scour; whereas, this is not the case in the field. Due to the relatively small scale of a laboratory flume, floodplain flow intercepted by the embankment and returned to the main channel is directly related to the length of the obstruction. However, in the field case the embankment length is not a good measure of the discharge returned to the main channel. This results in "ineffective flow" on the floodplain which can be even more pronounced on wide heavily vegetated floodplains. In order to properly apply laboratory derived abutment scour equations to the field case, an assessment must be made of the location of the boundary between "live flow" and "ineffective flow." The location of this boundary should then be used to establish the length of the abutment or embankment for abutment scour computations (see Chapter 8).

5. Pier length has no appreciable effect on local scour depth as long as the pier is aligned with the flow. When the pier is skewed to the flow, the pier length has a significant influence on scour depth. For example, doubling the length of the pier increases scour depth from 30 to 60 percent (depending on the angle of attack).

6. Bed material characteristics such as size, gradation, and cohesion can affect local scour. Bed material in the sand-size range has little effect on local scour depth. Likewise, larger size bed material that can be moved by the flow or by the vortices and turbulence created by the pier or abutment will not affect the maximum scour, but only the time it takes to attain it. Very large particles in the bed material, such as coarse gravels, cobbles or boulders, may armor the scour hole. The size of the bed material also determines whether the scour at a pier or abutment is clear-water or live-bed scour. This topic is discussed in Section 3.4.

    Fine bed material (silts and clays) will have scour depths as deep as sand-bed streams. This is true even if bonded together by cohesion. The effect of cohesion is to influence the time it takes to reach maximum scour. With sand-bed material the time to reach maximum depth of scour is measured in hours and can result from a single flood event. With cohesive bed materials it may take much longer to reach the maximum scour depth, the result of many flood events.

7. Angle of attack of the flow to the pier or abutment has a significant effect on local scour, as was pointed out in the discussion of pier length. Abutment scour is reduced when embankments are angled downstream and increased when embankments are angled upstream. According to the work of Ahmad (1953), the maximum depth of scour at an embankment inclined 45 degrees downstream is reduced by 20 percent; whereas, the maximum scour at an embankment inclined 45 degrees upstream is increased about 10 percent.

8. Shape of the nose of a pier can have up to a 20 percent influence on scour depth. Streamlining the front end of a pier reduces the strength of the horseshoe vortex, thereby reducing scour depth. Streamlining the downstream end of piers reduces the strength of the wake vortices. A square-nose pier will have maximum scour depths about 20 percent greater than a sharp-nose pier and 10 percent greater than either a cylindrical or round-nose pier. The shape effect is negligible for flow angles in excess of five degrees. Full retaining abutments with vertical walls on the stream side (parallel to the flow) and vertical walls parallel to the roadway will produce scour depths about double that of spill-through (sloping) abutments.

9. Bed configuration of sand-bed channels affects the magnitude of local scour. In streams with sand-bed material, the shape of the bed (bed configuration) may be ripples, dunes, plane bed, or antidunes. The bed configuration depends on the size distribution of the sand-bed material, hydraulic characteristics, and fluid viscosity. The bed configuration may change from dunes to plane bed or antidunes during an increase in flow for a single flood event. It may change back with a decrease in flow. The bed configuration may also change with a change in water temperature or suspended sediment concentration of silts and clays. The type of bed configuration and change in bed configuration will affect flow velocity, sediment transport, and scour. HDS 6 discusses bed configuration in detail (FHWA 2001).

10. Potentially, ice and debris can increase the width of the piers, change the shape of piers and abutments, increase the projected length of an abutment, and cause the flow to plunge downward against the bed. This can increase both local and contraction scour. For pier scour, debris can be taken into account in the scour equations by estimating how much the debris will increase the effective width of a pier (see Chapter 7). Debris and ice effects on contraction scour can also be accounted for by estimating the amount of flow blockage (decrease in width of the bridge opening) in the equations for contraction scour. Limited field measurements of scour at ice jams indicate the scour can be as much as 10 to 30 ft (3 to 10 m).

3.6.2 Bridge Pier Flow Field

Since 1990, major progress has been made with numerical modeling of flow at piers. The three-dimensional Computational Fluid Dynamic (CFD) models available today can resolve all the main flow features and their unsteady interactions. Continued examination of the bridge pier flow field with data and flow visualization from hydraulic (laboratory) models has also shed light on the complexities of the bridge pier flow field. The observations and figures in this section were derived, primarily, from an investigation and evaluation of the results and applicability of recent bridge pier scour research on scour processes and estimation (NCHRP 2011a).

To understand pier scour, it is necessary to understand the flow field at a pier, and how the flow field varies with pier size and shape, as well as with flow depth and foundation material. A difficulty in this respect, however, is that the flow field is a class of junction flow (i.e., flow at the junction of a structural form and a base plane), a notably three-dimensional unsteady flow field marked by interacting turbulence structures (Ettema 1980). The eroding forces exerted on the foundation material supporting the pier are generated by flow contraction around the pier, by a pronounced down-flow at the pier's leading edge, and by turbulence structures of a wide range of turbulence scales. Variations of pier width and shape, and flow depth, alter the flow field, enhancing or weakening these flow features.

In terms of ranges of pier width, a, and flow depth, y, commonly encountered in the field, it is convenient to consider three categories of pier flow field, which produce significantly different pier scour morphologies:

1. Narrow piers ($y/a > 1.4$), for which scour typically is deepest at the pier face
2. Transitional piers ($0.2 < y/a < 1.4$)
3. Wide piers ($y/a < 0.2$), for which scour typically is deepest at the pier flank

The values of y/a indicated for the flow-field categories are based on data trends delineating differences in the relationship between scour depth and y/a (e.g., Melville and Coleman 2000).

The pier flow field may become more complicated if the pier has a complex shape, such as a column supported on a pile cap underpinned by a pile cluster. Additionally, the close proximity of an abutment and/or a channel bank further complicates the flow field.

Narrow Piers. The main features of the flow field at narrow piers can be explained by viewing the flow field and scour at an isolated circular cylindrical pier in a relatively deep, wide channel. Figure 3.3 illustrates the main features of the flow field for a pier founded in sediment, and conveys a sense of the complexities of the flow field to be considered when attempting to understand scour at a simple pier.

An interacting and unsteady set of flow features entrains and transports sediment from the pier foundation. They include: flow impact against the pier face, producing a down-flow and an up-flow with roller; flow converging, contracting, then diverging; the generation, transport and dissipation of large-scale turbulence structures (macro-turbulence) at the base of the pier-foundation junction (commonly termed the horseshoe vortex as in Figure 3.2); detaching shear layer at each pier flank; and, wake vortices convected through the pier's wake. The features evolve as scour develops.

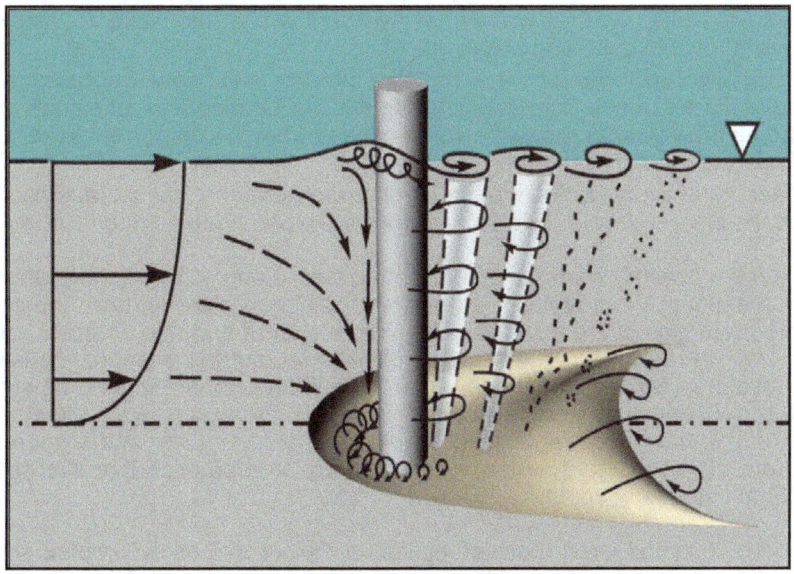

Figure 3.3. The main flow features forming the flow field at a narrow pier of circular cylindrical form (NCHRP 2011a).

Flow approaching the pier decelerates, impinges against the pier's centerline, and then strongly deflects both down and up the pier's face. These two vertical flows act almost as wall-attached jet-like flows along the pier's centerline, one directed up toward the free surface, and the other down toward the bed. The up-flow attains a height approximating a stagnation head, interacts with the free surface, and forms a surface roller or vortex. The stagnation pressure on the upstream face of the pier attains a maximum near the level where these two jet-like flows form. Also, at the stagnation line the deceleration is greatest. The deceleration decreases as the bed and, respectively, the free surface are approached. The down-flow is driven by the resulting downward gradient (below the still water level) of stagnation pressure along the pier's leading face. This downward gradient results largely because the velocity distribution of the approach flow is that of a fully turbulent shear flow; i.e., velocity generally decreases toward the bed. As the scour hole develops, the down-flow is augmented by the approach flow diverging into the scour hole (NCHRP 2011a).

In addition to the vertical component of flow at the pier's leading face, flow contracts as it passes around the sides of the pier and local values of flow velocity and bed shear stress increase. For many piers, the increases are such that scour begins at the sides of a pier. Once the scour region develops as a hole fully around the pier, the down-flow and the horseshoe vortices strengthen. Scour-hole formation draws flow into the hole.

The flow field, during all stages of scour development, is marked by the presence of organized, coherent turbulence structures, notably:

1. A horseshoe vortex system formed of several necklace vortices (standard term for junction flows) commonly termed the horseshoe vortex. It forms around the leading perimeter of the pier. These vortices wrap around the base of the pier such that the legs are oriented approximately parallel to the approaching flow. The legs break up and are shed intermittently;
2. Small but very energetic elongated eddies (vortex tubes whose main axis is approximately vertical relative to the bed) in the detached shear layers;
3. Large-scale rollers or wake vortices, which form behind the two flanks of the pier, and are shed into its wake. As they convect away from the pier, the wake vortices expand in diameter, then dissipate and break up;
4. A horizontal vortex formed by flow passing over the stationary, depositional mound formed at the exit slope from the scour hole. The location and amplitude of the mound depend on the power of the wake vortices shed from the pier (the weaker the vortices, the closer the mound to the pier); and,
5. A surface roller situated close to the junction between the free surface and the upstream face of the pier. The roller is akin to a bow wave of a boat moving through water.

In summary, the down-flow impingement on the bed, along with the wide range of turbulence structures present in the flow field, entrain and transport material from the scour hole. The details and interaction of the flow field vary with pier shape, angle of attack, and the stage of scour development between initiation and equilibrium, but the essential consideration is that these flow features are responsible for scour. Therefore, to understand how scour develops, to model scour, and to estimate scour depth it is necessary to understand the general structure of the flow field, and determine how flow entrains and transports foundation material from the scour hole. Also, it is important to recognize that the flow field evolves during different stages of scour.

The flow field becomes even more complicated if the pier has a complex shape, such as a column supported on a pile cap underpinned by a pile cluster. Additionally, the flow field can be complicated by debris or ice accumulation, the proximity of an abutment, and aspects of channel morphology.

Transition Piers. The main flow-field features described for narrow piers exist in the flow field of piers within the transition range of y/a, but the features now begin to alter in response to reductions of y and or increases in a. The closer proximity of the water surface to the foundation boundary (for constant pier width), or the increased width of a pier (for constant flow depth), partially disrupt the formation of the features, and thereby reduce their capacity to erode foundation material. Though further research is needed to systematically describe and document the flow field changes, ample data show that reducing y/a results in shallower scour depths for this transition category of flow field (NCHRP 2011a).

Figure 3.4 depicts a sequence of flow field adjustments commensurate with three values of y/a, indicating how the scour capacity of flow field reduces as the flow depth, y, and the value of y/a decrease. The down-flow at the pier face becomes less well developed because it has a shortened length over which to develop, whereas the up-flow associated with the (flow stagnation) bow wave remains essentially unchanged. The vorticity (circulation) of the large-scale turbulence structures (Horseshoe vortex) aligned more-or-less horizontally in the pier flow field weakens as the down-flow weakens, and the vertically aligned turbulence structures (wake vortices) also weaken due to the increased importance of bed friction in a shallower flow.

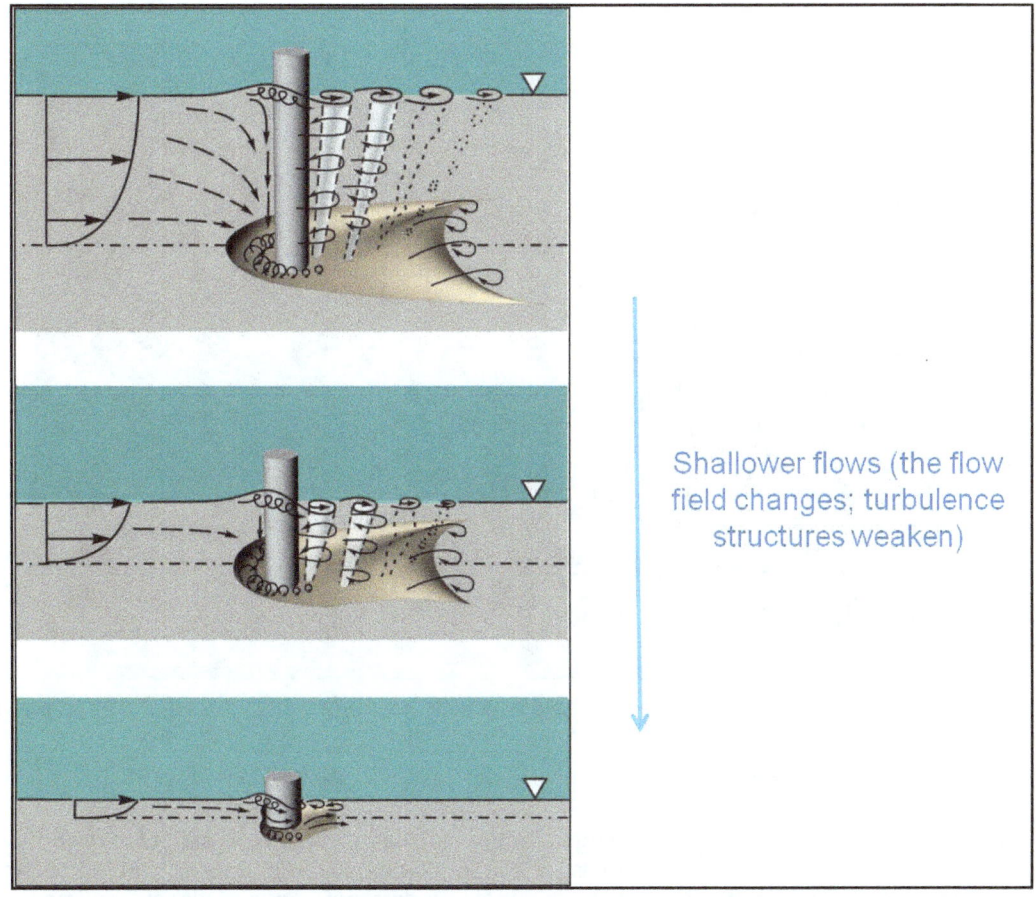

Figure 3.4. Variation of flow field with reducing approach flow depth; narrow to transitional pier of constant pier width (NCHRP 2011a).

Wide Piers. For wide piers, the flow approaching the pier decelerates, turns, and flows laterally along the pier face before contracting and passing around the sides of the pier. The down-flow at the pier face is weakly developed, and only slightly erodes the foundation at the pier centerplane. The circulation of the necklace vortices peaks at vertical sections situated around the flanks of the pier. Flow velocities near the pier are greatest where flow contracts around the pier's sides. Erosive turbulence structures now principally comprise wake vortices and the part of the horseshoe vortex system located in the scour region close to each flank of the pier. Deepest scour occurs at the pier flanks. Figure 3.5 illustrates schematically the flow field around a wide pier (NCHRP 2011a and c).

For a given flow depth, greater pier width increases flow blockage and therefore causes more of the approach flow to be swept laterally along the pier face than around the pier's flanks. Increased blockage modifies the lateral distribution of approach flow over a longer distance upstream of a pier.

The flow field around each side of a wide pier is essentially the same as that at an abutment built with a solid foundation extending with depth into the foundation material (also that at a long spur dike or coffer dam).

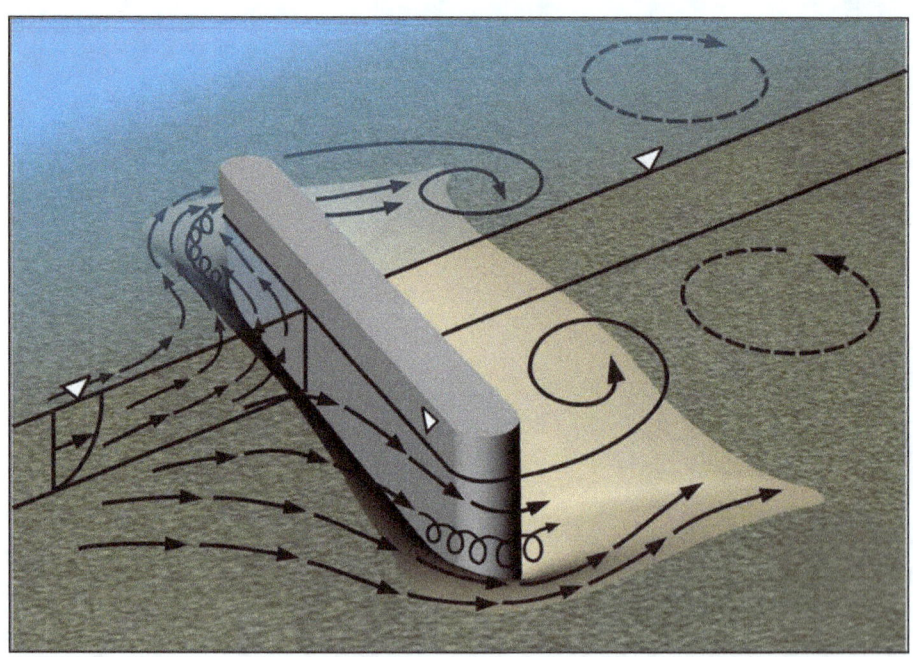

Figure 3.5. Main features of the flow field at a wide pier (y/a < 0.2) (NCHRP 2011a and c).

3.6.3 Bridge Abutment Flow Field

A typical bridge waterway crossing creates a highly complex set of flow/boundary interactions. The boundary materials of the main-channel, floodplain, and embankment components of a bridge-waterway usually constitute different zones of alluvial sediments and soil, as indicated in Figure 3.6. Abutment scour usually occurs within several zones of sediment and soil, leading to different erosion processes and varying rates of erosion. The observations and figures in this section were derived, primarily, from an investigation and evaluation of the results and applicability of bridge abutment scour research on scour processes and estimation (NCHRP 2011b).

Alluvial non-cohesive sediment (sands and gravels) most frequently forms the bed of the main channel, whereas the channel floodplain may be formed from considerably finer sediments (silts and clays), typically causing the floodplain soil to be more cohesive in character than the bed sediment of the main channel. The banks of the main channel usually are formed of the floodplain soils, and thus also may behave cohesively so as to stand at a fairly steep slope.

Most abutments have an earthfill approach embankment formed of compacted soils. The soils may have been excavated from the floodplain or have been brought to the bridge site from elsewhere. The earthfill embankment is placed and compacted to a specific value of shear strength so as to support the traffic load.

Flow through a bridge waterway narrowed by a bridge abutment and its embankment is essentially flow around a short streamwise contraction. Figure 3.7 illustrates schematically the characteristic flow features and the connection between the contraction and the formation of a complex flow field around the abutments. The flow width narrows and the flow accelerates through the contraction, generating macro-turbulence structures (eddies and various vortices spun from the contraction boundary) that shed and disperse within the flow. Flow contraction and turbulence at many bridge waterways, though, is complicated by the shape of the channel. It is common for waterways to traverse a compound channel formed of a deeper main channel flanked by floodplain channels, as shown in Figure 3.8. To a varying extent, all flow boundaries are erodible (NCHRP 2011b).

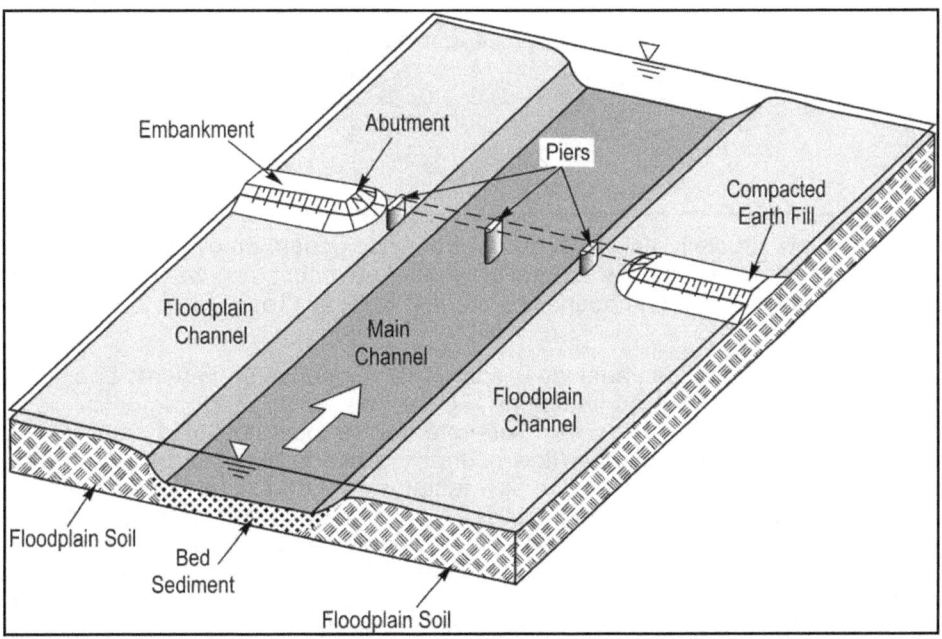

Figure 3.6. Variation of soil and sediment types at a bridge crossing (NCHRP 2011a and b).

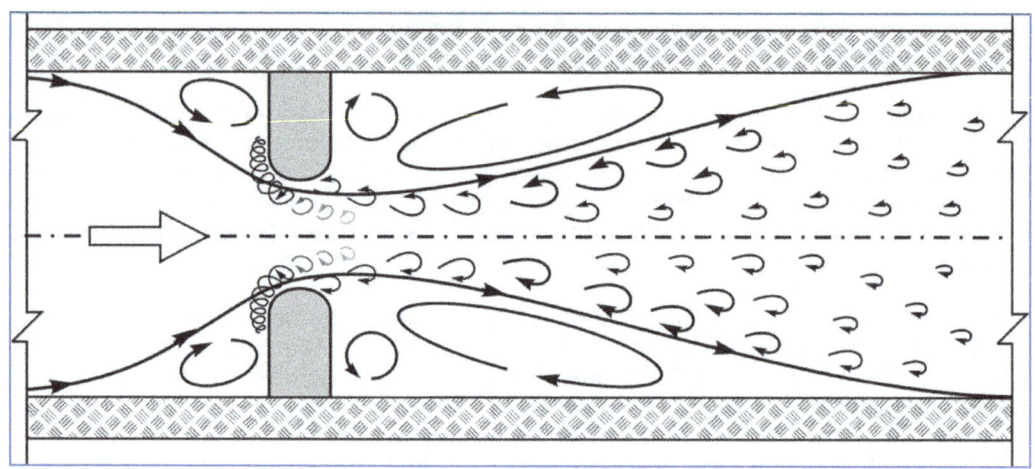

Figure 3.7. Flow structure including macro-turbulence generated by flow around abutments in a narrow main channel (NCHRP 2011b).

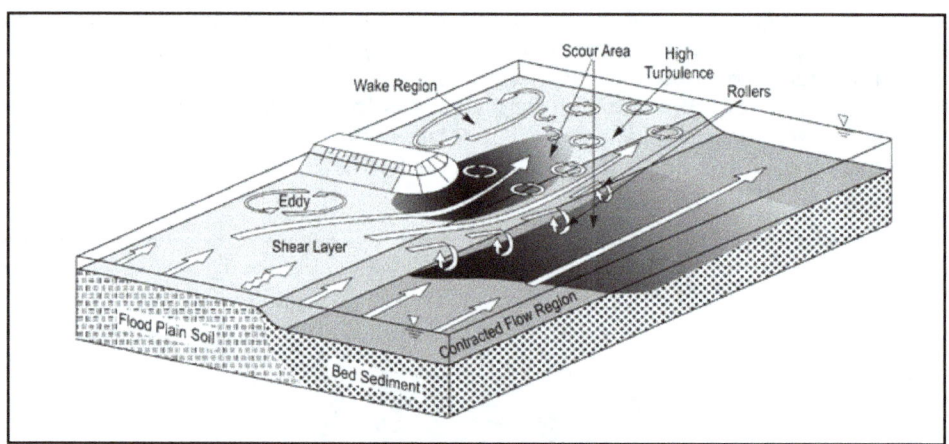

Figure 3.8. Flow structure including macro-turbulence generated by floodplain/main channel flow interaction, flow separation around abutment, and wake region on the floodplain of a compound channel (NCHRP 2011b).

Though the short-contraction analogy is somewhat simplistic, an important point to be made is that the flow field around an abutment, like the flow field through an orifice, is not readily delineated as a contraction flow field separate from a local flow field established near the abutment. The two flow features (flow contraction and large-scale turbulence) are related and difficult to separate. Either of the flow features may dominate, depending on the extent of flow contraction and the characteristics of the abutment and its foundation. When an abutment barely constricts flow through the waterway, scour at the abutment may develop largely due to the local flow field generated by the abutment. This flow field is characterized by a local contraction of flow and by generation of large-scale turbulence. For a severely contracted bridge waterway, flow contraction dominates the flow field and a substantial backwater occurs upstream of the bridge. In this situation, the approach flow slows as it approaches the upstream side of the bridge, and then accelerates to a higher velocity as it passes through the bridge waterway (NCHRP 2011b).

When the foundation of the end of an abutment consists of a solid contiguous form extending into the bed (floodplain or main channel), scour development may become similar to that at a wide pier where the flow becomes contracted and large-scale turbulence is produced. Such abutments include situations where a sheet-pile skirt is placed around the toe of the spill-slope of a spill-through abutment (to protect against spill-slope instability and failure), or when a wing-wall column is founded on sheet-piles.

In summary, abutments are essentially erodible short contractions. Higher flow velocities and large-scale turbulence around an abutment may erode the abutment boundary. As noted, the bed of the main channel is more erodible than the floodplain, because the bed is formed of loose sediment while the floodplain is formed of more cohesive soil often protected by a cover of vegetation. Field observations indicate that, typically, two prime scour regions develop, as indicated in Figure 3.9:

- One region is where the boundary is least resistant to hydraulic erosion. This could be the main bed if flow velocities (and unit discharges) are sufficiently large.
- The other region is where the flow velocities (and unit discharges) and turbulence are greatest. This usually is near the abutment.

For an abutment well set back on a floodplain, laboratory experiments indicate that deepest scour usually coincides with the region where flow contraction is greatest (NCHRP 2011b, c, and d).

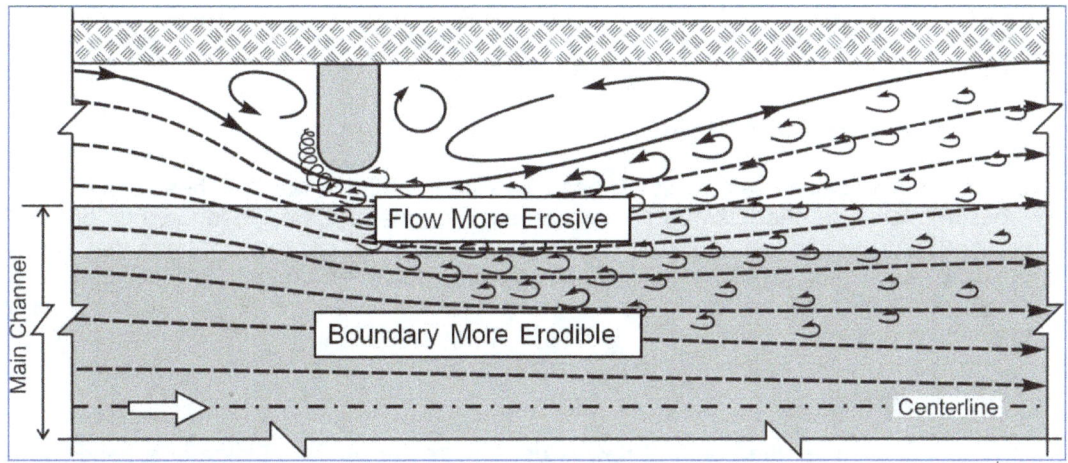

Figure 3.9. Interaction of flow features causing scour and erodibility of boundary (NCHRP 2011b).

3.7 LATERAL SHIFTING OF A STREAM

Streams are dynamic. Areas of flow concentration continually shift banklines, and in meandering streams having an "S-shaped" planform, the channel moves both laterally and downstream. A braided stream has numerous channels which are continually changing. In a braided stream, the deepest natural scour occurs when two channels come together or when the flow comes together downstream of an island or bar. This scour depth has been observed to be 1 to 2 times the average flow depth.

A bridge is static. It fixes the stream at one place in time and space. A meandering stream whose channel moves laterally and downstream into the bridge reach can erode the approach embankment and can affect contraction and local scour because of changes in flow direction. A braided stream can shift under a bridge and have two channels come together at a pier or abutment, increasing scour. Descriptions of stream morphology are given in HDS 6 and HEC-20 (FHWA 2001 and 2012b).

Factors that affect lateral shifting of a stream and the stability of a bridge are the geomorphology of the stream, location of the crossing on the stream, flood characteristics, the characteristics of the bed and bank material, and wash load. It is difficult to anticipate when a change in planform may occur. It may be gradual or the result of a single major flood event. Also, the direction and magnitude of the movement of the stream are not easily predicted. While it is difficult to evaluate the vulnerability of a bridge due to changes in planform, it is important to incorporate potential planform changes into the design of new bridges and design of countermeasures for existing bridges. These factors are discussed and analysis techniques are presented in HEC-20 (FHWA 2012b).

Countermeasures for lateral shifting and instability of the stream may include changes in the bridge design, construction of river control works, protection of abutments with riprap, or careful monitoring of the river in a bridge inspection program. Serious consideration should be given to placing footings/foundations located on floodplains at elevations the same as those located in the main channel. Control of lateral shifting requires river training works, bank stabilizing by riprap, and/or guide banks. The design of these works is beyond the scope of this document. Design methods are given by FHWA in HEC-23 (FHWA 2009), HDS 6 (FHWA 2001), and similar publications. The USACE and AASHTO provide additional guidance (USACE 1981, 1983, 1993a; AASHTO 1992b).

3.8 OTHER TYPES OF SCOUR

Other scour conditions can affect the bed elevation in a bridge reach. These can result from erosion related to the planform characteristics of the stream (meandering, braided or straight), variable downstream control, flow around a bend, or other changes that decrease the bed elevation. These scour conditions can occur at bridges located upstream or downstream of a confluence of two streams or the confluence of individual braids in a braided stream.

3.8.1 Discussion

In a natural channel, the depth of flow is usually greater on the outside of a bend. In fact, there may well be deposition on the inner portion of the bend at a point bar. If a bridge is located on or close to a bend, scour will generally be concentrated on the outer portion of the bend. Also, in bends, the thalweg (the part of the stream where the flow is deepest and, typically, the velocity is the greatest) may shift toward the inside of the bend as the flow increases. This can increase scour and nonuniform distribution of scour in the bridge opening. In some cases during high flow the point bar may have a channel (chute channel) eroded across it (see FHWA 2012b). This can further skew the distribution of scour in the bridge reach. Consequently, other scour conditions such as these are differentiated from contraction scour which involves removal of material from the bed across all or most of the channel width.

The relatively shallow straight reaches between bendway pools are called crossings. With changes in discharge and stage the patterns of scour and fill can also change in the crossing and pool sequence. These geomorphic processes are discussed in more detail in HEC-20 and HDS 6 (FHWA 2012b and 2001). These processes can contribute to scour in a bridge reach. They are cyclic and may be in equilibrium around some general bed elevation. There are no equations for predicting these changes in elevation. Generally, a study of the stream using aerial photographs and/or successive cross section surveys can determine trends. In this case, the long-term safety of the bridge may depend, primarily, on inspection.

Some unique scour conditions are associated with a particular channel morphology. Braided channels will have deep scour holes when two channels come together downstream from a bar or island (confluence scour). At other times a bar or island will move into the bridge opening concentrating the flow onto a pier or abutment or changing the angle of attack. In anabranching flow, where flow is in two or more channels around semi-permanent islands, there is a problem of determining the distribution of flow between the channels, and over time the distribution may change. The bridge could be designed for the anticipated worst case flow distribution or designed using the present distribution. In either case, inspection and maintenance personnel should be informed of the potential for the flow distribution and scour conditions to change.

Other scour conditions can be caused by short-term (daily, weekly, yearly, or seasonal) changes in the downstream water surface elevation that control backwater and hence, the velocity through the bridge opening. Similarly, a bridge located upstream or downstream of a confluence can experience scour caused by variable flow conditions on the main river and tributary. This scour is reversible and it is considered "scour" rather than long-term aggradation or degradation.

3.8.2 Determining Other Types of Scour

Scour at a bridge cross-section resulting from variable water surface elevation downstream of the bridge (e.g., tributary or downstream control) is analyzed by determining the lowest potential water-surface elevation downstream of the bridge insofar as scour processes are concerned. Then determine contraction and local scour depths using these worst-case conditions.

Scour in a channel bendway resulting from the flow through the bridge being concentrated toward the outside of the bend is analyzed by determining the superelevation of the water surface on the outside of the bend and estimating the resulting velocities and depths through the bridge. The maximum velocity in the outer part of the bend can be 1.5 to 2 times the mean velocity. A physical model study can also be used to determine the velocity and scour depth distribution through the bridge for this case.

Estimating scour in the bridge cross-section for unusual situations involves particular skills in the application of principles of river mechanics to the site-specific conditions. To determine these other types of scour in the bridge opening may require 2-dimensional (2-D) computer programs (for example, FST2DH (FHWA 2003b) or a physical model. Such studies should be undertaken by engineers experienced in the fields of hydraulics and river mechanics.

(page intentionally left blank)

CHAPTER 4 – SOILS, ROCK, AND GEOTECHNICAL CONSIDERATIONS

4.1 GENERAL

Because scour is caused by the erosive action of flowing water as it excavates and removes earth materials from the beds and banks of streams, it is useful for engineers to have a background and basic understanding of the properties of earth materials as they relate to scour and erosion. In this context, hydraulic forces can be considered a load, and the engineering properties of the soil characterize the resistance to that load. This chapter provides an introduction to the physical properties of soils and rock, and the behavior of these materials at the interface between the channel boundary and the flow field. Figure 4.1 provides examples of scour in soil and rock.

| a. Scour in soil (sand) | b. Scour in rock (sandstone) |

Figure 4.1. Photographs of scour in soil and rock.

In the most general sense, earth materials can be categorized as either soil or rock. In the engineering sense, soil can be defined as any unconsolidated geomaterial composed of discrete particles with gases and liquids in between. The maximum particle size that qualifies as soil is not fixed, but depends on the uses and functions to which the material is put, and the ease with which it can be moved and placed. As a rule of thumb, for trench and footing excavations and for the construction of fill in layers, an upper (limiting) particle size for soils can be considered to be about 12 inches in diameter, a practical limit on what a worker can lift by hand (Sowers and Sowers 1970).

For engineering purposes, rock may be defined as any indurated geomaterial that requires drilling, wedging, blasting, or other methods of applying force for excavation. The minimum degree of induration that qualifies as rock has sometimes been defined by an unconfined compressive strength of about 100 to 200 psi (700 to 1400 kPa). From an engineering or functional viewpoint, the definition of rock is complicated by structure and defects, such as jointing and fractures (discontinuities). For purposes of this discussion, indurated earth materials having discontinuity spacings greater than about 4 to 8 inches (0.1 to 0.2 m) can be considered to be more rock-like, whereas materials having smaller spacings may behave more like soils, especially where this spacing frequency is in multiple dimensions.

Intermediate geomaterials (IGMs) are transition materials between soils and rocks. The distinction of IGMs from soils or rocks for geotechnical engineering purposes is made purely on the basis of strength of the geomaterials.

4.2 SCOUR PROCESSES

4.2.1 Cohesionless Soils

Cohesionless soils are eroded particle by particle, and scour occurs relatively rapidly such that the maximum (equilibrium) scour depth is reached within a time period of a few hours to a few days, often within the duration of a single flood event. For this reason, current practice assumes that scour in cohesionless soils is not a function of time, and therefore occurs essentially instantaneously once the threshold hydraulic condition for particle motion is exceeded.

For situations where it is desirable to estimate the time rate of scour, NCHRP Report 682 (NCHRP 2011c) provides a method for estimating the depth of scour at a bridge pier as a function of time, called the scour evolution rate, at structures with simple geometries in cohesionless soils. In addition, that report presents equations to predict the equilibrium pier scour depth, including the cases where the piers are very wide with respect to the flow depth and/or skewed to the flow direction (see Chapter 7).

4.2.2 Cohesive Soils

In the case of cohesive soils, erosion can take place particle by particle but also block of particles by block of particles. The boundaries of these blocks are formed naturally in the soil matrix by micro-fissures which result from various phenomena, such as compression and extension. Resistance to erosion is influenced by a combination of weight and, more importantly, electromagnetic and electrostatic interparticle forces that are the source of the cohesive properties of these soils. Table 4.1 provides a list of factors that influence the erodibility of cohesive soils.

One major difference between cohesionless and cohesive soils is the rate of erosion beyond the critical (threshold) shear stress. In cohesive soils, this rate increases slowly and is typically measured in millimeters per hour. This slow rate makes it advantageous to consider that scour processes in cohesive soils are time dependent and to find ways to accumulate the effect of long-term hydraulic loading, including the effects of many flood events over many years, rather than to consider a single flood event for design.

There is a critical shear stress τ_c below which no erosion occurs and above which erosion starts. One can also define the initial slope $S_i = (dz\dot{}/d\tau)$ at the origin of the erosion rate vs. shear stress curve. Both τ_c and S_i are parameters that help describe the erosion function and, therefore, the erodibility of a material. This concept, while convenient, may not be theoretically simple. The fact that during NCHRP Project 24-15 (NCHRP 2004) no relationship could be found between the critical shear stress or the initial slope of the erosion function and common soil properties seems to be at odds with the accepted idea that different cohesive soils erode at different rates. If different clays erode at different rates, then the erosion function and therefore its parameters should be functions of the soil properties.

The likely explanation is that there is a relationship between erodibility and soil properties, but that this relationship is quite complicated and involves advanced understanding of soil properties or combination of soil properties and environmental conditions. For this reason, the direct testing and measurement of soil erodibility using devices such as those described in Section 4.3 was recommended by Briaud et al. (NCHRP 2004) as the preferred method for determining the erosion function for site-specific geomaterials.

Table 4.1. Factors Influencing the Erodibility of Cohesive Soils (NCHRP 2004).

When this Parameter Increases	Erodibility
Soil water content	*
Soil unit weight	decreases
Soil plasticity index	decreases
Soil undrained shear strength	increases
Soil void ratio	increases
Soil swell	increases
Soil mean grain size	*
Soil percent passing sieve #200	decreases
Soil clay minerals	*
Soil dispersion ratio	increases
Soil cation exchange capacity	*
Soil sodium absorption ratio	increases
Soil pH	*
Soil temperature	increases
Water temperature	increases
Water chemical composition	*
*unknown	

A time-dependent method for estimating scour in cohesive soils known as the Scour Rate in Cohesive Soils method (SRICOS) was developed by Briaud et al. (1999a and b) for estimating scour at bridge piers, and was later expanded to include contraction scour (NCHRP 2004). In general, the method considers the cumulative effect of many floods in order to estimate the scour vs. time relationship for bridge pier or contraction scour in cohesive soils. In practice, this means generating a long-term synthetic hydrograph of daily flow values, or assuming that a historically-observed flow series of 20, 30, or 40 years of daily data will repeat itself. The concept is illustrated schematically in Figure 4.2 where the scour from two consecutive floods is considered as a cumulative effect over time.

4.2.3 Rock

Prediction of scour in rock is a function of hydraulic loading conditions as well as rock resistance properties; it is not a function of rock properties alone. Rock scour can occur in four modes (NCHRP 2011e):

1. Dissolution of soluble rocks
2. Cavitation
3. Quarrying and plucking of durable, jointed rock
4. Abrasion and grain-scale plucking of degradable rock

Soluble rock formations suitable for support of bridge foundations do not dissolve in engineering time; however, these rock formations can produce complex deposits of rock blocks in a clayey soil matrix that respond to hydraulic forces in a complicated way with gradual wear of the matrix until rock blocks become susceptible to plucking. Flow conditions required for cavitation are not likely to occur in typical natural channels where bridge foundations are placed.

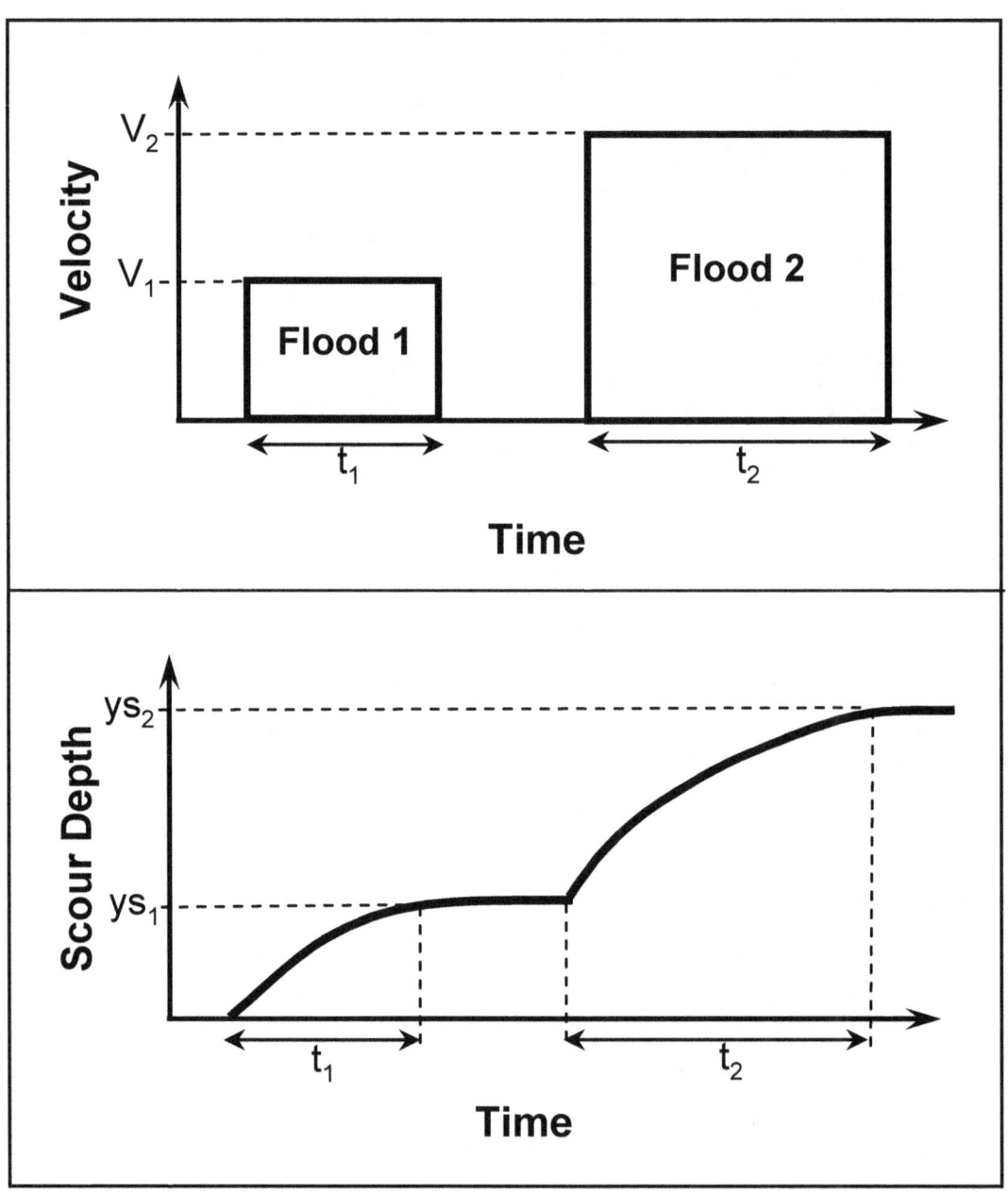

Figure 4.2. Method for accumulating the effects of scour resulting from multiple floods (modified from Briaud et al. 2011).

Jointed, durable rock-bed channels can scour rapidly in response to threshold flow conditions. Degradable rocks scour gradually and progressively during periods of time that a threshold stream power is exceeded. Stream power is the preferred hydraulic parameter for predicting scour in degradable rock because it can be accumulated over time, over the course of many individual flood events, small and large.

Key parameters associated with scour in rock materials as identified by Dickenson and Baillie (FHWA 1999b) are listed in Table 4.2. They note that their results are limited to rock units in the Coast Range of Oregon, and they selected reaches where the stream channels were straight and unobstructed. The extent of scour was calculated from current and earlier channel surveys. Scour depths and rates were computed and laboratory tests were performed on samples of rock to obtain relevant geotechnical index properties. Stream gage data were used to develop hydraulic parameters of the stream flows during the time intervals between the surveys.

Table 4.2. Parameters Influencing the Rate of Scour in Rock (modified from Dickenson and Baillie (FHWA 1999b).

Factors Contributing to Scour in Rock		
Geologic	Geotechnical	Hydraulic
Lithology	Rock density	Channel geometry
Frequency and character of discontinuities	Abrasion resistance	Year-round flow characteristics
Orientation of discontinuities	Slake durability	Energy gradient
Degree of weathering	Rock strength	Bedload characteristics
Degree of induration of sedimentary rock	--	Intensity and duration of flood events

A conceptual stream power model proposed by Costa and O'Connor (1995) is presented in Figure 4.3 for three hypothetical floods. Flood intensity is expressed as stream power in this model in a way which is comparable to the rate of energy dissipation developed by Annandale (1995). Flood A is a long duration, low peak-power flood that would cause insignificant scour because it does not exceed the threshold condition to begin eroding alluvial deposits. Floods B and C have the same peak power but different durations and serve to demonstrate the essence of Costa and O'Connor's (1995) model. Flood B is a short-duration, high-power flood which exceeds the threshold for eroding bedrock as well as alluvium; however, the area under the power curve is small indicating that Flood B has relatively small scour potential. Flood C, on the other hand, has the same peak power as Flood B but a longer duration, indicating that Flood C could cause significant scour in susceptible materials.

Dickenson and Baillie (FHWA 1999b) note that the term "slake durability" describes the behavior of samples that have been subjected to cycles of wetting and complete drying. Therefore, they developed a modification to the ASTM standard test procedure which excludes heated drying on the durability of the rock samples; therefore, the weight loss observed during their test reflects abrasion resistance, not loss caused by desiccation-induced slaking. They called this modified slake durability test a "continuous abrasion" test to distinguish the behavior of the specimens tested to the behavior of similar rocks tested by the ASTM D 4644 standard slake durability test procedure.

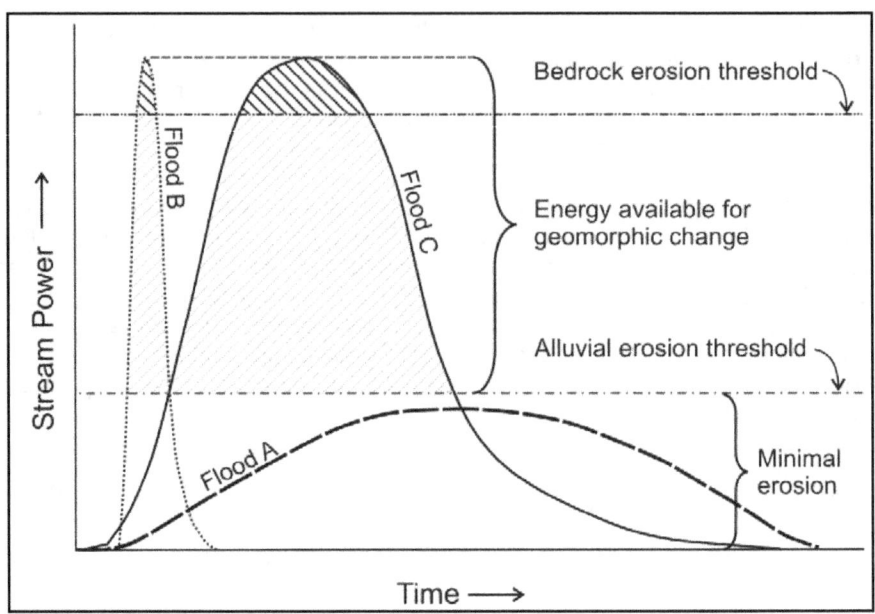

Figure 4.3. Conceptual stream power model for geomorphically effective floods (NCHRP 2011e).

The continuous abrasion test shows an initially high rate which diminishes with time. The relatively high rate of weight loss at the start of the test is caused by angular rock fragments becoming sub-rounded to rounded. The fragments become well rounded and exhibit a much lower and typically uniform rate of weight loss after 120 to 200 minutes which Dickenson and Baillie (FHWA 1999b) interpreted to represent the initiation of long-term abrasion loss. They plotted weight loss with time after 120 minutes and used the slope of the line as the basis for an index property they called the Abrasion Number. Larger abrasion number values are calculated for rock fragments that abrade quickly, whereas smaller abrasion number values indicate rock fragments whose edges do not chip easily and are more resistant to abrasion.

Using the abrasion number, a straightforward procedure for estimating scour was developed based on the scour resistance of the rock in the channel and the hydraulic parameters causing scour. The procedure uses the abrasion number to represent the abrasion resistance of the rock, and the stream power of the flow to represent hydraulic turbulence and uplift forces on rock particles, as well as the effects of bedload translating and saltating over the rock-bed channel. The results of Dickenson and Baillie's (FHWA 1999b) procedure are illustrated in Figure 4.4. In this figure, the cumulative (integrated) stream power over any time period of interest is represented on the x-axis. For rock materials, the time period typically considers the cumulative effect of many flood events over a period of years, for example, over the life of the bridge.

4.3 ERODIBILITY

Erodibility involves both the hydraulic conditions that create erosive forces, and the properties of the geomaterials to resist erosion when exposed to those conditions. This section presents the concepts and equations used to define and quantify erosive conditions of flow and the erosion resistance of soils and rock, the combination of which results in predictive methods for estimating erodibility under a wide range of flow conditions and geomaterials.

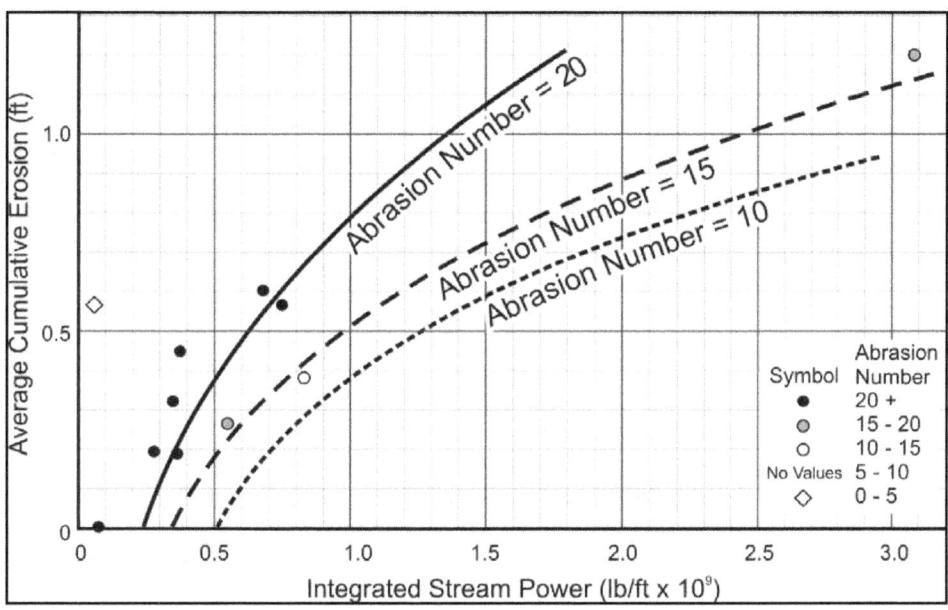

Figure 4.4. Average cumulative erosion related to integrated stream power and abrasion number (NCHRP 2011e).

4.3.1 Velocity

The velocity of flowing water is a vector quantity, i.e., it exhibits both a magnitude and direction. In open channels, velocity is a function of many variables, including the slope and cross-sectional geometry of the channel, the roughness of the channel boundaries (bed and banks), and the total amount of flow (discharge) in the channel at any given time. The average velocity V, expressed in ft/s (m/s) in a channel is determined by the continuity equation:

$$V = \frac{Q}{A} \tag{4.1}$$

where Q is the total discharge (ft^3/s or m^3/s) at a given time, and A is the cross-sectional area of flow in ft^2 (m^2). However, in rivers and streams, flow and velocity are never uniform across the cross section; velocity is greatest near the deepest section of the channel, and decreases near the channel margins. Figure 4.5 presents a diagram of a typical river cross section and illustrates the velocity distribution as a series of isovels (lines of equal velocity).

Standard 1-dimensional hydraulic models such as HEC-RAS are used to quantify hydraulic conditions at selected channel cross sections in a river reach, including depth-averaged (stream tube) velocity distributions across the channel (USACE 2010a).

In erosion studies, one of the most important characteristics of geomaterials is the erosion threshold. Below this threshold, hydraulic conditions are mild enough such that erosion does not occur, whereas above this threshold, erosion occurs at rates that increase as the hydraulic conditions become more and more severe. In terms of velocity, this threshold is referred to as the critical velocity V_c.

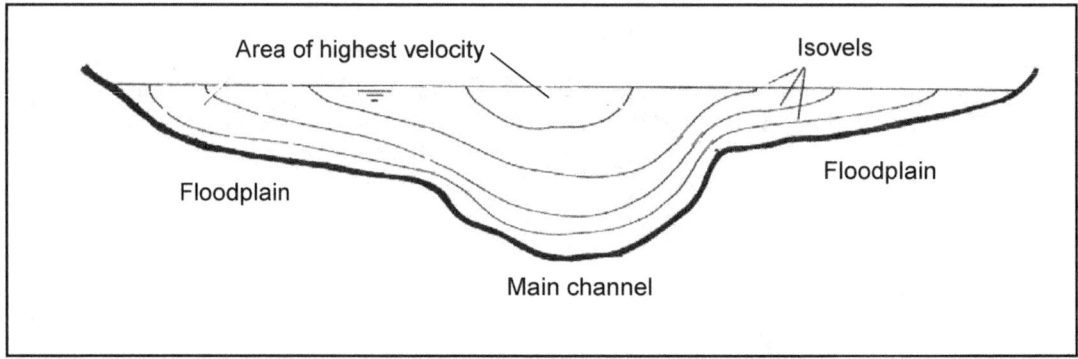

Figure 4.5. Typical velocity distribution in an open channel

While velocity in and of itself is not a force, it is often used as a surrogate or index value that is related to erosive potential of the flow. For example, the velocity at which cohesionless particles (e.g., sands and gravels) begin to move can be determined by the critical velocity equation:

$$V_c = K_u y^{1/6} d^{1/3} \tag{4.2}$$

where V_c is the critical velocity at which particles begin to move, y is the flow depth, and d is the grain size of the particle. K_u is a unit correction factor equal to 11.17 for U.S. customary units (ft-lb-s), and 6.19 for SI units (m-kg-s).

4.3.2 Shear Stress

Shear stress is the shear force per unit area exerted on the channel boundary by flowing water. The average shear stress on the channel boundary across the entire cross section is calculated as follows:

$$\tau = K_b \gamma R S_f \tag{4.3}$$

where:

τ	=	Design shear stress, lb/ft² (N/m²)
K_b	=	Bend coefficient (dimensionless)
γ	=	Unit weight of water, lb/ft³
R	=	Hydraulic radius (area divided by wetted perimeter), ft (m)
S_f	=	Slope of the energy grade line, ft/ft (m/m)

Figure 4.6 presents a plot of the critical shear stress as a function of the mean grain size. The data in Figure 4.6 come from measurements in a laboratory erosion device as well as measurements published in the literature. As can be seen from this figure, the relationship between the critical value and the grain size has a "V" shape indicating that the most erodible soils are fine sands with a mean grain size in the range of 0.1 to 0.5 mm. This V shape also illustrates that particle size controls the erosion threshold of coarse-grained soils, **while particle size does not correlate with the erosion threshold of fine-grained soils**. This effect is due to the cohesive nature of fine-grained soils (i.e., silts and clays).

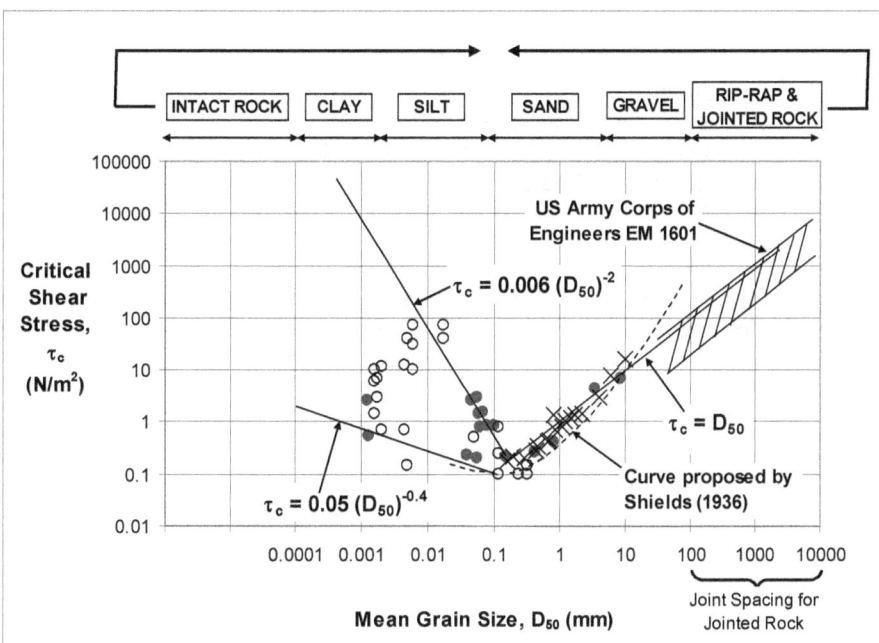

Figure 4.6. Critical shear stress vs. particle grain size (Briaud et al. 2011).

The bend coefficient K_b is used to calculate the increased shear stress on the outside of a bend. This coefficient ranges from 1.05 to 2.0, depending on the severity of the bend. The bend coefficient is a function of the radius of curvature R_c divided by the top width of the channel T, as follows:

$K_b = 2.0$ for $2 \geq R_c/T$

$$K_b = 2.38 - 0.206\left(\frac{R_c}{T}\right) + 0.0073\left(\frac{R_c}{T}\right)^2 \quad \text{for } 10 > R_c/T > 2 \tag{4.4}$$

$K_b = 1.05$ for $R_c/T \geq 10$

The shear stress on the bed of the channel at any point in the cross section is determined by substituting the depth of flow y for the hydraulic radius R. Alternatively, the local shear stress, for example in the vicinity of a bridge pier or abutment, can be calculated as:

$$\tau_{local} = \left(\frac{nV_{local}}{K_u}\right)^2 \frac{\gamma_w}{y^{1/3}} \tag{4.5}$$

where:

τ_{local}	=	Local shear stress, lb/ft² (N/m²)
n	=	Manning "n" value
V_{local}	=	Local velocity, ft/s (m/s)
γ_w	=	Density of water, 62.4 lb/ft³ (9,800 N/m³) for fresh water
y	=	Local depth of flow, ft (m)
K_u	=	1.486 for U.S. customary units, 1 for SI units

At bridge piers, the maximum stream-tube velocity V_{max} in the cross section, multiplied by a shape factor K_p to account for local acceleration around the pier, will provide a more suitable representation of local conditions at the pier itself. Shape factors K_p are typically taken as 1.5 for round-nose piers and 1.7 for blunt (or square-nosed) piers.

Critical shear stress τ_c for the initiation of motion for cohesionless soil particles can be estimated using the Shields relation:

$$\tau_c = K_s(\rho_s - \rho_w)gd \tag{4.6}$$

where:

τ_c	=	Critical shear stress for a particle of size "d," lb/ft² (N/m²)
K_s	=	Dimensionless Shields parameter, 0.047 for sand and 0.030 for gravel
ρ_s	=	Particle mass density, slugs/ft³ (kg/m³)
ρ_w	=	Mass density of water, 1.94 slugs/ft³ (1,000 kg/m³) for fresh water
g	=	Gravitational acceleration, 32.2 ft/s² (9.8 m/s²)
d	=	Particle size, ft (m)

4.3.3 Stream Power

Power is defined as a rate of doing work or a rate of expending energy. In open channel flow, instantaneous stream power (the stream power at any particular moment) is defined as:

$$P = \gamma q S_f = \gamma q (\Delta E) \tag{4.7}$$

where:

P	=	Instantaneous stream power, lb-ft/s per square foot (kW/m²)
γ	=	Unit weight of water, 62.4 lb/ft³ (9,800 N/m³)
q	=	Unit discharge, ft³/s per foot width (m³/s per meter width)
S_f	=	Slope of the energy grade line, ft/ft (m/m)
ΔE	=	Energy loss per unit distance in direction of flow, ft/ft (m/m)

Stream power is conveniently expressed in terms of shear stress and velocity as:

$$P = \tau V$$

where:

τ	=	Representative shear stress, lb/ft² (N/m²)
V	=	Representative velocity, ft/s (m/s)

The stream power calculated by the above equations must be representative of the conditions for which the scour is being evaluated. For example, if long-term scour across the entire cross section is of interest, the cross-sectional average velocity and shear stress will be satisfactory. However, if the scour at a specific location in the cross section is of interest, for example at a bridge pier, it is more appropriate to use local values for these variables.

4.3.4 Erosion Rates

When hydraulic conditions exceed the critical (threshold) value, the soil erodes at a rate which increases with increasing hydraulic load. Figure 4.7 shows erosion rate relationships as a function of the Unified Soil Classification System and other factors. A degree of uncertainty is associated with such a chart. For advanced studies and when the economy of the project warrants, it is preferable to test the soil in an appropriate erosion testing device.

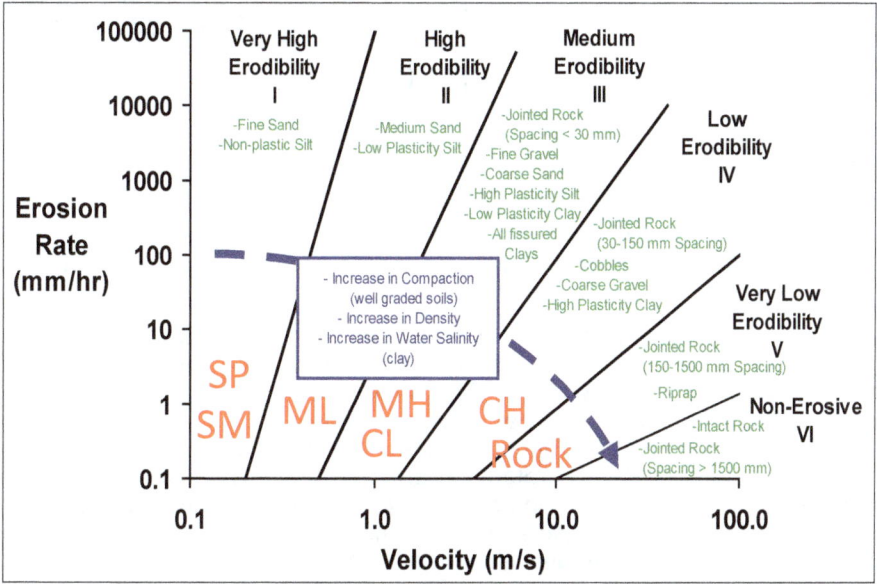

Figure 4.7. Erosion rate vs. velocity for a wide range of geomaterials (Briaud et al. 2011).

4.3.5 Devices to Measure Erodibility

<u>Piston-Type Devices</u>: Piston-type devices are laboratory devices that measure the erosion of a sample of soil or rock exposed to flowing water in a relatively small flume under controlled conditions. A tube filled with soil or rock is placed through the bottom of the flume where water flows at a constant velocity (Figure 4.8). The soil or rock is gradually pushed out of the sampling tube only as fast as it is eroded by the water flowing over it. For each velocity, an erosion rate is measured and a corresponding shear stress is calculated. After a series of gradually increasing velocities, the data are compiled to develop a relationship between erosion rate and velocity (or shear stress).

<u>Rotating-Type Devices</u>: Rotating-type devices measure the rate at which geomaterials erode as a function of the shear stress applied by the flowing water. Stiff clays, sandstones, limestones and other geomaterials that have sufficient strength to stand under their own weight can be tested with these devices. Cohesionless materials such as sands and gravels are not suitable for testing in this type of device.

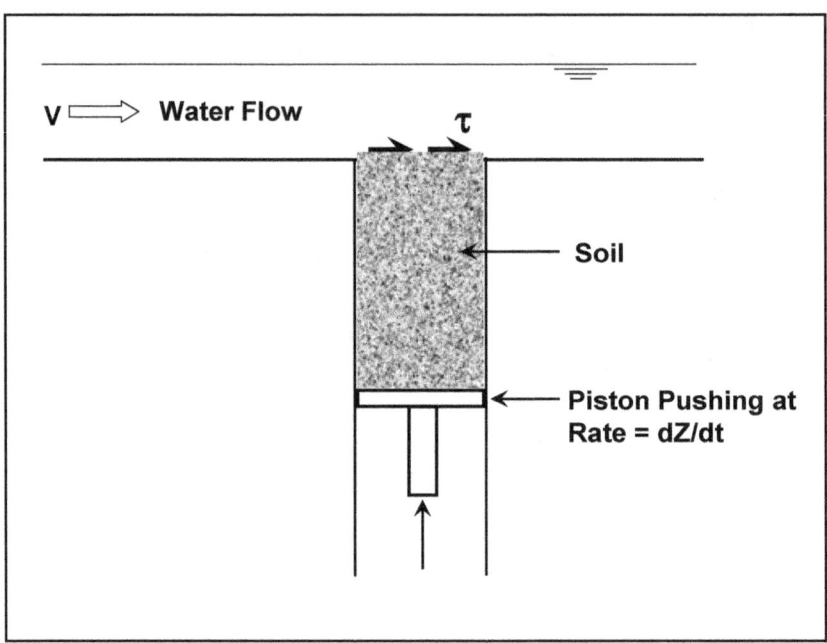

Figure 4.8. Schematic diagram of piston-type erosion rate device.

A cylindrical sample is taken from the site and sent to the laboratory where it is prepared for testing. A small diameter hole is drilled through the axis of the sample and a support shaft is inserted and attached to the sample. The other end of the shaft is attached to a torque-measuring sensor through a clutch as shown in Figure 4.9. The sample is lowered into a slightly larger cylinder and the annulus between the sample and the outer cylinder is filled with water. The outer cylinder is rotated in order to create a shear stress on the surface of the sample. The shear stress increases with increasing rotational speed of the outer cylinder. Erosion rates are measured over a range of applied shear stress values and used to develop the erodibility rate relationship for the tested material.

Submerged Jet-Type Devices: An apparatus that can be used in the laboratory or field to evaluate soil erodibility is the submerged jet erosion device. It applies a water jet to a submerged soil surface and the scour depth beneath the jet is measured over time. The device is composed of three parts as shown in Figure 4.10 - a source of water for the jet, a container to emplace around the sample and submerge it, and a device to measure the depth of scour at intervals during the test.

The erosive stress applied by the jet is adjusted by varying the pressure delivered at the nozzle, and varying the initial distance of the nozzle to the soil-water interface. Shear stresses applied during a test are estimated from an analysis of the jet hydrodynamics at the centerline of the jet. Diffusion of the jet causes the water velocity V at the soil-water interface to be inversely proportional to the distance from the nozzle, where the nozzle velocity is V_0. At intervals during the test, the water velocity V is typically related to the shear stress using the Chezy equation as the depth of scour increases. These data are then used to develop an erodibility rate relationship for the tested material.

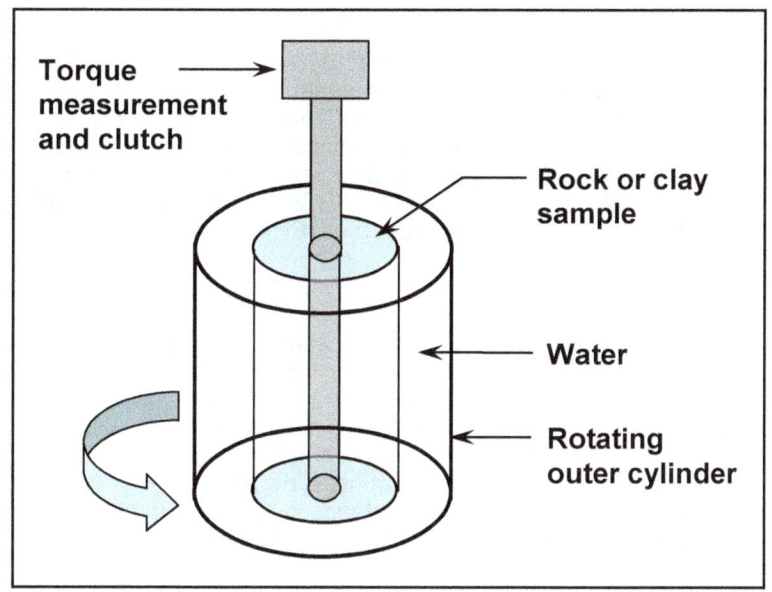

Figure 4.9. Schematic diagram of rotating-type erosion rate device.

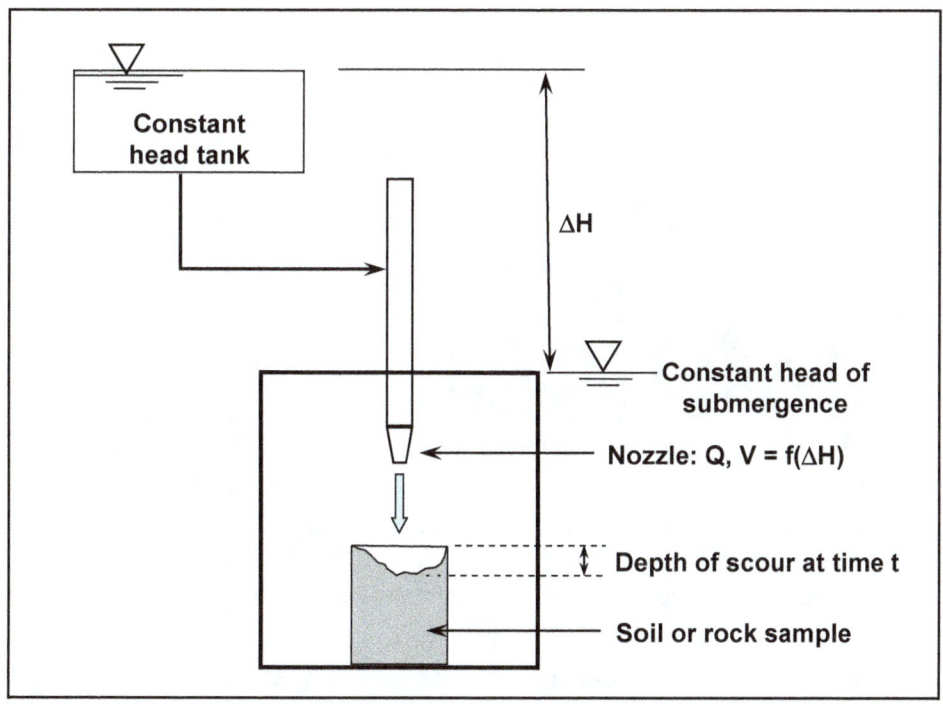

Figure 4.10. Schematic diagram of jet-type erosion rate device.

4.4 SOIL PROPERTIES

Scour and erosion of soils is a soil-water interaction phenomenon. This section presents common methods used to characterize the engineering properties of soils that are relevant to physical processes involved in this interaction. These properties are grain size distribution, plasticity, density, strength, and hydraulic conductivity. Finally, common methods of classification are summarized with emphasis on terminology. Soil properties are discussed in greater detail in Soils and Foundations (FHWA 2006) and Geotechnical Engineering Circular 5, Evaluation of Soil and Rock Properties (FHWA 2002).

4.4.1 Particle Size

The range of sizes of soil particles and grains spans many orders of magnitude, from the largest size that can be moved by hand, down to the smallest size that can be detected using electron microscopes or X-ray diffraction. Two methods are commonly used to determine the grain sizes present in a soil: sieving and sedimentation. Calibrated sieves having openings as large as 4 inches (200 mm) and as small as 0.074 mm (U.S. Standard Sieve no. 200) are used for separating the coarser grains. The amount (dry weight) retained on each sieve is measured to determine the percentages by weight of each retained sieve (grain) size. The grain size distribution curve is plotted using percentages passing for the different sieve sizes.

The portion of the grains smaller than 0.074 mm are measured by a hydrometer based on their effect on the density of a water-sediment mixture, using the principle that the smaller the grain, the more slowly it will settle in a column of still water.

Typical sieves and hydrometers used for soil grain size analyses are shown in Figure 4.11. Figure 4.12 provides a typical graph illustrating the result of a grain size analysis using both sieve and hydrometer techniques for two different soils.

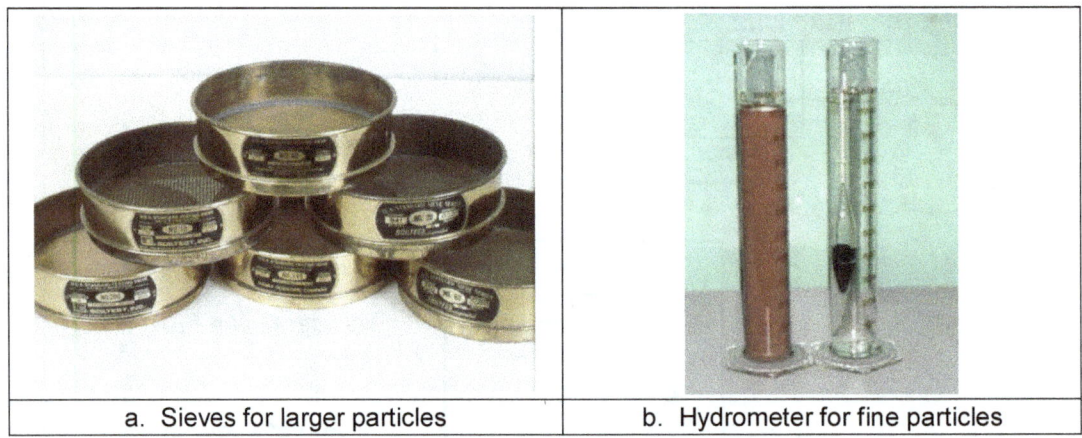

| a. Sieves for larger particles | b. Hydrometer for fine particles |

Figure 4.11. Typical sieves and hydrometers used for grain size analyses (from NRCS website).

The shape of the grain-size distribution (GSD) curve or "gradation curve" as it is frequently called, is one of the more important aspects in a soil classification system for coarse-grained soils. The shape of the gradation curve can be characterized by two coefficients, called the Coefficient of Uniformity, C_u and the Coefficient of Curvature, C_c, to which numerical values may be assigned, as follows:

$$C_u = \frac{d_{60}}{d_{10}} \tag{4.8}$$

and

$$C_c = \frac{(d_{30})^2}{(d_{60})(d_{10})} \tag{4.9}$$

In equations 4.8 and 4.9, d_{10}, d_{30}, and d_{60} are the particle sizes for which 10, 30, and 60 percent of the soil by weight is finer, respectively. The Coefficient of Uniformity is used to determine whether a soil is comprised of grains that are more or less uniform in size, or whether the soil is comprised of a wide range of grain sizes. In general, a C_u value of 5 or less is characteristic of a soil that is uniform with respect to grain size. The Coefficient of Curvature is a measure of the curvature of the grain size curve between the d_{60} and d_{10} particle sizes, and is useful in identifying gap-graded soils, such as Soil 2 shown in Figure 4.12.

By use of the two coefficients, C_u and C_c, the uniformity of the coarse-grained soil (gravel and sand) can be classified as well-graded (non-uniform), poorly graded (uniform), or gap graded. Table 4.3 presents criteria for such classifications.

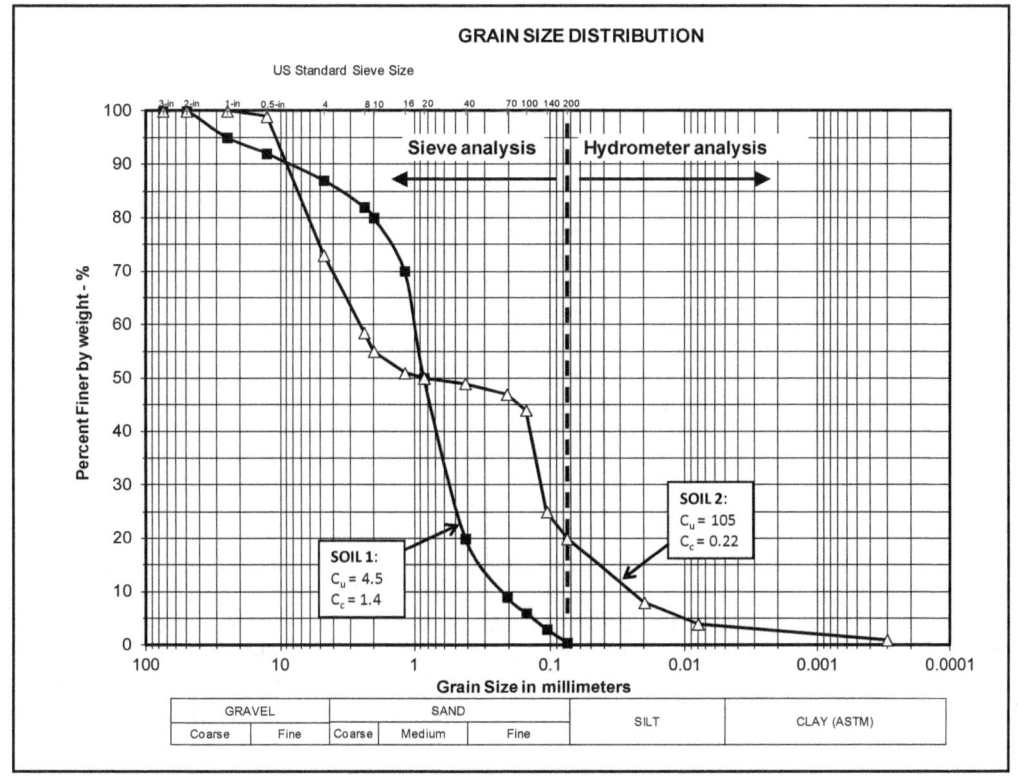

Figure 4.12. Typical grain size curves for two different soils.

Table 4.3. Gradation Based on C_u and C_c Parameters (FHWA 2006).		
Gradation	Gravels	Sands
Well-graded	$C_u \geq 4$ and $1 < C_c < 3$	$C_u \geq 6$ and $1 < C_c < 3$
Poorly graded	$C_u < 4$ and $1 < C_c < 3$	$C_u < 6$ and $1 < C_c < 3$
Gap graded*	C_c not between 1 and 3	C_c not between 1 and 3
*Gap-graded soils may be well-graded or poorly graded. In addition to the C_c value, it is recommended that the shape of the GSD be the basis for definition of gap-graded.		

Some key points to remember with respect to grain size distribution are as follows:

1. Grain size can vary from smaller than the eye can see to large boulders.

2. Some soils have one or two dominant grain sizes and others have a wide range of particle sizes.

3. Many engineering characteristics and properties are related to the grain size distribution of a soil.

4.4.2 Plasticity and the Atterberg Limits

Soils with appreciable amounts of fine grains (less than 0.074 mm, which pass the #200 sieve) may adsorb water due to electrostatic attraction to clay minerals comprising the fine grains. When wetted, these soils result in a mass which holds together and deforms plastically at varying water contents. These soils are known as cohesive soils. Many soils are mixtures of larger bulky grains and finer particles, and they exhibit some degree of varying consistency with changes in moisture. There is no sharp dividing line between cohesive and cohesionless soils, but it is useful to divide soils into these two main groups for engineering purposes. Because fine grains soil particles tend to coat larger grains they have a dominant effect on soil behavior at relatively low percentages. As a rule of thumb, a soil with as little as 10% fines will exhibit some cohesion and fine grained soil behavior, and a soil with more than 35% fines will be dominated by it.

The water content of a soil, w, is defined as the weight of water W_w within a soil mass divided by the weight of the solid particles W_s:

$$w = \frac{W_w}{W_s} \tag{4.10}$$

The Swedish soil scientist Atterberg developed a method for quantifying the effect of varying water content on the consistency of fine-grained soils. Arbitrary but very well-defined limits based on this definition of water content have been established to describe soil consistency as a function of water content. AASHTO test methods T 89, "Determining the Liquid Limit of Soils," and T 90, "Determining the Plastic Limit and Plasticity Index of Soils," provide detailed standardized procedures for quantifying the consistency of soils using the method developed by Atterberg.

The liquid limit (LL) is defined as the water content at which a trapezoidal groove of specified shape, cut in moist soil held in a special cup, is closed after 25 taps on a hard rubber plate using a calibrated apparatus for this purpose. The plastic limit (PL) is the water content at which the soil begins to break apart and crumble when rolled by hand into threads 1/8 inch in diameter. Figure 4.13 illustrates these tests.

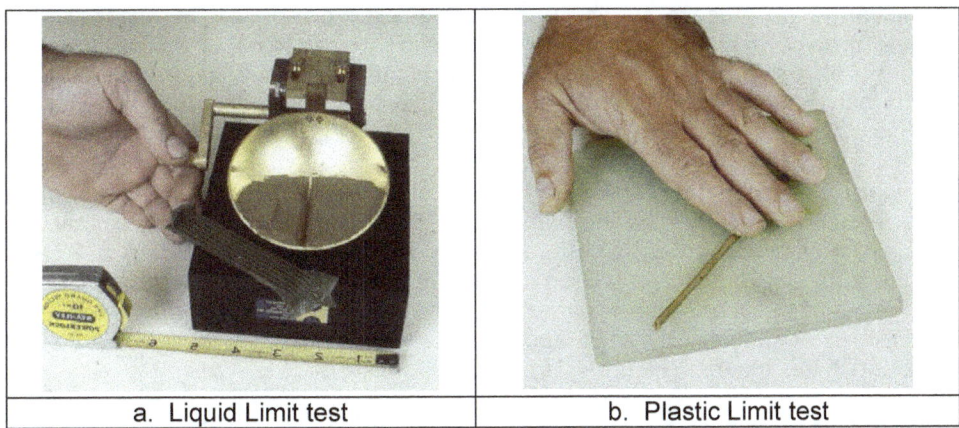

| a. Liquid Limit test | b. Plastic Limit test |

Figure 4.13. Atterberg Limit tests for Liquid Limit and Plastic Limit (from NRCS website).

As indexes of cohesive properties of a soil, these so-called Atterberg Limits are very useful. For example, liquid limit has been found to be directly proportional to compressibility of the soil. The difference between liquid limit and plastic limit is termed the Plasticity Index (PI) and represents the range of water contents for which soil is in the plastic state. Atterberg Limits are also useful in classification of soils, as described in Section 4.5.

A few key things to remember about plasticity are as follows:

1. The smallest soil particles are almost always clay minerals and they attract water because of electrochemical charges. This attraction causes the particles to want to stay together and is responsible for the plasticity these soils exhibit.

2. This property is called plasticity because over a range of water content these soils deform plastically, not like a brittle solid or a liquid.

3. Soils that are plastic also exhibit cohesion which is a strength that is independent of externally applied confining pressure.

4.4.3 Density and Compaction

The density of a soil is simply its weight per volume, and is typically expressed in terms of pounds per cubic foot (kilograms per cubic meter). Density is a function of the relative amounts of soil, water, and air in the sample, and therefore to eliminate the variability introduced by a variable water content, it is usual to express the density in terms of the dry density, i.e., the oven-dry density for which no water content is present. Since the weight of air is negligible, the dry density γ_d is:

$$\gamma_d = \frac{W_s}{V} \tag{4.11}$$

where W_s is the weight of the soil solids, and V is the total volume of solids and the air in the voids between the grains.

Compaction is the process by which soils are made more dense, by either reorienting the particles to achieve closer packing, or by bending or distortion of the particles. In either case, the net result is an expulsion of air, such that there is a greater proportion of solid particles occupying a given volume of soil. Achieving a more dense state requires energy, which is known as compactive effort.

A standard test method developed by soil scientist R. R. Proctor uses a calibrated compactive effort to determine the compaction characteristics of a given soil, as detailed in AASHTO test method T 99, "Moisture-Density Relations of Soils Using a 2.5-kg (5.5-lb) Rammer and 305-mm (12-in.) Drop." The test is conducted using a 4-inch diameter cylindrical mold, and a 2-inch rammer weighing 5.5 pounds. The rammer is dropped 25 times from a height of 12 inches above the soil layer, with 3 layers being successively compacted in this manner (Figure 4.14).

For any soil, there is a unique relationship between the dry density that can be achieved and the water content of the soil during compaction. This relationship can be graphically expressed by a moisture-density curve, as shown in Figure 4.15. Note that there is a specific water content for which compaction is maximized, known as the optimum water content. The optimum water content will be different for different compactive efforts, as shown in the figure. The optimum moisture represents a compromise whereby there is enough water to permit the grains to distort and reposition themselves, but not so much water that the voids are filled. Providing more compactive energy will result in a more dense soil. Many current highway projects use a higher compactive effort than test method T 99, which is called Modified effort (AASHTO test method T 180, "Moisture-Density Relations of Soils Using a 4.54-kg (10-lb) Rammer and a 457-mm (18-in.) Drop") and Figure 4.15 shows an example of how these two efforts typically compare.

For purely cohesionless soils, the most effective compactive effort is achieved through vibration, which reduces the friction between grains and allows repositioning. With all other soils, compaction is best achieved with a combination of static and kneading pressures that bend and reorient the grain structure. Theoretically, given enough compactive effort, the maximum dry density that can be achieved at any given water content is described by the zero air voids curve, as shown in Figure 4.15.

For purely cohesionless soils, relative density, rather than relative compaction, is used. Relative density is based on void ratio. The highest void ratio for a given soil is denoted the maximum void ratio e_{max}, and the smallest void ratio, upon combined tamping and vibration until no further densification is possible, is the minimum void ratio e_{min}. The relative density D_r of a soil is found by its actual void ratio compared to the maximum and minimum values:

$$D_r = \frac{(e_{max} - e)}{(e_{max} - e_{min})} \tag{4.12}$$

Figure 4.14. Standard laboratory compaction test (from the University of British Columbia web site).

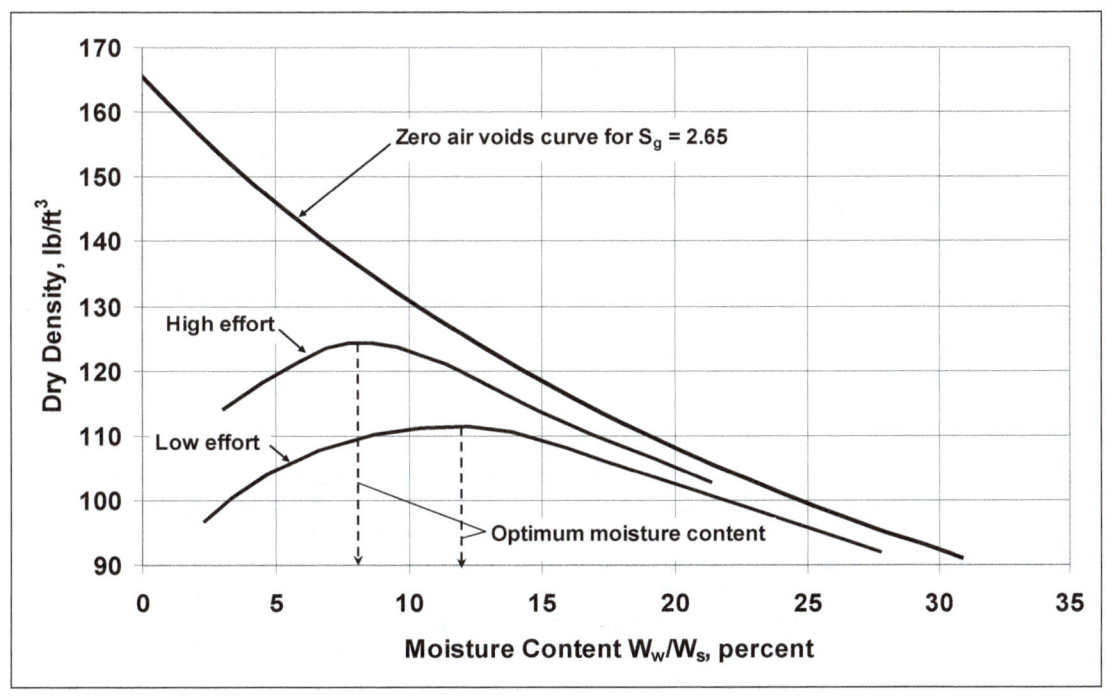

Figure 4.15. Moisture-density curves for different compactive efforts.

where e is the actual void ratio of the soil, defined as the ratio of the volume of voids V_v to the volume of solids V_s:

$$e = \frac{V_v}{V_s} \tag{4.13}$$

Qualitative descriptions of relative density are provided in Table 4.4.

Table 4.4. Soil Density (Sowers and Sowers 1970).		
Term	Relative Density, percent	Field test
Loose	0 – 50	Easily penetrated by ½-inch rebar pushed by hand
Firm	50 - 70	Easily penetrated by ½-inch rebar driven with a 5-lb hammer
Dense	70 - 90	Penetrated 12 inches by ½-inch rebar driven with a 5-lb hammer
Very dense	90 - 100	Penetrated only a few inches by ½-inch rebar driven with a 5-lb hammer

A few things to remember with respect to density and compaction are as follows:

1. All soils exhibit a range of density. The range for coarse grained soils tends to be larger when soils have multiple grain sizes and smaller when soils have more uniform grain size.

2. Uniformly graded soils get about as dense as they are going to get with little effort and well graded soils take considerable effort to densify.

3. Relative density and relative compaction are not the same. A soil with a relative density of 90% is about as dense as can practically be achieved in the field. A soil with relative compaction of 90% (with respect to Standard or Modified effort) can be made much denser and is generally less dense than acceptable in transportation related construction.

4. Increasing the density improves strength and stiffness, and reduces hydraulic conductivity.

4.4.4 Shear Strength

Shear strength is a term used in soil mechanics to describe the magnitude of the shear stress that a soil can sustain. The shear resistance of a saturated soil is a result of friction and interlocking of particles, and possibly cementation or bonding at particle contacts. Due to interlocking, particulate material may attempt to expand or contract in volume as it is subject to shear strains. If soil expands its volume, the density of particles will decrease and the strength will decrease; in this case, the peak strength would be followed by a reduction of shear stress. The stress-strain relationship levels off when the material stops expanding or contracting, and when interparticle bonds are broken. The theoretical state at which the shear stress and density remain constant while the shear strain increases is often referred to as the residual strength. However, soils with high clay content will continue to lose strength with even larger strains because the clay particles, which are platy in shape, become aligned with one another and form shear surfaces.

The shear strength of soil is always a function of the effective stress acting to confine the soil, not the total stress. Effective stress represents the intergranular forces between particles that contributes to the frictional strength and it is calculated by subtracting water pressure from the total confining stress.

The stress-strain relationship of soils, and therefore the shear strength, is affected by:

1. Soil composition (basic soil material): Mineralogy, grain size and grain size distribution, shape of particles, pore fluid type and content, ions on grain and in pore fluid.
2. State (initial): Defined by the initial void ratio, effective normal stress and shear stress (stress history). State can be described by terms such as: loose, dense, overconsolidated, normally consolidated, stiff, soft, contractive, dilative, etc.
3. Structure: Refers to the arrangement of particles within the soil mass; the manner the particles are packed or distributed. Features such as layers, joints, fissures, slickensides, voids, pockets, cementation, etc., are part of the structure. Structure of soils is described by terms such as: undisturbed, disturbed, remolded, compacted, cemented; flocculent, honey-combed, single-grained; flocculated, deflocculated; stratified, layered, laminated; isotropic and anisotropic.
4. Loading conditions: Effective stress path, i.e., drained vs. undrained; and type of loading, i.e., magnitude, rate (static, dynamic), and time history (monotonic, cyclic).

If water is not allowed to flow in or out of the soil, the stress path is called an *undrained stress path*. During undrained shear, if the particles are surrounded by a nearly incompressible fluid such as water, then the density of the particles cannot change without drainage, but the water pressure and effective stress will change. On the other hand, if the fluids are allowed to freely drain out of the pores, then the pore pressures will remain constant and the stress path is called a *drained stress path*. The soil is free to dilate or contract during shear if the soil is drained. In reality, soil is partially drained, somewhere between the perfectly undrained and drained idealized conditions. Exactly where is a function of the rate of load application and the hydraulic conductivity of the soil. For example, sandy and gravelly soils require the rapid loading of an earthquake to behave as an undrained material, whereas clayey soils can behave as undrained material at common excavation or embankment construction rates.

Probably the two most common tests for determining the shear strength of soils are (1) the direct shear test (AASHTO standard test T 236), and (2) the triaxial shear test (AASHTO standard tests T 296 and T 297). In the direct shear test, a sample of soil is placed in a rectangular box, the top half of which is free to slide over the bottom half. The lid of the box is free to move vertically, and a normal stress σ_n is applied to the lid. A horizontal shearing stress τ is applied to the top half of the box, gradually increasing in strength until the soil begins to shear.

In the triaxial shear test, a cylindrical soil sample is encased in a rubber membrane with rigid caps on top and bottom. The sample is then placed in a closed chamber and subjected to a confining pressure σ_3 on all sides using air or water as the confining medium. An axial stress σ_1 is applied to the ends of the cylinder. The axial stress is either increased, or the confining stress decreased, until the sample fails in shear, which happens along a diagonal plane or number of planes. A special case of the triaxial test is when the confining stress σ_3 is zero, which leads to a shear strength value known as the unconfined compressive strength. Table

4.5 provides a qualitative description of soils as defined by the unconfined compressive strength.

Unconfined compressive strength is a special case and is primarily useful for classifying soils. More generally, shear strength is a function of confining stress and the direct shear and triaxial tests, and other tests, are used to develop shear strength parameters of c and Φ (with respect to total stress) or c' and Φ' (with respect to effective stress). These parameters represent the coefficients of a straight line plotted through through the results of similar tests on the same soil, with the only variable being changes in confining stress. The tangent of Φ (or Φ') is the coefficient of friction and represents the frictional component of soil strength and c (or c') is the value of the intercept of the line, representing strength with no confining pressure (no friction). Under certain conditions soils with significant fines content (especially clay) exhibit a significant c intercept and this is the source of their label as 'cohesive' soils.

Table 4.5. Soil Strength (Sowers and Sowers 1970).

Term	Unconfined compressive strength (kips per square foot)	Field test
Very soft	0 – 0.5	Squeezes between fingers when fist is closed
Soft	0.5 – 1.0	Easily molded by fingers
Firm	1.0 – 2.0	Molded by strong pressure of fingers
Stiff	2.0 – 3.0	Dented by strong pressure of fingers
Very stiff	3.0 – 4.0	Dented only slightly by finger pressure
Hard	4.0 or greater	Dented only slightly by pencil point

A few important things to remember with respect to soil strength are as follows:

1. Soils do not have intrinsic strength properties.
2. Soil strength is proportional to the effective stress and it is possible to develop strength parameters c or c' and Φ or Φ' relative to a range of effective stress and other variables as listed in bullets 1 through 4 above.
3. Some strength parameters such as unconfined compressive strength and the parameters c and Φ do not refer to the effective stress. This only because assumptions have been made with respect to effective stress and this observation is useful to emphasize that strength parameters should only be used for engineering problems for which they are applicable, or as indices for empirical approaches, such as classification.
4. When a soil is partially saturated the water pressure in the pore spaces is actually negative and this causes an increase in effective stress. This is difficult to measure and is not often measured. Instead, the increased strength caused by the increase in effective stress is recognized as an apparent cohesion and it can be quite significant in fine grained soils. Partially saturated soil strength is higher than saturated soil strength and is not often relied upon in practice because of the likelihood that saturation will occur at some point and that stability needs to be ensured at that time.

4.4.5 Hydraulic Conductivity

Hydraulic conductivity, sometimes referred to as permeability, is a measure of the ability of soil to transmit water. Water moves through soil from high hydraulic head to lower hydraulic head. If the difference in hydraulic head between two locations in a saturated soil is denoted ΔH, and the length of the flow path between those locations is L, then the hydraulic gradient is defined as:

$$i = \frac{\Delta H}{L} \qquad (4.14)$$

where i is the hydraulic gradient in feet per foot. Darcy's law is an empirical relationship that relates the volumetric rate of flow Q through the soil to the hydraulic gradient i, cross sectional area of flow A, and the hydraulic conductivity of the porous material K:

$$Q = KiA \qquad (4.15)$$

Two standard laboratory test methods for determining hydraulic conductivity are AASHTO standard test T215, "Permeability of Granular Soils (Constant Head)" and ASTM D5084, "Measurement of Hydraulic Conductivity of Saturated Porous Materials Using a Flexible Wall Permeameter." In these tests, the amount of water passing through a saturated soil sample is measured over a specified time interval, along with the sample's cross-sectional area and the hydraulic head at specific locations. The soil's hydraulic conductivity is then calculated from these measured values. Figure 4.16 provides schematic diagrams of constant-head and falling-head test devices.

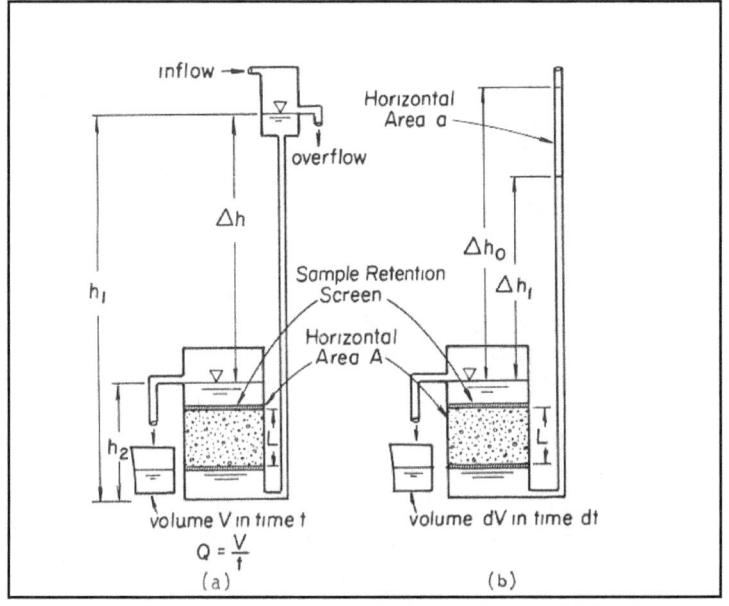

Figure 4.16. Permeameters: (a) Constant head, (b) Falling head. (McWhorter and Sunada 1977)

Hydraulic conductivity is related more to particle size distribution than to porosity, as water moves through large and interconnected voids more easily than small or isolated voids. Various equations are available to estimate hydraulic conductivity based on the grain size distribution, and the practitioner is encouraged to consult with geotechnical and materials engineers on estimating this property.

Table 4.6 lists representative values of void ratio, porosity and hydraulic conductivity for typical alluvial soils and common sedimentary rocks. Note that the hydraulic conductivity of common geomaterials can range over some seven orders of magnitude.

Table 4.6. Typical Void Ratio, Porosity, and Hydraulic Conductivity of Geomaterials (after McWhorter and Sunada 1977).

Type of material	Representative void ratio e (V_v/V_s)	Representative porosity n (V_v/V_T)	Representative hydraulic conductivity (cm/s)
Gravel	0.39	0.28	4×10^{-1}
Sand, coarse	0.64	0.39	5×10^{-2}
Sand, medium	0.69	0.41	1×10^{-2}
Sand, fine	0.75	0.43	3×10^{-3}
Silt	0.85	0.46	3×10^{-5}
Clay	0.72	0.42	9×10^{-8}
Sandstone (fine-grained)	0.52	0.34	3×10^{-4}
Siltstone	0.54	0.35	2×10^{-7}

4.5 CLASSIFICATION OF SOILS

A soil classification system is an arrangement of different soils into groups having similar properties. The purpose is to make it possible to estimate soil properties by association with soils of the same class whose properties are known, and to provide the engineer with a more or less accurate method of soil description. There are so many different soil properties of interest, and so many different combinations of these properties in any natural soil deposit, that any attempt at universal classification seems impractical.

However, classification systems are useful for identifying soil properties which are most important to that area of engineering for which the classification system was developed. In this section, the two most widely used systems in the United States by highway, bridge, and hydraulic engineers are presented.

4.5.1 Unified Soil Classification System

The Unified Soil Classification System (USCS) is a rapid method for identifying and grouping soils, originally developed by Casagrande for military construction purposes. The USCS is based on grain size and plasticity. In this system, soils are first divided into one of two classes: coarse-grained (more than 50% of the particles by weight are larger than 0.074 mm, or the #200 sieve), or fine-grained (more than 50% of the particles are finer than this size).

Coarse-grained soils are then identified by a letter designation "G" if more than half of the coarse fraction is greater than 4.76 mm (#4 sieve), or "S" if more than half are finer. The G or S is followed by a second letter:

- W if the soil is well-graded (i.e., a wide range of particle sizes) with little or no fines
- P if the soil is poorly graded, uniform, or gap-graded with little or no fines
- M if the soil contains appreciable amounts of silt (as assumed based on measured plasticity)
- C if the soil contains appreciable amounts of clay (as assumed based on measured plasticity)

Fine-grained soils are divided into 3 types: "C" for clays, "M" for silts and silty clays, and "O" for organic silts and clays. These symbols are followed by a second letter denoting the liquid limit as defined by the Atterberg method presented in Section 4.2: "L" for soils with a liquid limit less than 50, and "H" for a liquid limit exceeding 50. Fine-grained soils are divided using the plasticity chart of Casagrande as shown in Figure 4.17. Table 4.7 presents a summary of the Unified Soil Classification System.

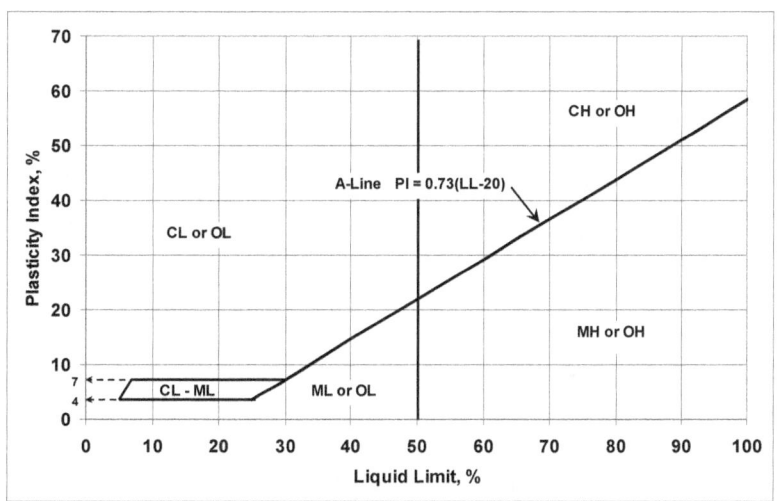

Figure 4.17. Plasticity chart for the Unified Soil Classification System for fine-grained soils.

4.5.2 AASHTO Classification System

The AASHTO classification system is commonly used for highway projects, and groups soils into categories having similar load carrying capacity and service characteristics for pavement subgrade design. It is useful in determining the relative quality of the soil material for use in earthwork structures, particularly embankments, subgrades, subbases and bases.

According to this system, soil is classified into seven major groups, A-1 through A-7. Soils classified under groups A-1, A-2, and A-3 are granular materials where 35% or less of the particles pass through the No. 200 sieve (0.075 mm). Soils where more than 35% pass the No. 200 sieve (0.075 mm) are classified under groups A-4, A-5, A-6, and A-7. Soils where more than 35% pass the No. 200 sieve (0.075 mm) have behavior, including plasticity and cohesion, that is characteristic of the fine grains, not the coarse grains.

Table 4.7. Unified Soil Classification System (after FHWA 2006).

Criteria for Assigning Group Symbols and Group Names Using Laboratory Tests
COARSE-GRAINED SOILS (Sands and Gravels): more than 50% retained on No. 200 (0.074 mm) sieve
FINE-GRAINED SOILS (Silts and Clays): 50% or more passes the No. 200 (0.074 mm) sieve

General Classification	Major Division	Minor Division	Soil Classification Group Symbol	Soil Classification Group Name
GRAVELS More than 50% of coarse fraction retained on No. 4 sieve	CLEAN GRAVELS < 5% fines	$C_u \geq 4$, and $1 \leq C_c \leq 3$	GW	Well-graded gravel
		$C_u < 4$, and $1 > C_c > 3$	GP	Poorly-graded gravel
	GRAVELS WITH FINES > 12% fines	Fines classify as ML or MH	GM	Silty gravel
		Fines classify as CL or CH	GC	Clayey gravel
SANDS More than 50% of coarse fraction passes No. 4 sieve	CLEAN SANDS < 5% fines	$C_u \geq 6$, and $1 \leq C_c \leq 3$	SW	Well-graded sand
		$C_u < 6$, and $1 > C_c > 3$	SP	Poorly-graded sand
	SANDS WITH FINES > 12% fines	Fines classify as ML or MH	SM	Silty sand
		Fines classify as CL or CH	SC	Clayey sand
SILTS AND CLAYS Liquid limit less than 50	Inorganic	PI > 7 and plots on or above "A" line	CL	Lean clay
		PI < 4 or plots below "A" line	ML	Silt
	Organic	Liquid limit – overdried / Liquid limit – not dried < 0.75	OL	Organic clay, organic silt
SILTS AND CLAYS Liquid limit 50 or more	Inorganic	PI plots on or above "A" line	CH	Fat clay
		PI plots below "A" line	MH	Elastic silt
	Organic	Liquid limit – overdried / Liquid limit – not dried < 0.75	OH	Organic clay, organic silt
Highly fibrous organic soils	Primarily organic matter, dark in color, and organic odor		Pt	Peat

The AASHTO classification system is based on the following criteria:

1. *Grain Size*: The grain size terminology for this classification system is as follows:

    Gravel: Fraction passing the 3 in (75 mm) sieve and retained on the No. 10 (2 mm) sieve.

    Sand: Fraction passing the No. 10 (2 mm) sieve and retained on the No. 200 (0.075 mm) sieve.

    Silt and clay: Fraction passing the No. 200 (0.075 mm) sieve.

2. *Plasticity*: The term *silty* and *clayey* are used as follows:

    Silty: Use when the fine fractions of the soil have a plasticity index of 10 or less.

    Clayey: Use when the fine fractions have a plasticity index of 11 or more.

3. If cobbles and boulders (size larger than 3 inches (75 mm)) are encountered they are excluded from the portion of the soil sample on which the classification is made. However, the percentage of material is recorded.

To evaluate the quality of a soil as a highway subgrade material, a number called the *group index* (GI) is also incorporated along with the groups and subgroups of the soil. The group index is written in parenthesis after the group or subgroup designation. The group index is given by Equation 4.16 where F is the percent passing the No. 200 (0.075 mm) sieve, LL is the liquid limit, and PI is the plasticity index.

$$GI = (F-35)[0.2+0.005(LL-40)] + 0.01(F-15)(PI-10) \tag{4.16}$$

In general, the quality of performance of a soil as a subgrade material is inversely proportional to the group index. The relationship of the Group Index to plasticity is shown in Figure 4.18. Table 4.8 presents a summary of the AASHTO classification system.

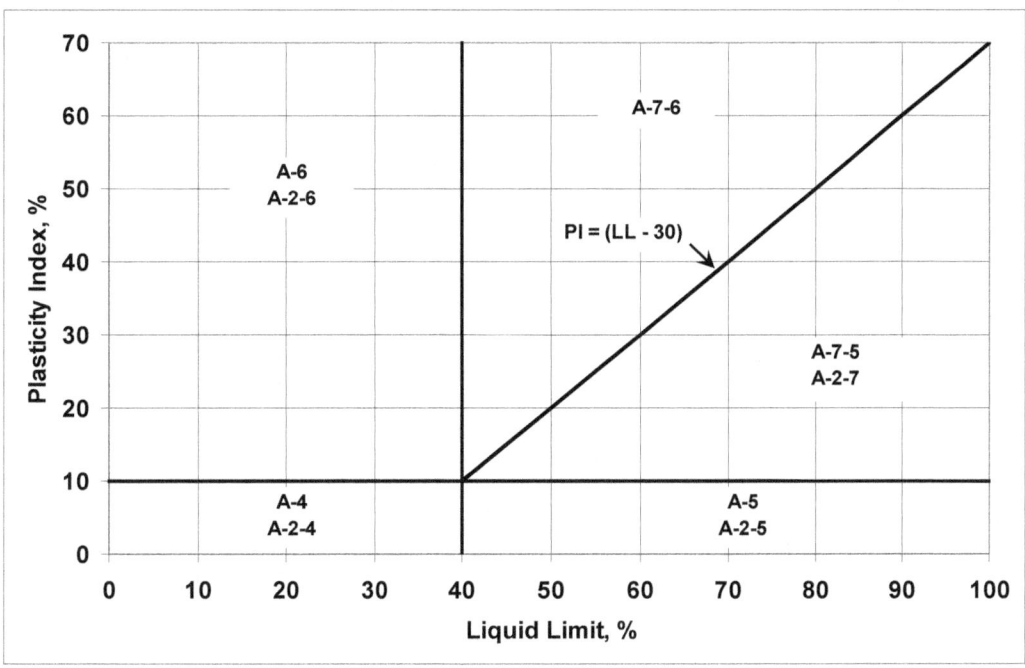

Figure 4.18. Plasticity chart for the AASHTO classification system for fine-grained soils.

4.6 ROCK PROPERTIES

4.6.1 Igneous, Sedimentary, and Metamorphic Rocks

Rocks are classified according to their origin into three major divisions: igneous, sedimentary, and metamorphic. These three groups are subdivided into types according to mineral and chemical composition, texture, and internal structure. Table 4.9 presents a summary of predominant rock types within these three categories.

4.6.2 Rock Mass Descriptions and Characteristics

When providing rock descriptions, geotechnical specialists should use technically correct geological terms. Local terms in common use may be acceptable if they help describe distinctive characteristics. Rock cores should be wetted before and during logging for consistency of color description and greater visibility of rock features such as hairline fractures. The guidelines presented in the International Society for Rock Mechanics (ISRM 1981) and Geotechnical Engineering Circular No. 5 (FHWA 2002) should be reviewed for additional information regarding logging procedures for core drilling.

Table 4.8. AASHTO Soil Classification System (FHWA 2006).

GENERAL CLASSIFICATION	GRANULAR MATERIALS (35 percent or less of total sample passing No. 200 sieve (0.075 mm)							SILT-CLAY MATERIALS (More than 35 percent of total sample passing No. 200 sieve (0.075 mm)			
GROUP CLASSIFICATION	A-1		A-3	A-2				A-4	A-5	A-6	A-7 A-7-5, A-7-6
	A-1-a	A-1-b		A-2-4	A-2-5	A-2-6	A-2-7				
Sieve analysis, percent passing:											
No. 10 (2 mm)	50 max.										
No. 40 (0.425 mm)	30 max.	50 max.	51 min.								
No. 200 (0.075 mm)	15 max.	25 max.	10 max.	35 max	35 max	35 max	35 max.	36 min.	36 min.	36 min.	36 min.
Characteristics of fraction passing No 40 (0.425 mm)											
Liquid limit				40 max.	41 min.	40 max.	41 min.	40 max.	41 min.	40 max.	41 min.
Plasticity index	6 max.		NP	10 max.	10 max.	11 min.	11 min.	10 max.	10 max.	11 min.	11 min.*
Usual significant constituent materials	Stone fragments, gravel and sand		Fine sand	Silty or clayey gravel and sand				Silty soils		Clayey soils	
Group Index**	0		0	0		4 max.		8 max.	12 max.	16 max.	20 max.

Classification procedure:

With required test data available, proceed from left to right on chart; correct group will be found by process of elimination. The first group from left into which the test data will fit is the correct classification.

*Plasticity Index of A-7-5 subgroup is equal to or less than LL minus 30. Plasticity Index of A-7-6 subgroup is greater than LL minus 30 (see Fig 4.18).

**See group index formula (Eq. 4-16 Group index should be shown in parentheses after group symbol as: A-2-6(3), A-4(5), A-7-5(17), etc.

Table 4.9. Rock Groups and Types (FHWA 2006).		
Igneous	Intrusive (Coarse Grained)	Granite, Syenite, Diorite, Diabase, Gabbro, Peridotite, Pegmatite
Igneous	Extrusive (Fine Grained)	Rhyolite, Trachyte, Andesite, Basalt
Igneous	Pyroclastic	Obsidian, Pumice, Tuff
Sedimentary	Clastic (Sediment)	Shale, Mudstone, Claystone, Siltstone, Sandstone, Conglomerate, Limestone, oolitic
Sedimentary	Chemically Formed	Limestone, Dolomite, Gypsum, Halite
Sedimentary	Organic Remains	Chalk, Coquina, Lignite, Coal
Metamorphic	Foliated	Slate, Phyllite, Schist, Gneiss
Metamorphic	Non-foliated	Quartzite, Amphibolite, Marble, Hornfel

The rock's lithologic description should include as a minimum the following items:

- Rock type
- Mineral composition
- Color
- Grain size and shape
- Texture (stratification/foliation)
- Weathering and alteration
- Strength
- Hardness
- Discontinuities, joints, and fractures
- Other relevant notes

Rock types are classified according to their origin into three major divisions: igneous, sedimentary, and metamorphic. These three groups are subdivided into types according to mineral and chemical composition, texture, and internal structure. Predominant rock types within these categories are presented in Table 4.9. For some projects a library of hand samples and photographs representing lithologic rock types present in the project area should be maintained.

The mineral composition of rock should be identified by a geologist based on experience and the use of appropriate references. The most abundant mineral should be listed first, followed by minerals in decreasing order of abundance. For some common rock types, the mineral composition need not be specified (e.g., dolomite, limestone). Color used to describe rock material should be consistent with a Munsell Color Chart (USDA 1993) and recorded for both wet and dry conditions as appropriate.

Grain size should be classified according to the terms presented in Table 4.10. The grain size descriptions in Table 4.10 are consistent with those used in the USCS for soil particles. Table 4.11 is used to classify the grain shape.

Table 4.10. Terms Used to Describe Grain Size (Typically for Sedimentary Rocks) (FHWA 2006).

Description	Grain Size, sieve (mm)	Characteristic of Individual Grains
Very coarse grained	#4 (> 4.75)	Can be easily distinguished by eye
Coarse grained	#10 to #4 (2.00 - 4.75)	Can be easily distinguished by eye
Medium grained	#40 to #10 (0.425 - 2.00)	Can be distinguished by eye
Fine grained	#200 to #40 (0.075 - 0.425)	Can be distinguished by eye with difficulty
Very fine grained	< #200 (< 0.075)	Cannot be distinguished by unaided eye

Table 4.11. Terms Used to Describe Grain Shape (for Sedimentary Rocks) (FHWA 2006).

Description	Characteristic
Angular	Showing very little evidence of wear. Grain edges and corners are sharp. Secondary corners are numerous and sharp.
Subangular	Showing some evidence of wear. Grain edges and corners are slightly rounded off. Secondary corners are slightly less numerous and slightly less sharp than in angular grains.
Subrounded	Showing considerable wear. Grain edges and corners are rounded to smooth curves. Secondary corners are reduced greatly in number and highly rounded.
Rounded	Showing extreme wear. Grain edges and corners are smoothed off to broad curves. Secondary corners are few in number and rounded.
Well-rounded	Completely worn. Grain edges or corners are not present. No secondary edges or corners are present.

The texture of rock in terms of significant non-fracture structural features should be described. Rock textures are separated into foliated and non-foliated categories. Foliated rock is a product of differential stress that deforms the rock in one plane, sometimes creating a plane of cleavage. For example, slate is a foliated metamorphic rock, originating from shale. Non-foliated rock does not exhibit planar patterns of strain. The thickness of bedding/foliation should be described by using the terms in Table 4.12. The orientation of the bedding/ foliation should be measured from the horizontal or with respect to the core axis.

Table 4.12. Terms Used to Describe Stratum Thickness (FHWA 2006).	
Descriptive Term	Stratum Thickness (m or mm)
Very thickly bedded	(> 1 m)
Thickly bedded	(0.5 to 1.0 m)
Thinly bedded	(50 mm to 500 mm)
Very thinly bedded	(10 mm to 50 mm)
Laminated	(2.5 mm to 10 mm)
Thinly laminated	(< 2.5 mm)

<u>Weathering</u> is the breaking down of the earth's rocks, soils and minerals through direct contact with the planet's atmosphere and biosphere. Weathering occurs *in situ*, or "with no movement", and thus should not be confused with erosion, which involves the movement of rocks and minerals by agents such as water, ice, wind, and gravity. Weathering and the associated alteration of rock can be caused by physical, chemical, biological, and thermal mechanisms. Terms used to describe weathering and alteration are presented in Table 4.13.

The <u>strength</u> of rock materials can be conveniently classified using the point load test. This test is an index test that may also be used to predict other strength parameters with which it is correlated, such as the uniaxial tension and compressive strengths, and is therefore recommended for the measurement of sample strength. The point-load index, I_s, obtained from the point load test should be converted to uniaxial compressive strength. The test apparatus can accommodate irregularly-shaped rock samples as well as rectangular (saw-cut) or cylindrical samples (for example, samples obtained from a core barrel).

Categories and terminology for describing rock strength based on the uniaxial compressive strength are presented in Table 4.14. The table also presents guidelines for common qualitative assessments of strength that can be performed with the aid of a geologist's hammer and a pocket knife. The field estimates of strength properties should be confirmed where appropriate by comparison with selected laboratory tests.

Table 4.13. Terms Used to Describe Rock Weathering and Alteration (ISRM 1981).	
Grade (Term)	**Description**
I (Fresh)	Rock shows no discoloration, loss of strength, or other effects of weathering/alteration
II (Slightly Weathered/Altered)	Rock is slightly discolored, but not noticeably lower in strength than fresh rock
III (Moderately Weathered/Altered)	Rock is discolored and noticeably weakened, but less than half is decomposed; a minimum 2 in (50 mm) diameter sample cannot be broken readily by hand across the rock fabric
IV (Highly Weathered/Altered)	More than half of the rock is decomposed; rock is weathered so that a minimum 2 in (50 mm) diameter sample can be broken readily by hand across the rock fabric
V (Completely Weathered/Altered)	Original minerals of rock have been almost entirely decomposed to secondary minerals even though the original fabric may be intact; material can be granulated by hand
VI (Residual Soil)	Original minerals of rock have been entirely decomposed to secondary minerals, and original rock fabric is not apparent; material can be easily broken by hand

Table 4.14. Terms Used to Describe the Strength of Rock (ISRM 1981).		
Grade (Description)	Field Identification	Approximate Range of Uniaxial Compressive Strength, psi (kPa)
R0 (Extremely Weak Rock)	Can be indented by thumbnail	35 to 150 (250 to 1,000)
R1 (Very Weak Rock)	Can be peeled by pocket knife	150 to 725 (1,000 to 5,000)
R2 (Weak Rock)	Can be peeled with difficulty by pocket knife	725 to 3,500 (5,000 to 25,000)
R3 (Medium Strong Rock)	Can be indented 3/16 in (5 mm) with sharp end of pick	3,500 to 7,000 (25,000 to 50,000)
R4 (Strong Rock)	Requires one blow of geologist's hammer to fracture	7,000 to 15,000 (50,000 to 100,000)
R5 (Very Strong Rock)	Requires many blows of geologist's hammer to fracture	15,000 to 36,000 (100,000 to 250,000)
R6 (Extremely Strong Rock)	Can only be chipped with blows of geologist's hammer	> 36,000 (>250,000)

Hardness of rock is commonly assessed by the scratch test. Descriptions and abbreviations used to describe rock hardness are presented in Table 4.15.

Discontinuity is the general term for any mechanical break in a rock mass that results in zero or low tensile strength. Discontinuity is the collective term used for most types of joints, weak bedding planes, weak schistosity planes, weakness zones, and faults. The spacing between discontinuities is defined as the perpendicular distance between adjacent discontinuities. The spacing should be measured perpendicular to the planes in the set. Table 4.16 presents guidelines to describe discontinuity spacing.

Table 4.15. Terms Used to Describe Rock Hardness (FHWA 2002).	
Description (Abbreviation)	Characteristic
Soft (S)	Reserved for plastic material alone.
Friable (F)	Easily crumbled by hand, pulverized or reduced to powder.
Low Hardness (LH)	Can be gouged deeply or carved with a pocket knife.
Moderately Hard (MH)	Can be readily scratched by a knife blade; scratch leaves a heavy trace of dust and scratch is readily visible after the powder has been blown away.
Hard (H)	Can be scratched with difficulty; scratch produces little powder and is often faintly visible; traces of the knife steel may be visible.
Very Hard (VH)	Cannot be scratched with pocket knife. Leaves knife steel marks on surface.

Discontinuities should be described as closed, open, or filled. *Aperture* is the term used to describe the perpendicular distance separating the adjacent rock walls of an open discontinuity in which the intervening space is filled with air or water. *Width* is the term used to describe the distance separating the adjacent rock walls of discontinuities filled with broken rock material or soil. The terms presented in Table 4.17 should be used to describe apertures. Terms such as "wide," "narrow" and "tight" are used to describe the width of discontinuities such as the thickness of veins, fault gouge filling, or joints openings as indicated in Tables 4.16 and 4.17.

For faults or shears that are not thick enough to be represented on the boring log, the measured thickness is recorded numerically in millimeters. Discontinuities are further characterized by the surface shape of the joint and the roughness of its surface in addition to the fill material separating the adjacent rock walls of the discontinuities. Filling is characterized by its type, amount, width (i.e., perpendicular distance between adjacent rock walls) and strength. If non-cohesive fillings are identified, then the filling should be identified qualitatively, e.g., fine sand. Refer to Table 4.16 for guidelines to characterize these features.

Table 4.16. Terms to Describe Discontinuities (after ISRM 1981).

Discontinuity Type	Amount of Infilling	Type of Infilling
F – Fault J - Joint Sh - Shear Fo - Foliation V - Vein B - Bedding	Su – Surface stain Sp - Spotty Pa – Partially filled Fi - Filled No - None	Cl – Clay Ca - Calcite Ch - Chlorite Fe – Iron Oxide Gy – Gypsum/Talc H - Healed No - None Py - Pyrite Qz - Quartz Sd – Sand

Discontinuity Width (mm)	Discontinuity Spacing (m)	Surface Shape of Joint
W – Wide (12.5-5.0) MW - Moderately Wide (2.5-12.5) N - Narrow (1.25-2.5) VN - Very Narrow (<1.25) T - Tight (~ 0)	EW - Extremely Wide (>6) VW - Very Wide (2-6) W - Wide (0.6-2) M - Moderate (0.2-0.6) C - Close (0.06-0.2) VC - Very Close (0.02-0.06) EC - Extremely close (<0.02)	Wa – Wavy Pl - Planar St - Stepped Ir - Irregular

Roughness of Surface

Slk - Slickensided (surface has smooth, glassy finish with visual evidence of striations)
S - Smooth (surface appears smooth and feels so to the touch)
SR - Slightly Rough (asperities on the discontinuity surface are distinguishable and can felt)
R - Rough (some ridges and side-angle steps are evident; asperities are clearly visible, and discontinuity surface feels very abrasive)
VR - Very Rough (near-vertical steps and ridges occur on the discontinuity surface

Table 4.17. Terms to Classify Discontinuities Based on Aperture Size (ISRM 1981).		
Aperture (mm)	Description	Description
<0.1 0.1 - 0.25 0.25 – 0.5	Very tight Tight Partly open	"Closed Features"
0.5 – 2.5 2.5 - 10 > 10	Open Moderately open Wide	"Gapped Features"
1-100 100-1000 >1 m	Very wide Extremely wide Cavernous	"Open Features"

Naturally occurring fractures are numbered and described using the same terminology that is used for discontinuities (Table 4.16). During core drilling and recovery, the number of naturally occurring fractures observed in each 1 ft (0.3 m) of core should be recorded as the fracture frequency. Mechanical breaks, thought to have occurred during drilling, are not counted. The following criteria can be used to identify natural breaks:

1. A rough brittle surface with fresh cleavage planes in individual rock minerals suggests an artificial fracture.
2. A generally smooth or somewhat weathered surface with soft coating or infilling materials, such as talc, gypsum, chlorite, mica, or calcite indicates a natural discontinuity.
3. In rocks showing foliation, cleavage or bedding it may be difficult to distinguish between natural discontinuities and artificial fractures when the discontinuities are parallel with the incipient weakness planes. If drilling has been carried out carefully, then the questionable breaks should be counted as natural features to be on the conservative side.
4. Depending upon the drilling equipment, part of the length of core being drilled may occasionally rotate with the inner barrels in such a way that grinding of the surfaces of discontinuities and fractures occurs. In weak rock types it may be very difficult to decide if the resulting rounded surfaces represent natural or artificial features. When in doubt, conservatively assume that they are natural.

The description of fractures can be strongly time dependent and moisture-content dependent in the case of certain varieties of shales and mudstones that have relatively weakly developed diagenetic bonds. A diagenetic bond is the bond that is formed in a deposited sediment by chemical and physical processes during its conversion to rock. A frequent problem is "discing," in which an initially intact core separates into discs on incipient planes. The process generally becomes noticeable perhaps within a few minutes of core recovery. This phenomenon is experienced in several different forms:

1. Stress relief cracking and swelling by the initially rapid release of strain energy in cores recovered from areas of high stress, especially in the case of shaley rocks.
2. Dehydration cracking experienced in the weaker mudstones and shales that may reduce Rock Quality Designation (RQD) values from 100% to 0% in a matter of minutes. The initial integrity might possibly have been due to negative pore water pressure.
3. Slaking and cracking experienced by some of the weaker mudstones and shales when they are subjected to wetting and drying.

Any of these forms of "discing" may make logging of fracture frequency unreliable. Whenever such conditions are anticipated, core should be logged by a geotechnical specialist as it is being recovered and at subsequent intervals until the phenomenon is predictable.

Recommended values for allowable bearing pressure of bridge footings founded in different rock types as a function of the Rock Quality Designation (RQD) are provided in Table 4.18 (FHWA 1991).

Table 4.18. Recommended Allowable Bearing Pressure for Footings on Rock (FHWA 1991).		
MATERIAL	ALLOWABLE CONTACT PRESSURE	
	Kpa	(tsf)
Such igneous and sedimentary rock as crystalline bedrock, including granite, diorite, gneiss, traprock; and hard limestone, and dolomite, in sound condition:		
RQD = 75 to 100 percent	11491	(120 tsf)
RQD = 50 to 75 percent	6224	(65 tsf)
RQD = 25 to 50 percent	2873	(30 tsf)
RQD = 0 to 25 percent	958	(10 tsf)
Such metamorphic rock as foliated rocks, such as schist or slate; and bedded limestone, in sound condition:		
RQD > 50 percent	3830	(40 tsf)
RQD < 50 percent	958	(10 tsf)
Sedimentary rocks, including hard shales and sandstones, in sound condition:		
RQD > 50 percent	2394	(25 tsf)
RQD < 50 percent	958	(10 tsf)
Soft or broken bedrock (excluding shale), and soft limestone:		
RQD > 50 percent	1149	(12 tsf)
RQD < 50 percent	766	(8 tsf)
Soft shale	383	(4 tsf)

4.7 CLASSIFICATION OF ROCK

For engineering purposes, rock classification is not based on origin or age, for example, but on how it is expected to behave. Some common classification systems are the Rock Mass Rating system (Bieniawski 1989) and the Erodibility Index method (Annandale 1995).

4.7.1 Rock Mass Rating System

In determining the rock strength for transportation facilities constructed in, on, or of rock, it is most important to account for the presence of discontinuities, such as joints, fractures, faults or bedding planes. Therefore, for most conditions, the rock mass strength properties, rather than the intact rock properties must be determined for use in design. The rock mass is the in-situ, fractured rock that will almost always have significantly lower strength than the intact rock because of discontinuities that divide the rock mass into blocks. Therefore, the strength of the rock mass will depend on such factors as the shear strength of the surfaces of the blocks, the spacing and continuous length of the discontinuities, and their alignment relative to the direction of loading.

Using these factors, Bieniawski (1989) proposed a method for estimating rock mass properties from an index that characterizes the overall properties of the rock mass quality. This index is known as the rock mass rating (RMR). Originally developed for tunnel support design, the RMR has been adopted by AASHTO (2010) because the RMR is determined from readily measurable parameters. Table 4.19 identifies the following five measurable parameters and assigns relative ratings to each parameter:

1. Strength of intact rock material
2. Drill core quality as expressed by the Rock Quality Designation (RQD)
3. Spacing of joints
4. Condition of joints
5. Ground water conditions

The Rock Quality Designation (RQD) (see bullet 2) is a standard parameter in drill core logging and is determined as the ratio between the sum of the lengths of pieces of rock that are longer than 0.1 m and the total core run length (usually 1.5 m), expressed as a percent. RQD values range between 5 and 100. A RQD of 5 represents very poor quality rock, and a RQD of 100 represents very good quality rock.

The Rock Mass Rating (RMR) is a numerical value determined as the sum of the five relative ratings. The RMR should be adjusted in accordance with the criteria in Table 4.20. The rock classification should be determined in accordance with Table 4.21 where RMR refers to the adjusted value. Note that the use of the RMR has been superseded by the Geological Strength Index (GSI) in some FHWA technical guidance (FHWA 2010) as it provides a more direct correlation to the Hoek-Brown strength parameters. The GSI is essentially an extension of the RMR system.

4.7.2 Erodibility Index Method

Another rock classification system uses rock mass properties to classify rock materials with respect to their erosion potential when subjected to flowing water. The Erodibility Index method developed by Annandale (1995) defines an erodibility threshold for a wide variety of earth materials and hydraulic flow conditions by relating the erosive power of water to a geomechanical index. Erosive power is expressed in terms of stream power, also known as the rate of energy dissipation. The Erodibility Index quantifies the ability of earth materials to resist erosion.

Table 4.19. Geomechanics Classification of Rock Masses (AASHTO 2010).

PARAMETER		RANGES OF VALUES						
1. Strength of intact rock material	Point load strength index	>1,200 psi	600 to 1,200 psi	300 to 600 psi	150 to 300 psi	For this low range – uniaxial compressive test is preferred		
	Uniaxial compressive strength	>30,000 psi	15,000 to 30,000 psi	7,500 to 15,000 psi	3,600 to 7,500 psi	1,500 to 3,600 psi	500 to 1,500 psi	150 to 500 psi
Relative Rating		15	12	7	4	2	1	0
2. Drill core quality RQD		90% to 100%	75% to 90%	50% to 75%	25% to 50%	<25%		
Relative Rating		20	17	13	8	3		
3. Spacing of joints		>10 ft	3 to 10 ft	1 to 3 ft	2 in. to 1 foot	<2 in.		
Relative Rating		30	25	20	10	5		
4. Condition of joints		• Very rough surfaces • Not continuous • No separation • Hard joint wall rock	• Slightly rough surfaces • Separation <0.05" • Hard joint wall rock	• Slightly rough surfaces • Separation <0.05" • Soft joint wall rock	• Slickensided surfaces - or - • Gouge <0.2 in thick –or– • Joints open 0.05-0.2" • Continuous joints	• Soft gouge >0.2" thick - or - • Joints open >0.2" • Continuous joints		
Relative Rating		25	20	12	6	0		
5. Ground water conditions (use one of the three evaluation criteria as appropriate to the method of exploration)	Inflow per 30 ft tunnel length	None	<400 gallons/hr	400 to 2,000 gallons/hr	>2,000 gallons/hr			
	Ratio= joint water pressure/ major principal stress	0	0.0 to 0.2	0.2 to 0.5	>0.5			
	General Conditions	Completely Dry	Moist only (interstitial water)	Water under moderate pressure	Severe water problems			
Relative Rating		10	7	4	0			

Table 4.20. Geomechanics Rating Adjustment for Joint Orientations (after AASHTO 2010).

Orientations of Joints		Very Favorable	Favorable	Fair	Unfavorable	Very Unfavorable
Ratings	Tunnels	0	-2	-5	-10	-12
	Foundations	0	-2	-7	-15	-25
	Slopes	0	-5	-25	-50	-60

Table 4.21. Geomechanics Rock Mass Classes Determined From Total Ratings (AASHTO 2010).

RMR (Note 1)	100 to 81	80 to 61	60 to 41	40 to 21	<20
Class No.	I	II	III	IV	V
Description	Very good rock	Good rock	Fair rock	Poor rock	Very poor rock

Note 1: RMR is adjusted for structural application and rock joint orientation as per Table 4.20 prior to evaluating the Class No.

The Erodibility Index is identical to Kirsten's excavatability index which is used to characterize rock for determining the power requirements of earth-moving equipment that can rip the subject material. The index is expressed as the product of four parameters:

$$K = (M_s)(K_b)(K_d)(J_s) \tag{4.17}$$

where:

K = Erodibility Index
M_s = Intact rock mass strength parameter
K_b = Block size parameter
K_d = Shear strength parameter
J_s = Relative orientation parameter

The values of the parameters are determined by making use of tables and equations published by Annandale (1995) and Kirsten (1982) as provided in Tables 4.22 through 4.26 below. The intact rock mass strength parameter M_s is related to the unconfined compressive strength as shown in Table 4.22.

Joint spacing and the number of joint sets within a rock mass determines the value of K_b for rock. Joint spacing is estimated from borehole data by means of the rock quality designation (RQD) and the number of joint sets is represented by the joint set number (J_n). The values of the joint set numbers (J_n) are found in Table 4.23. As seen in the table, J_n is a function of the number of joint sets, ranging from rock with no or few joints (essentially intact rock), to rock formations consisting of one to more than four joint sets. The classification accounts for rock that displays random discontinuities in addition to regular joint sets. Random joint discontinuities are discontinuities that do not form regular patterns. For example, rock with two joint sets and random discontinuities is classified as having two joint sets plus random. Having determined the values of RQD and J_n, K_b is calculated as:

$$K_b = \frac{RQD}{J_n} \tag{4.18}$$

Table 4.22. Values of the Rock Mass Strength Parameter M_s.			
Hardness	Identification in Profile	Unconfined Compressive Strength (MPa)	Mass Strength Number (M_s)
Very soft rock	Material crumbles under firm (moderate) blows with sharp end of geological pick and	Less than 1.7	0.87
	can be peeled off with a knife; is too hard to cut triaxial sample by hand.	1.7 – 3.3	1.86
Soft rock	Can just be scraped and peeled with a knife; indentations 1 mm to 3-mm show in the	3.3 – 6.6	3.95
	specimen with firm (moderate) blows of the pick point.	6.6 – 13.2	8.39
Hard rock	Cannot be scraped or peeled with a knife; hand-held specimen can be broken with hammer end of geological pick with a single firm (moderate) blow.	13.2 – 26.4	17.70
Very hard rock	Hand-held specimen breaks with hammer end of pick under more than one blow.	26.4 – 53.0
53.00 – 106.0	35.0
70.0
Extremely hard rock	Specimen requires many blows with geological pick to break through intact material.	Larger than 212.0	280.0

With the values of RQD ranging between 5 and 100, and those of J_n ranging between 1 and 5, the value of K_b ranges between 1 and 100 for rock.

Table 4.23. Rock Joint Set Number J_n.	
Number of Joint Sets	Joint Set Number (J_n)
Intact, no or few joints/fissures	1.00
One joint/fissure set	1.22
One joint/fissure set plus random	1.50
Two joint/fissure sets	1.83
Two joint/fissure sets plus random	2.24
Three joint/fissure sets	2.73
Three joint/fissure sets plus random	3.34
Four joint/fissure sets	4.09
Multiple joint/fissure sets	5.00

The discontinuity or shear strength number (K_d) is the parameter that represents the relative strength of discontinuities in rock. In rock, it is determined as the ratio between joint wall roughness (J_r) and joint wall alteration (J_a), where J_r represents the degree of roughness of opposing faces of a rock discontinuity, and J_a represents the degree of alteration of the materials that form the faces of the discontinuity. Alteration relates to amendments of the rock surfaces, for example weathering or the presence of cohesive material between the opposing faces of a joint. Values of J_r and J_a can be found in Tables 4.24 and 4.25. The values of K_d calculated with the information in these tables change with the relative degree of resistance offered by the joints. Increases in resistance are characterized by increases in the value of K_d. The shear strength of a discontinuity is directly proportional to the degree of roughness of opposing joint faces and inversely proportional to the degree of alteration.

$$K_d = \frac{J_r}{J_a} \tag{4.19}$$

Table 4.24. Joint Roughness Number J_r.

Condition of Joint	Joint Roughness Number J_r
Stepped joints/fissures	4.0
Rough or irregular, undulating	3.0
Smooth undulating	2.0
Slickensided undulating	1.5
Rough or irregular, planar	1.5
Smooth planar	1.0
Slickensided planar	0.5
Joints/fissures either open or containing relatively soft gouge of sufficient thickness to prevent joint/fissure wall contact upon excavation	1.0
Shattered or micro-shattered clays	1.0

Table 4.25. Joint Alteration Number J_a.

Description of Gouge	Joint Alteration Number (J_a) for Joint Separation (mm)		
	1.0 [1]	1.0 –5.0 [2]	5.0 [3]
Tightly healed, hard, non-softening impermeable filling	0.75	-	-
Unaltered joint walls, surface staining only	1.0	-	-
Slightly altered, non-softening, non-cohesive rock mineral or crushed rock filling	2.0	2.0	4.0
Non-softening, slightly clayey non-cohesive filling	3.0	6.0	10.0
Non-softening, strongly over-consolidated clay mineral filling, with or without crushed rock	3.0	6.0**	10.0
Softening or low friction clay mineral coatings and small quantities of swelling clays	4.0	8.0	13.0
Softening moderately over-consolidated clay mineral filling, with or without crushed rock	4.0	8.00**	13.0
Shattered or micro-shattered (swelling) clay gouge, with or without crushed rock	5.0	10.0**	18.0

Note:
(1) Joint walls effectively in contact.
(2) Joint walls come into contact after approximately 100-mm shear.
(3) Joint walls do not come into contact at all upon shear.
**Also applies when crushed rock occurs in clay gouge without rock wall contact.

Relative orientation, in the case of rock, is a function of the relative shape of the rock and its dip and dip direction relative to the direction of flow. The relative orientation parameter J_s represents the relative ability of earth material to resist erosion due to the structure of the ground. This parameter is a function of the dip and dip direction of the least favorable discontinuity (most easily eroded) in the rock with respect to the direction of flow, and the shape of the material units. These two variables (orientation and shape) affect the ease by which the stream can penetrate the ground and dislodge individual material units.

Conceptually, the function of the relative orientation parameter J_s incorporating shape and orientation is as follows. If rock is dipped against the direction flow, it will be more difficult to scour the rock than when it is dipped in the direction of flow. When it is dipped in the direction of flow, it is easier for the flow to lift the rock, penetrate underneath and remove it. Rock that is dipped against the direction of flow will be more difficult to dislodge. The shape of the rock, represented by the length to width ratio r, impacts the erodibility of rock in the following manner. Elongated rock will be more difficult to remove than equi-sided blocks of rock. Therefore, large ratios of r represent rock that is more difficult to remove because it represents elongated rock shapes. Values of the relative orientation parameter J_s are provided in Table 4.26.

The material characteristics to quantify the Erodibility Index parameters are generally obtained from borehole data, field observation and testing, and laboratory testing (to obtain the unconfined compressive strength). Depending on the importance of the project, it is also possible to obtain parameter values by making use of geologic descriptions of the material [see tables of Annandale (1995)]. Larger values of the Erodibility Index value K indicate greater resistance to erosion (see Section 7.13).

4.8 SUMMARY

An understanding of soil and rock property classification is important because it provides a basis for describing common engineering properties of geomaterials and how different materials may be expected to behave under various environmental conditions and loads. As noted in Sections 4.2 and 4.3, the physical processes causing erosion of different types of soils and rock vary based on the nature of the material. Various methods for estimating and/or measuring erodibility characteristics also depend of the nature of the material being considered.

The characteristics of soils and rock (the resisting materials) are important to estimating scour and erosion under different combinations of geotechnical and hydraulic conditions. While the most widely used equations for scour assume cohesionless materials such as sand or gravel (see Chapters 6, 7, and 8), some guidance is available for estimating scour components in cohesive soils and rock. Reference is suggested to the following sections:

Section 6.7	Contraction Scour in Cohesive Materials
Section 6.8	Contraction Scour in Erodible Rock
Section 7.12	Pier Scour in Cohesive Materials
Section 7.13	Pier Scour in Erodible Rock

| Table 4.26. Relative Orientation Parameter J_s. ||||||
| Dip Direction of Closer Spaced Joint Set (degrees) | Dip Angle of Closer Spaced Joint Set (degrees) | Ratio of Joint Spacing, r ||||
Dip Direction	Dip Angle	Ratio 1:1	Ratio 1:2	Ratio 1:4	Ratio 1:8
180/0	90	1.14	1.20	1.24	1.26
In direction of stream flow	89	0.78	0.71	0.65	0.61
In direction of stream flow	85	0.73	0.66	0.61	0.57
In direction of stream flow	80	0.67	0.60	0.55	0.52
In direction of stream flow	70	0.56	0.50	0.46	0.43
In direction of stream flow	60	0.50	0.46	0.42	0.40
In direction of stream flow	50	0.49	0.46	0.43	0.41
In direction of stream flow	40	0.53	0.49	0.46	0.45
In direction of stream flow	30	0.63	0.59	0.55	0.53
In direction of stream flow	20	0.84	0.77	0.71	0.67
In direction of stream flow	10	1.25	1.10	0.98	0.90
In direction of stream flow	5	1.39	1.23	1.09	1.01
In direction of stream flow	1	1.50	1.33	1.19	1.10
0/180	0	1.14	1.09	1.05	1.02
Against direction of stream flow	-1	0.78	0.85	0.90	0.94
Against direction of stream flow	-5	0.73	0.79	0.84	0.88
Against direction of stream flow	-10	0.67	0.72	0.78	0.81
Against direction of stream flow	-20	0.56	0.62	0.66	0.69
Against direction of stream flow	-30	0.50	0.55	0.58	0.60
Against direction of stream flow	-40	0.49	0.52	0.55	0.57
Against direction of stream flow	-50	0.53	0.56	0.59	0.61
Against direction of stream flow	-60	0.63	0.68	0.71	0.73
Against direction of stream flow	-70	0.84	0.91	0.97	1.01
Against direction of stream flow	-80	1.26	1.41	1.53	1.61
Against direction of stream flow	-85	1.39	1.55	1.69	1.77
Against direction of stream flow	-89	1.50	1.68	1.82	1.91
180/0	-90	1.14	1.20	1.24	1.26

Notes:

1. For intact material take J_s = 1.0.
2. For values of r greater than 8 take J_s as for r = 8.
3. If the flow direction FD is not in the direction of the true dip TD, the effective dip ED is determined by adding the ground slope to the apparent dip AD: ED = AD + GS

CHAPTER 5

LONG-TERM AGGRADATION AND DEGRADATION

5.1 INTRODUCTION

This chapter discusses the factors affecting long-term bed elevation changes, methods available for estimating these changes, and the role of sediment transport computer models to complement HEC-20 procedures. This chapter links long-term degradation, which is fundamentally a stream instability problem, to the other components of scour at a bridge site. In following chapters methods and equations are given for determining the other components of total scour. Procedures for estimating long-term aggradation and degradation at a bridge are presented in HEC-20 (FHWA 2012b).

5.2 LONG-TERM BED ELEVATION CHANGES

Long-term bed elevation changes may be the natural trend of the stream or may be the result of some modification to the stream or watershed. The streambed may be aggrading, degrading, or in relative equilibrium in the vicinity of the bridge crossing. In this section, long-term trends are considered. Long-term aggradation and degradation do not include the cutting and filling of the streambed at a bridge that might occur during a runoff event (contraction and local scour). A stream may cut and fill at specific locations during a runoff event and also have a long-term trend of an increase or decrease in bed elevation over a longer reach of a stream. The problem for the engineer is to estimate the long-term bed elevation changes that will occur during the life of the structure.

A long-term trend may change during the life of the bridge. These long-term changes are the result of modifications to the stream or watershed. Such changes may be the result of natural processes or human activities. The engineer must assess the present state of the stream and watershed and then evaluate potential future changes in the river system. From this assessment, the long-term streambed changes must be estimated.

Factors that affect long-term bed elevation changes are dams and reservoirs (up- or downstream of the bridge), changes in watershed land use (urbanization, deforestation, etc.), channelization, cutoffs of meander bends (natural or man-made), changes in the downstream channel base level (control), gravel mining from the streambed, diversion of water into or out of the stream, natural lowering of the fluvial system, movement of a bend and bridge location with respect to stream planform, and stream movement in relation to the crossing. Tidal ebb and flood may degrade a coastal stream; whereas, littoral drift may result in aggradation. The elevation of the bed under bridges which cross streams tributary to a larger stream will follow the trend of the larger stream unless there are controls. Controls could be bed rock, dams, culverts or other structures.

Data from the USACE, USGS, and other Federal and State agencies should be considered when evaluating long-term streambed variations. If no data exist or if such data require further evaluation, an assessment of long-term streambed elevation changes for riverine streams should be made using the principles of river mechanics (see HDS 6 (FHWA 2001)). Such an assessment requires the consideration of all influences upon the bridge crossing, i.e., runoff from the watershed to a stream (hydrology), sediment delivery to the channel (watershed erosion), sediment transport capacity of a stream (hydraulics), and response of a stream to these factors (geomorphology and river mechanics).

With coastal streams, the principles of both river and coastal engineering mechanics are needed. In coastal streams, estuaries or inlets, in addition to the above, consideration must be given to tidal conditions, i.e., the magnitude and period of the storm surge, sediment delivery to the channel by the ebb and flow of the tide, littoral drift, sediment transport capacity of the tidal flows, and response of the stream, estuary, or inlet to these tidal and coastal engineering factors (see Chapter 9).

Significant morphologic impacts can result from human activities. The assessment of the impact of human activities requires a study of the history of the river, estuary, or tidal inlet, as well as a study of present water and land use and stream control activities. All agencies involved with the river or coastal area should be contacted to determine possible future changes.

5.3 ESTIMATING LONG-TERM AGGRADATION AND DEGRADATION

To organize an assessment of long-term aggradation and degradation, a three-level fluvial system approach can be used. The three level approach consists of (1) a qualitative determination based on general geomorphic and river mechanics relationships, (2) an engineering geomorphic analysis using established qualitative and quantitative relationships to estimate the probable behavior of the stream system to various scenarios or future conditions, and (3) physical models or physical process computer modeling using mathematical models such as the USACE HEC-RAS (USACE 2010a) to make predictions of quantitative changes in streambed elevation due to changes in the stream and watershed. Methods to be used in Levels (1) and (2) are presented in HEC-20 and HDS 6 (FHWA 2012b, 2001).

For coastal areas, where highway crossings (bridges) and/or longitudinal stream encroachments are subject to tidal influences, the three-level approach used in fluvial systems is also appropriate (Chapter 9). The following sections outline procedures that can assist in identifying long-term trends in vertical stability.

5.3.1 Bridge Inspection Records

The biannual bridge inspection reports for bridges on the stream where a new or replacement bridge is being designed are an excellent source of data on long-term aggradation or degradation trends. Also, inspection reports for bridges crossing streams in the same area or region should be studied. In most states the biannual inspection includes taking the elevation and/or cross section of the streambed under the bridge. These elevations are usually referenced to the bridge, but these relative bed elevations will show trends and can be referenced to sea level elevations. Successive cross sections from a series of bridges in a stream reach can be used to construct longitudinal streambed profiles through the reach.

5.3.2 Gaging Station Records

The USGS and many State Water Resource and Environmental agencies maintain gaging stations to measure stream flow. In the process they maintain records from which the aggradation or degradation of the streambed can be determined. Gaging station records at the bridge site, on the stream to be bridged and in the area or region can be used.

Where an extended historical record is available, one approach to using gaging station records to determine long-term bed elevation change is to plot the change in stage through time for a selected discharge. This approach is often referred to as establishing a "specific gage" record.

Figure 5.1 shows a plot of specific gage data for a discharge of 400 cfs (14 m^3/sec) from about 1910 to 1980 for Cache Creek in California. Cache Creek has experienced significant gravel mining with records of gravel extraction quantities available since about 1940. When the historical record of cumulative gravel mining is compared to the specific gage plot, the potential impacts are apparent. The specific gage record shows more than 10 ft (3 m) of long-term degradation in a 70-year period.

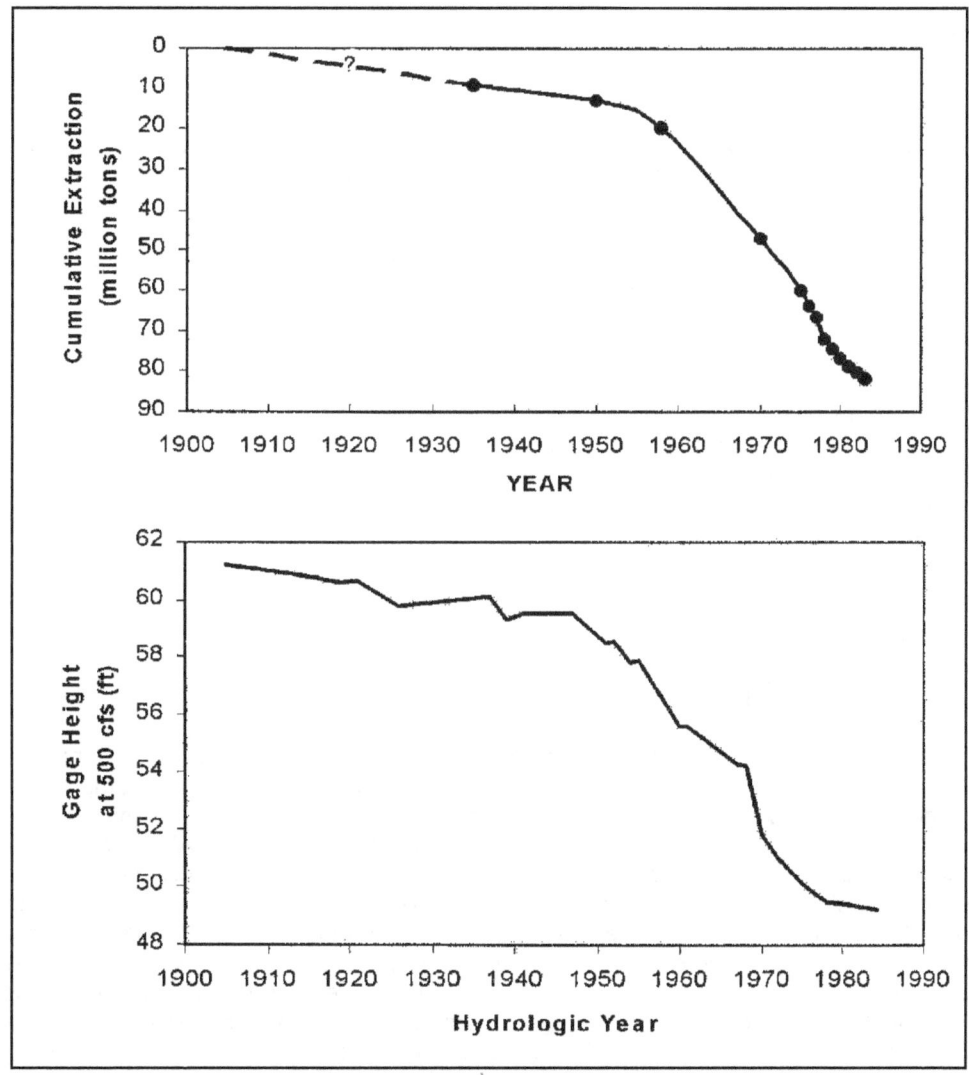

Figure 5.1. Specific gage data for Cache Creek, California.

5.3.3 Geology and Stream Geomorphology

The geology and geomorphology of the site needs to be studied to determine the potential for long-term bed elevation changes at the bridge site. Quantitative techniques for streambed aggradation and degradation analyses are covered in detail in HEC-20 (FHWA 2012b). These techniques include:

- Incipient motion analysis
- Analysis of armoring potential
- Equilibrium slope analysis
- Sediment continuity analysis

Sediment transport concepts and equations are discussed in detail in HDS 6 (FHWA 2001), HDS 7 (FHWA 2012a), and HEC-20 (FHWA 2012b).

5.3.4 Computer Models

Sediment transport computer models can be used to determine long-term aggradation or degradation trends. These computer models route sediment down a channel and adjust the channel geometry to reflect imbalances in sediment supply and transport capacity. The USACE HEC-RAS (USACE 2010a) model is an example of a sediment transport model that can be used for single event or long-term estimates of changes in bed elevation. The information needed to run these models includes:

- Channel and floodplain geometry
- Structure geometry
- Roughness
- Geologic or structural vertical controls
- Downstream water surface relationship
- Event or long-term inflow hydrographs
- Tributary inflow hydrographs
- Bed material gradations
- Upstream sediment supply
- Tributary sediment supply
- Selection of appropriate sediment transport relationship
- Depth of alluvium

In sediment routing, the sediment transport capacity is used to update cross section geometry, which is then used to update the hydraulic calculations. The geometry is updated for individual cross sections, though the hydraulic variables can be weighted with up- and downstream cross sections. A flood hydrograph or long-term flow hydrograph is entered as a series of constant flows. Within each flow time step, many sediment transport and cross section updating time steps are often required. The model does not assume that transport capacity is reached at every cross section, but limits erosion based on potential entrainment rates and limits deposition based on fall velocity, flow velocity and water depth. Sediment layer depths, as well as lateral limits for erosion and deposition are also input. Sediment transport modeling generally requires greater model upstream and downstream extent than a hydraulic flow model, as well as careful consideration of all boundary conditions (hydraulic and sediment).

5.3.5 Aggradation, Degradation, and Total Scour

Using all the information available estimate the long-term bed elevation change at the bridge site for the design life of the bridge. Usually, the design life is 100 years. **If the estimate indicates that the stream will degrade, use the elevation after degradation as the base elevation for contraction and local scour. That is, total scour must include the estimated long-term degradation**. If the estimate indicates that the stream will aggrade, then (1) make note of this fact to inspection and maintenance personnel, and (2) use existing ground elevation as the base for contraction and local scour.

5.3.6 Inspection, Maintenance, and Countermeasures

The estimate of long-term aggradation or degradation in the final design should be communicated to inspection and maintenance personnel. This information will aid them in tracking long-term trends and provide feedback for future design and evaluation. HEC-23 (FHWA 2009) outlines techniques for controlling long-term bed elevation changes and provides design guidance for countermeasures commonly used for vertical stability problems.

(page intentionally left blank)

CHAPTER 6

CONTRACTION SCOUR

6.1 INTRODUCTION

Contraction scour occurs when the flow area of a stream at flood stage is reduced, either by a natural contraction of the stream channel or by a bridge. It also occurs when overbank flow is forced back to the channel by roadway embankments at the approaches to a bridge. In most cases, contraction scour results in a decrease in the elevation of the bed across the bridge opening. It does not include localized scour at the foundations (local scour) or the long-term changes in the stream bed elevation (aggradation or degradation), and contraction scour may not have a uniform depth across the bridge opening. Live-bed contraction scour can be cyclic, that is, there can be an increase and decrease of the stream bed elevation (cutting and filling) during the passage of a flood. In this chapter, methods and equations will be presented to estimate contraction scour.

6.2 CONTRACTION SCOUR

6.2.1 Contraction Scour Conditions

Contraction scour equations are based on the principle of conservation of sediment transport (continuity). In the case of **live-bed scour**, the fully developed scour in the bridge cross section reaches equilibrium when sediment transported into the contracted section equals sediment transported out. As scour develops, the shear stress in the contracted section decreases as a result of a larger flow area and decreasing average velocity. For **live-bed** scour, maximum scour occurs when the shear stress reduces to the point that sediment transported in equals the bed sediment transported out and the conditions for sediment continuity are in balance. For **clear-water** scour, the sediment transport into the contracted section is essentially zero and maximum scour occurs when the shear stress reduces to the critical shear stress of the bed material in the section. Normally, for both live-bed and clear-water scour the width of the contracted section is constrained and depth increases until the limiting conditions are reached.

Live-bed contraction scour occurs at a bridge when there is transport of bed material in the upstream reach into the bridge cross section. With live-bed contraction scour the area of the contracted section increases until, in the limit, the transport of sediment out of the contracted section equals the sediment transported in.

Clear-water contraction scour occurs when (1) there is no bed material transport from the upstream reach into the downstream reach, or (2) the material being transported in the upstream reach is transported through the downstream reach mostly in suspension and at less than capacity of the flow. With clear-water contraction scour the area of the contracted section increases until, in the limit, the velocity of the flow (V) or the shear stress (τ_o) on the bed is equal to the critical velocity (V_c) or the critical shear stress (τ_c) of a certain particle size (D) in the bed material.

There are four conditions (cases) of contraction scour at bridge sites depending on the type of contraction, and whether there is overbank flow or relief bridges. Regardless of the case, contraction scour can be evaluated using two basic equations: (1) the **live-bed** scour equation, and (2) the **clear-water** scour equation. For any case or condition, it is only necessary to determine if the flow in the main channel or overbank area upstream of the bridge, or approaching a relief bridge, is transporting bed material (live-bed) or is not (clear-water), and then apply the appropriate equation with the variables defined according to the location of contraction scour (channel or overbank).

To determine if the flow upstream of the bridge is transporting bed material, calculate the critical velocity for beginning of motion V_c of the D_{50} size of the bed material being considered for movement and compare it with the mean velocity V of the flow in the main channel or overbank area upstream of the bridge opening. If the critical velocity of the bed material is larger than the mean velocity ($V_c > V$), then clear-water contraction scour will exist. If the critical velocity is less than the mean velocity ($V_c < V$), then live-bed contraction scour will exist. To calculate the critical velocity use the equation derived in the Appendix C. This equation is:

$$V_c = K_u y^{1/6} D^{1/3} \tag{6.1}$$

where:

V_c = Critical velocity above which bed material of size D and smaller will be transported, ft/s (m/s)
y = Average depth of flow upstream of the bridge, ft (m)
D = Particle size for V_c, ft (m)
D_{50} = Particle size in a mixture of which 50 percent are smaller, ft (m)
K_u = 6.19   SI units
K_u = 11.17   English units

The D_{50} is taken as an average of the bed material size in the reach of the stream upstream of the bridge. It is a characteristic size of the material that will be transported by the stream. Normally this would be the bed material size in the upper 1 ft (0.3 m) of the stream bed. As discussed in Section 6.7, a reasonable lower limit of D_{50} equal to 0.2 mm can be applied to this equation. For smaller sizes, cohesion tends to increase critical velocity.

Live-bed contraction scour depths may be limited by armoring of the bed by large sediment particles in the bed material or by sediment transport of the bed material into the bridge cross-section. Under these conditions, live-bed contraction scour at a bridge can be determined by calculating the scour depths using both the clear-water and live-bed contraction scour equations and using the smaller of the two depths.

6.2.2 Contraction Scour Cases

Four conditions (cases) of contraction scour are commonly encountered:

Case 1. Involves overbank flow on a floodplain being forced back to the main channel by the approaches to the bridge. Case 1 conditions include:
   a. The river channel width becomes narrower either due to the bridge abutments projecting into the channel or the bridge being located at a narrowing reach of the river (Figure 6.1);
   b. No contraction of the main channel, but the overbank flow area is completely obstructed by an embankment (Figure 6.2); or
   c. Abutments are set back from the stream channel (Figure 6.3).

Case 2. Flow is confined to the main channel (i.e., there is no overbank flow). The normal river channel width becomes narrower due to the bridge itself or the bridge site is located at a narrower reach of the river (Figures 6.4 and 6.5).

Case 3. A relief bridge in the overbank area with little or no bed material transport in the overbank area (i.e., clear-water scour) (Figure 6.6).

Case 4. A relief bridge over a secondary stream in the overbank area with bed material transport (similar to Case 1) (Figure 6.7).

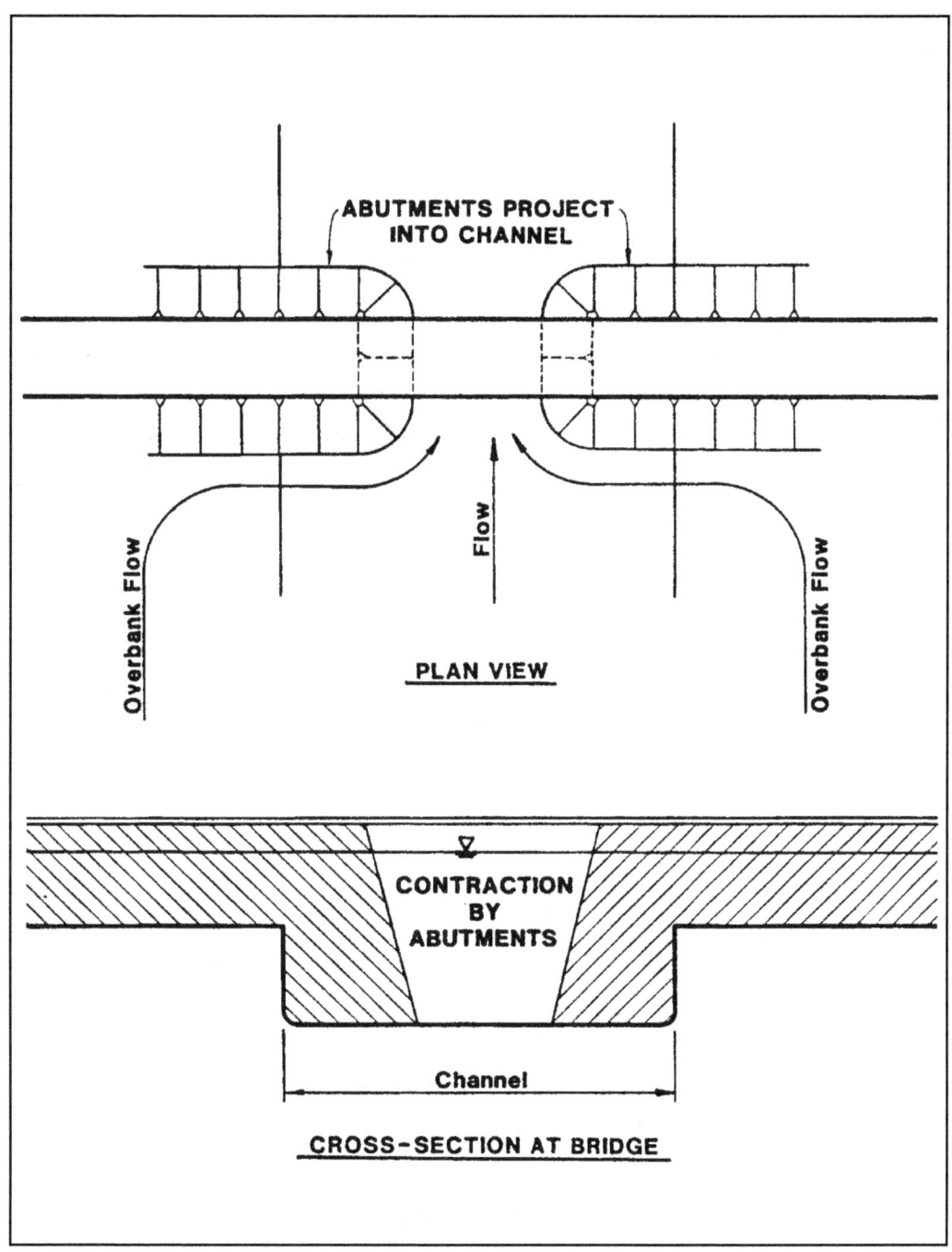

Figure 6.1. Case 1a: Abutments project into channel.

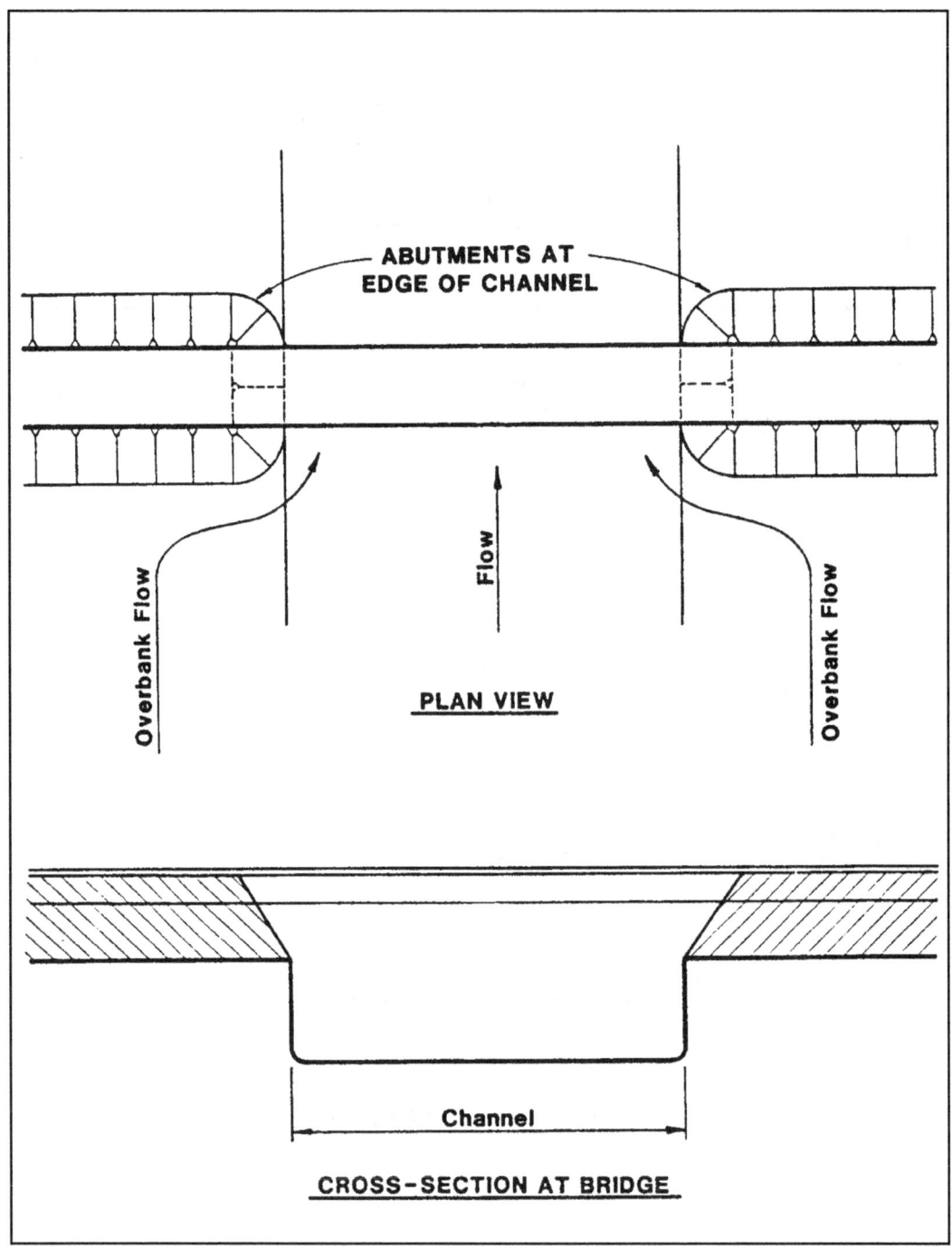

Figure 6.2. Case 1b: Abutments at edge of channel.

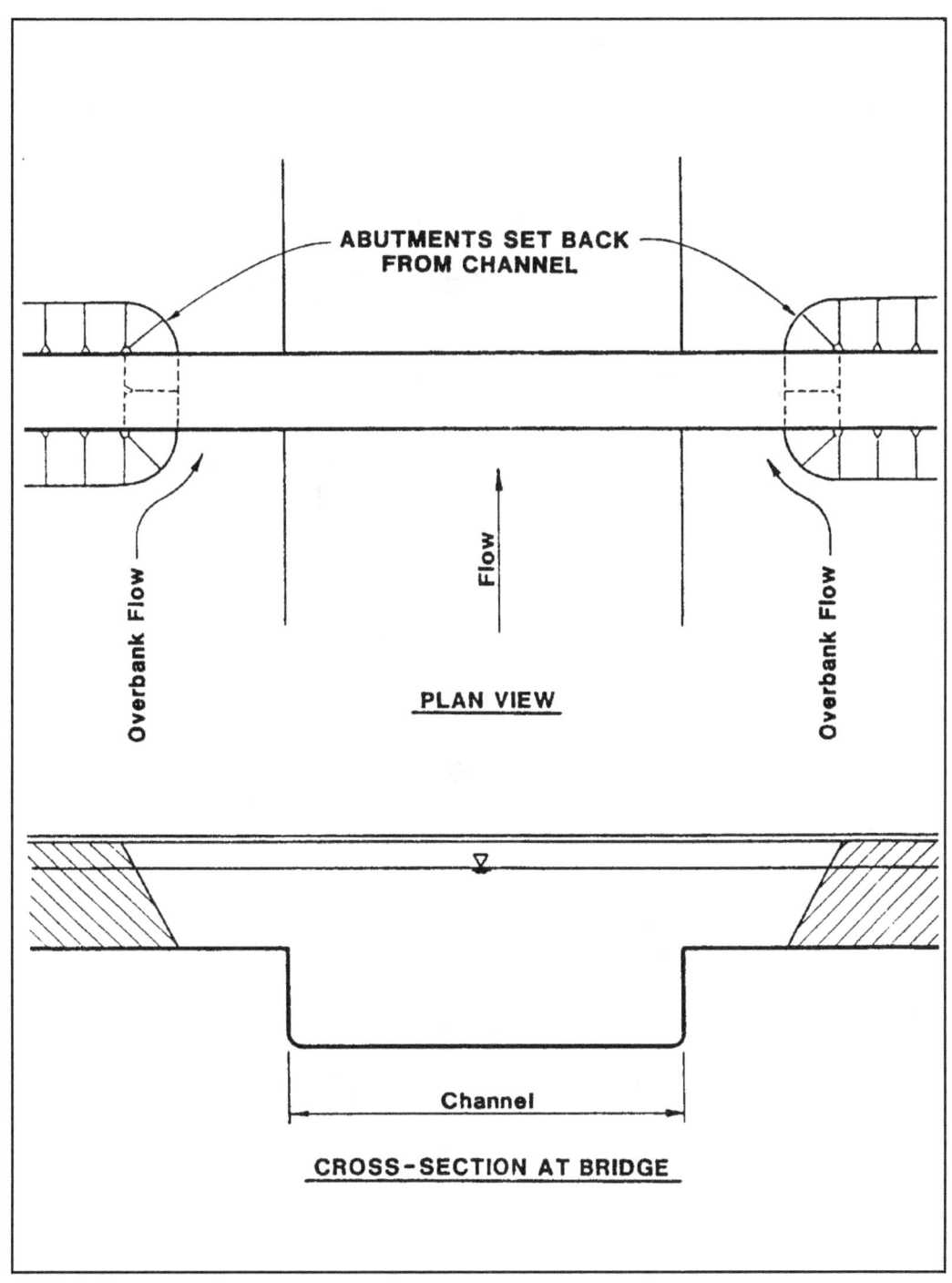

Figure 6.3. Case 1c: Abutments set back from channel.

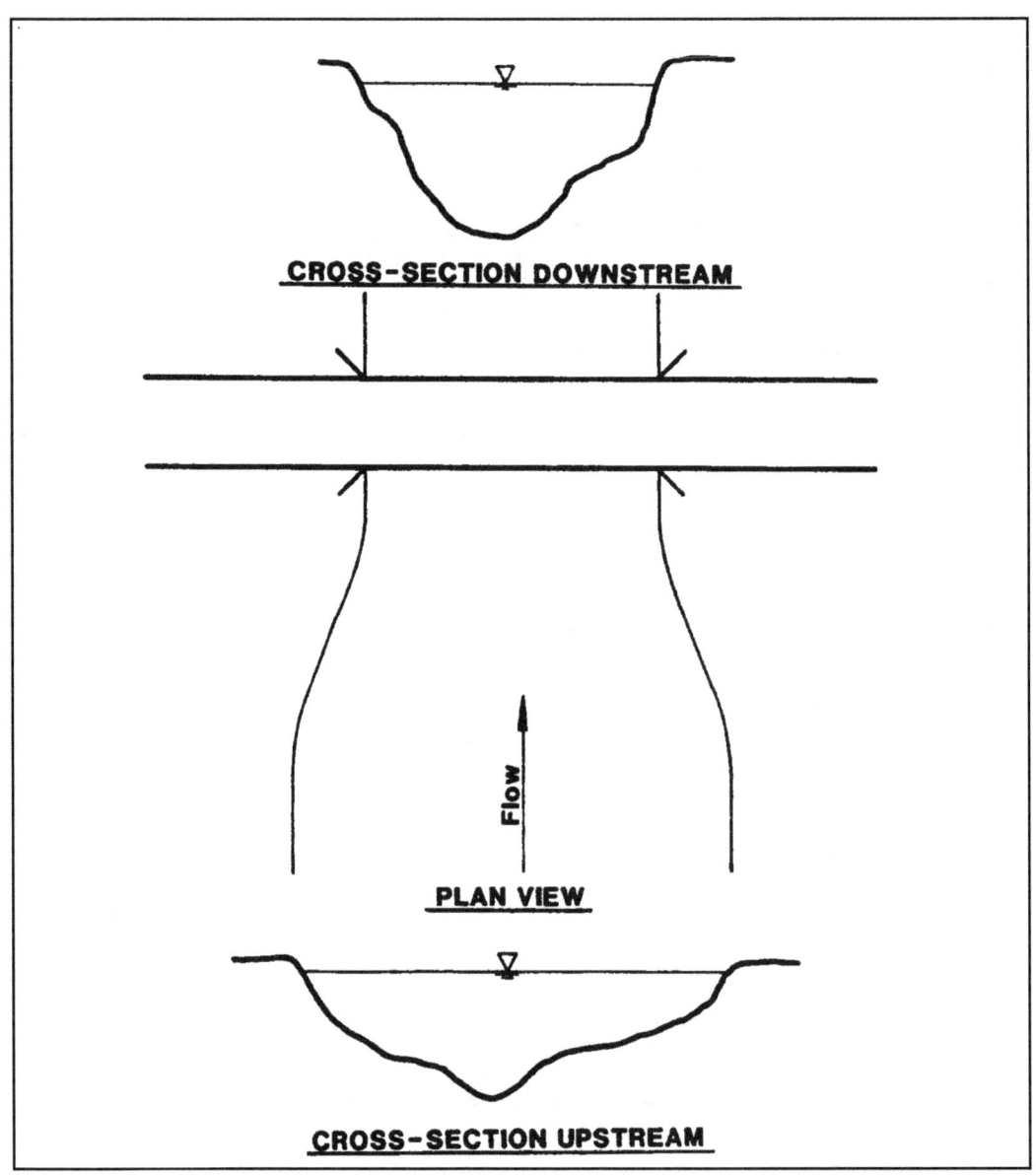

Figure 6.4. Case 2a: River narrows.

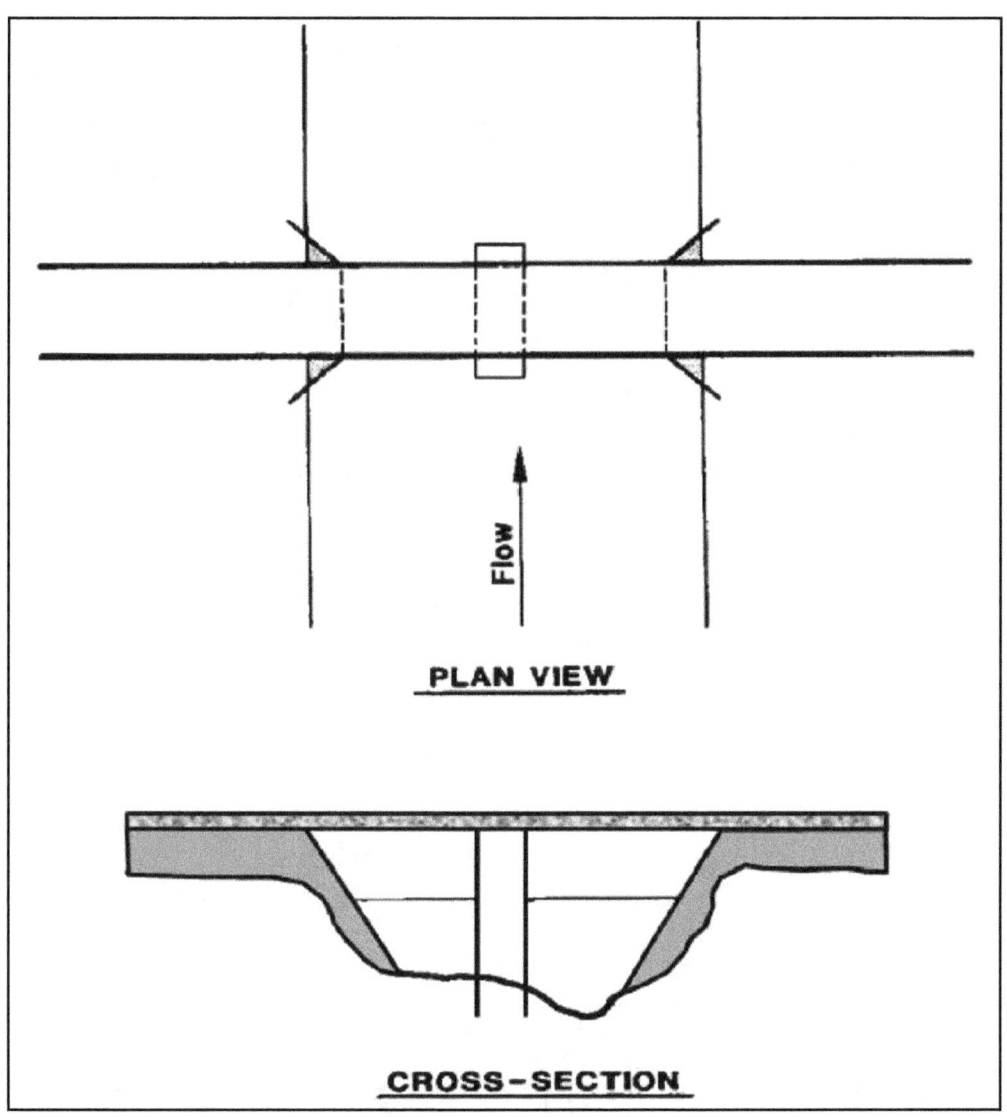

Figure 6.5. Case 2b: Bridge abutments and/or piers constrict flow.

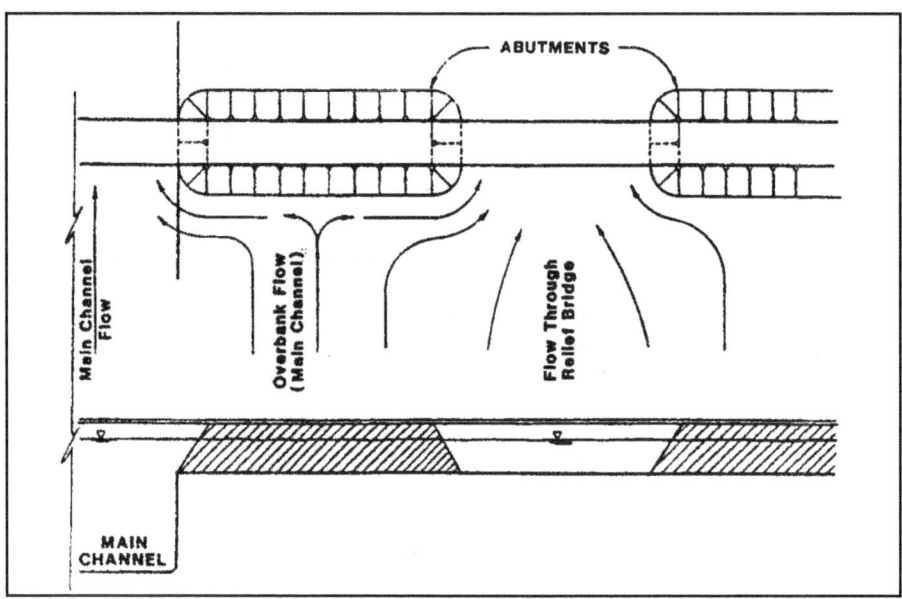

Figure 6.6. Case 3: Relief bridge over floodplain.

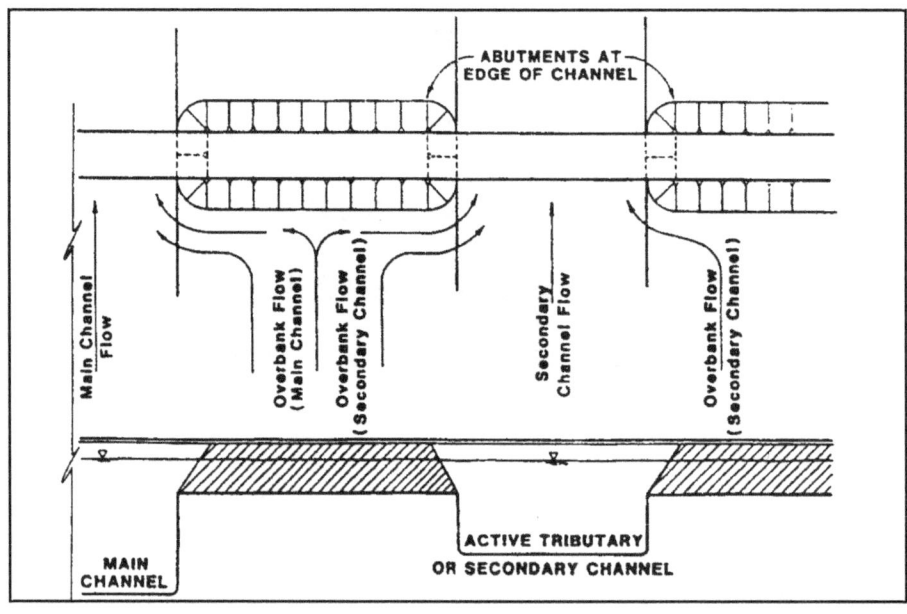

Figure 6.7. Case 4: Relief bridge over secondary stream.

Notes:

1. **Cases 1, 2, and 4** may either be live-bed or clear-water scour depending on whether there is bed material transport from the upstream reach into the bridge reach during flood flows. To determine if there is bed material transport compute the critical velocity at the approach section for the D_{50} of the bed material using the equation given above and compare to the mean velocity at the approach section. To determine if the bed material will be washed through the contraction determine the ratio of the shear velocity (V_*) in the contracted section to the fall velocity (ω) of the D_{50} of the bed material being transported from the upstream reach (see the definition of V_* in the live-bed contraction scour equation). If the ratio is much larger than 2, then the bed material from the upstream reach will be mostly suspended bed material discharge and may wash through the contracted reach (clear-water scour).

2. **Case 1c is very complex.** The depth of contraction scour depends on factors such as (1) how far back from the bankline the abutment is set, (2) the condition of the overbank (is it easily eroded, are there trees on the bank, is it a high bank, etc.), (3) whether the stream is narrower or wider at the bridge than at the upstream section, (4) the magnitude of the overbank flow that is returned to the bridge opening, and (5) the distribution of the flow in the bridge section, and (6) other factors.

   The main channel under the bridge may be live-bed scour; whereas, the set-back overbank area may be clear-water scour.

   HEC-RAS (USACE 2010a) can be used to determine the distribution of flow between the main channel and the set-back overbank areas in the contracted bridge opening. However, the distribution of flow needs to be done with care. Studies by Chang and Davis (1999a) and Sturm (1999) have shown that conveyance calculations do not properly account for the flow distribution under the bridge.

   If the abutment is set back only a small distance from the bank (less than 3 to 5 times the average depth of flow through the bridge), there is the possibility that the combination of contraction scour and abutment scour may destroy the bank. Also, the two scour mechanisms are not independent. Consideration should be given to using a guide bank and/or protecting the bank and bed under the bridge in the overflow area with rock riprap. See HEC-23 (FHWA 2009) for guidance on designing guide banks and rock riprap.

3. **Case 3** may be clear-water scour even though the floodplain bed material is composed of sediments with a critical velocity that is less than the flow velocity in the overbank area. The reasons for this are (1) there may be vegetation growing part of the year, and (2) if the bed material is fine sediments, the bed material discharge may go into suspension (wash load) at the bridge and not influence contraction scour.

4. **Case 4** is similar to Case 3, but there is sediment transport into the relief bridge opening (live-bed scour). This case can occur when a relief bridge is over a secondary channel on the floodplain. Hydraulically this is no different from case 1, but analysis is required to determine the floodplain discharge associated with the relief opening and the flow distribution going to and through the relief bridge. This information could be obtained from HEC-RAS (USACE 2010a).

6.3 LIVE-BED CONTRACTION SCOUR

A modified version of Laursen's 1960 equation for live-bed scour at a long contraction is recommended to predict the depth of scour in a contracted section (Laursen 1960). The original equation is given in Appendix C. The modification is to eliminate the ratio of Manning n (see the following Note #3). The equation assumes that bed material is being transported from the upstream section.

$$\frac{y_2}{y_1} = \left(\frac{Q_2}{Q_1}\right)^{6/7} \left(\frac{W_1}{W_2}\right)^{k_1} \tag{6.2}$$

$y_s = y_2 - y_o$ = (average contraction scour depth) \hfill (6.3)

where:

- y_1 = Average depth in the upstream main channel, ft (m)
- y_2 = Average depth in the contracted section, ft (m)
- y_o = Existing depth in the contracted section before scour, ft (m) (see Note 7)
- Q_1 = Flow in the upstream channel transporting sediment, ft³/s (m³/s)
- Q_2 = Flow in the contracted channel, ft³/s (m³/s)
- W_1 = Bottom width of the upstream main channel that is transporting bed material, ft (m)
- W_2 = Bottom width of main channel in contracted section less pier width(s), ft (m)
- k_1 = Exponent determined below

V_*/ω	k_1	Mode of Bed Material Transport
<0.50	0.59	Mostly contact bed material discharge
0.50 to 2.0	0.64	Some suspended bed material discharge
>2.0	0.69	Mostly suspended bed material discharge

- V_* = $(\tau_o/\rho)^{1/2}$ = $(gy_1 S_1)^{1/2}$, shear velocity in the upstream section, ft/s (m/s)
- ω = Fall velocity of bed material based on the D_{50}, m/s (Figure 6.8)
  For fall velocity in English units (ft/s) multiply ω in m/s by 3.28
- g = Acceleration of gravity (32.2 ft/s²) (9.81 m/s²)
- S_1 = Slope of energy grade line of main channel, ft/ft (m/m)
- τ_o = Shear stress on the bed, (lb/ft²) (Pa (N/m²))
- ρ = Density of water (1.94 slugs/ft³) (1000 kg/m³)

Notes:

1. Q_2 may be the total flow going through the bridge opening as in cases 1a and 1b. **It is not the total flow for Case 1c**. For Case 1c contraction scour must be computed separately for the main channel and the left and/or right overbank areas.
2. Q_1 is the flow in the main channel upstream of the bridge, not including overbank flows.
3. The Manning n ratio is eliminated in Laursen live-bed equation to obtain Equation 6.2 (Appendix C). This was done for the following reasons. The ratio can be significant for a condition of dune bed in the upstream channel and a corresponding plane bed, washed out dunes or antidunes in the contracted channel. However, Laursen's equation does not

correctly account for the increase in transport that will occur as the result of the bed planing out (which decreases resistance to flow, increases the velocity and the transport of bed material at the bridge). That is, Laursen's equation indicates a decrease in scour for this case, whereas in reality, there would be an increase in scour depth. In addition, at flood flows, a plane bedform will usually exist upstream and through the bridge waterway, and the values of Manning n will be equal. Consequently, the n value ratio is not recommended or presented in Equation 6.2.

4. W_1 and W_2 are not always easily defined. In some cases, it is acceptable to use the topwidth of the main channel to define these widths. Whether topwidth or bottom width is used, it is important to be consistent so that W_1 and W_2 refer to either bottom widths or top widths.

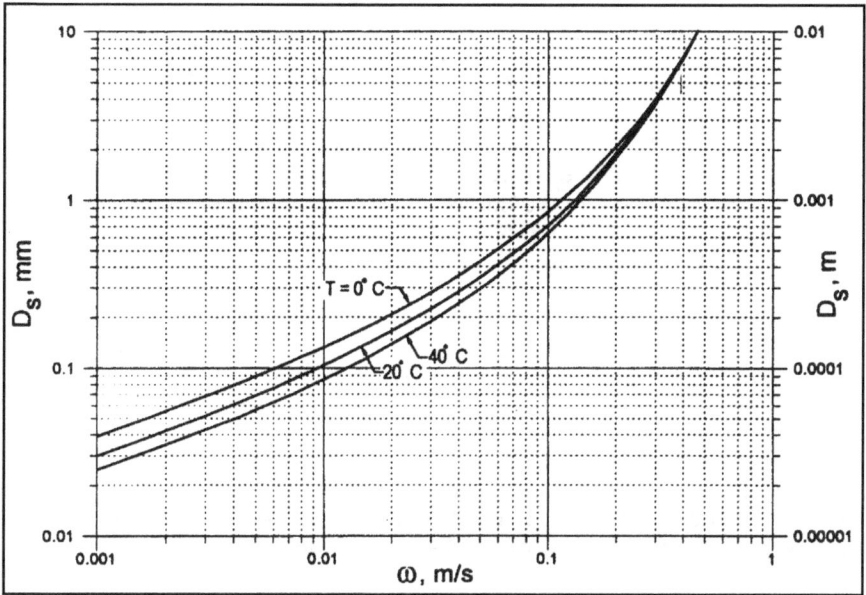

Figure 6.8. Fall velocity of sand-sized particles with specific gravity of 2.65 in metric units.

5. The average width of the bridge opening (W_2) is normally taken as the bottom width, with the width of the piers subtracted.

6. Laursen's equation will overestimate the depth of scour at the bridge if the bridge is located at the upstream end of a natural contraction or if the contraction is the result of the bridge abutments and piers. At this time, however, it is the best equation available.

7. In sand channel streams where the contraction scour hole is filled in on the falling stage, the y_0 depth may be approximated by y_1. Sketches or surveys through the bridge can help in determining the existing bed elevation.

8. **Scour depths with live-bed contraction scour may be limited by coarse sediments in the bed material armoring the bed. Where coarse sediments are present, it is recommended that scour depths be calculated for live-bed scour conditions using the clear-water scour equation (given in the next section) in addition to the live-bed equation, and that the smaller calculated scour depth be used.**

6.4 CLEAR-WATER CONTRACTION SCOUR

The recommended clear-water contraction scour equation is based on a development suggested by Laursen (1963) (presented in the Appendix C). The equation is:

$$y_2 = \left[\frac{K_u Q^2}{D_m^{2/3} W^2} \right]^{3/7} \tag{6.4}$$

$y_s = y_2 - y_o$ = (average contraction scour depth) (6.5)

where:

y_2	=	Average equilibrium depth in the contracted section after contraction scour, ft (m)
Q	=	Discharge through the bridge or on the set-back overbank area at the bridge associated with the width W, ft³/s (m³/s)
D_m	=	Diameter of the smallest nontransportable particle in the bed material (1.25 D_{50}) in the contracted section, ft (m)
D_{50}	=	Median diameter of bed material, ft (m)
W	=	Bottom width of the contracted section less pier widths, ft (m)
y_o	=	Average existing depth in the contracted section, ft (m)
K_u	=	0.0077 English units
K_u	=	0.025 SI units

Equation 6.4 is a rearranged version of Equation 6.1. As discussed in Section 6.7, a reasonable lower limit of D_{50} equal to 0.2 mm can be applied to this equation. Using a size smaller than 0.2 mm will over-estimate clear-water contraction scour.

Because D_{50} is not the largest particle in the bed material, the scoured section can be slightly armored. Therefore, the D_m is assumed to be 1.25 D_{50}. For stratified bed material the depth of scour can be determined by using the clear-water scour equation sequentially with successive D_m of the bed material layers.

6.5 CONTRACTION SCOUR WITH BACKWATER

The **live-bed** contraction scour equation is derived assuming a uniform reach upstream and a long contraction into a uniform reach downstream of the bridge. With live-bed scour the equation computes a depth after the long contraction where the sediment transport into the downstream reach is equal to the sediment transport out. The **clear-water** contraction scour equations are derived assuming that the depth at the bridge increases until the shear-stress and velocity are decreased so that there is no longer any sediment transport. With the clear-water equations it is assumed that flow goes from one uniform flow condition to another. Both equations calculate contraction scour depth assuming a level water surface ($y_s = y_2 - y_o$). A more consistent computation would be to write an energy balance before and after the scour. For live-bed the energy balance would be between the approach section (1) and the contracted section (2). Whereas, for clear-water scour it would be the energy at the same section before (1) and after (2) the contraction scour.

Backwater, in extreme cases, can decrease the velocity, shear stress and the sediment transport in the upstream section. This will increase the scour at the contracted section. The backwater can, by storing sediment in the upstream section, change live-bed scour to clear-water scour.

6.6 CONTRACTION SCOUR EXAMPLE PROBLEMS

6.6.1 Example Problem 1 - Live-Bed Contraction Scour

Given:
　　The upstream channel width = 322 ft (98.2 m); depth = 8.6 ft (2.62 m)
　　The discharge is 27,300 cfs (773 m³/s) and is all contained within the channel. Channel slope = 0.004 (ft/ft) (m/m)
　　The bridge abutments consist of vertical walls with wing walls, width = 122 ft (37.2 m); with 3 sets of piers consisting of 3 columns 15 inches (0.38 m) in diameter
　　The bed material size: from 0 to 3 ft (0 to 0.9 m) the D_{50} is 0.0010 ft (0.31 mm) and below 3 ft (0.9 m) the D_{50} is 0.0023 ft (0.70 mm) with a fall velocity of 0.33 ft/sec (0.10 m/s)
　　Original depth at bridge is estimated as 7.1 ft (2.16 m)

Determine:
　　The magnitude of the contraction scour depth.

Solution:
　　1. Determine if it is live-bed or clear-water scour.

Average velocity in the upstream reach

V = 27,300/(8.6 x 322) = 9.86 ft/s (3.0 m/s)

For velocities this large and bed material this fine **live-bed** scour will occur. Check by calculating V_c for 0.7 mm bed material size. If live-bed scour occurs for 0.7 mm it would also be live-bed for 0.3 mm.

$V_c = 11.17 (8.6)^{1/6} (0.0023)^{1/3}$ = 2.11 ft/s (0.65 m/s)

Live-bed contraction scour is verified

　　2. Calculate contraction scour.

　　a. Determine K_1 for mode of bed material transport

$V_* = (32.2 \times 8.6 \times 0.004)^{0.5}$ = 1.05 ft/s (0.32 m/s)

ω = 0.33;　　V_*/ω = 3.2;　　K_1 = 0.69

　　b. Live-bed contraction scour. Equation 6.2

$Q_1 = Q_2$

$$\frac{y_2}{8.6} = \left[\frac{322}{118.25}\right]^{0.69} = 2.00$$

y_2 = 8.6 x 2.00 = 17.2 ft (5.24 m) from water surface.
y_s = 17.2 - 7.1 = 10.1 ft (3.08 m) from original bed surface

6.6.2 Example Problem 2 - Alternate Method

An alternative approach is demonstrated to calculating y_s in Problem 1 to determine if scour is clear-water or live-bed. In this method calculate the scour depth using both the clear-water and the live-bed equation and take the smaller scour depth.

a. Live-bed scour depth is 10.1 ft (3.08 m) from Problem 1.

b. Clear-water scour depth (Equation 6.4)

$D_m = 1.25\ D_{50} = 1.25\ (0.0023) = 0.0030$ ft (0.0009 m)

$$y_2 = \left[\frac{0.0077\ (27,300)^2}{0.0030^{2/3}\ (118.25)^2} \right]^{3/7} = 69.31\text{ft (21.12m)}$$

$y_s = 69.31 - 7.1 = 62.2$ ft (18.96 m) from original bed surface

c. Live-bed scour (10.1 ft (3.08 m) < 62.2 ft (18.96 m)). The sediment transport limits the contraction scour depth rather than the size of the bed material.

6.6.3 Example Problem 3 - Relief Bridge Contraction Scour

The 1952 flood on the Missouri River destroyed several relief bridges on Highway 2 in Iowa near Nebraska City, Nebraska. The USGS made continuous measurements during the period April 2 through April 29, 1952. This data set is from the April 21, 1952 measurement (measurement #1013). The discharge in the relief bridge was 13,012 cfs (368 m³/s). The measurement was made on the upstream side of Cooper Creek ditch using a boat and tag line.

Given:

Q = 13,012 cfs (368 m³/s); Bridge width (minus piers) = 300 ft (91.4 m)
$V_{average}$ = 1.71 ft/s; y_0 = 6.4 ft (1.95 m)
D_{50} = 0.24 mm (D_m = 1.25 x 0.24 = 0.3 mm)

Determine:

The magnitude of the contraction scour depth.

Solution:

1. Clear- water scour because of low velocity flow on the floodplain (use Equation 6.4)
2. Calculate contraction scour.

$$y_2 = \left[\frac{0.0077\ (13,012)^2}{(0.0010)^{2/3}\ (300)^2} \right]^{3/7} = 22.6 \text{ ft (6.89m)}$$

y_2 = 22.6 ft (6.89 m) from the water surface, this compares to 25.3 ft (7.71 m) measured at the site.

$y_s = y_2 - y_0 = 22.6 - 6.4 = 16.2$ ft (4.94 m)

6.6.4 Comprehensive Example

Additional contraction scour problems are included in the Comprehensive Example in Appendix D.

6.7 CONTRACTION SCOUR IN COHESIVE MATERIALS

The live-bed and clear-water contraction scour equations presented in Sections 6.3 and 6.4 are developed for cohesionless sediments and provide estimates of scour for a hydraulic conditions sufficient to produce ultimate scour. As illustrated in Figure 6.9 (Briaud et al. 2011), for silts and clays the critical shear stress (τ_c) increases due to cohesion. The standard clear-water contraction scour equation (Equation 6.4 and Appendix C) is based on Shields relationship for critical shear. Figure 6.9 shows that grain size and critical shear are well correlated for sand and gravels sizes. Because the critical shear stress reaches a minimum (Figure 6.9) at a D_{50} of approximately 0.2 mm, it is reasonable to use this size as a lower limit when applying Equation 6.4. Although the 0.2 mm limit avoids extremely overly conservative clear-water contraction scour estimates, it is also reasonable to apply a critical shear value based on the lower-bound equation shown in Figure 6.9. However, silt and clay materials may have critical shear values orders of magnitude greater than that lower-bound. Therefore, there are silt and clay materials that may experience little or no contraction scour, even for extreme flood conditions.

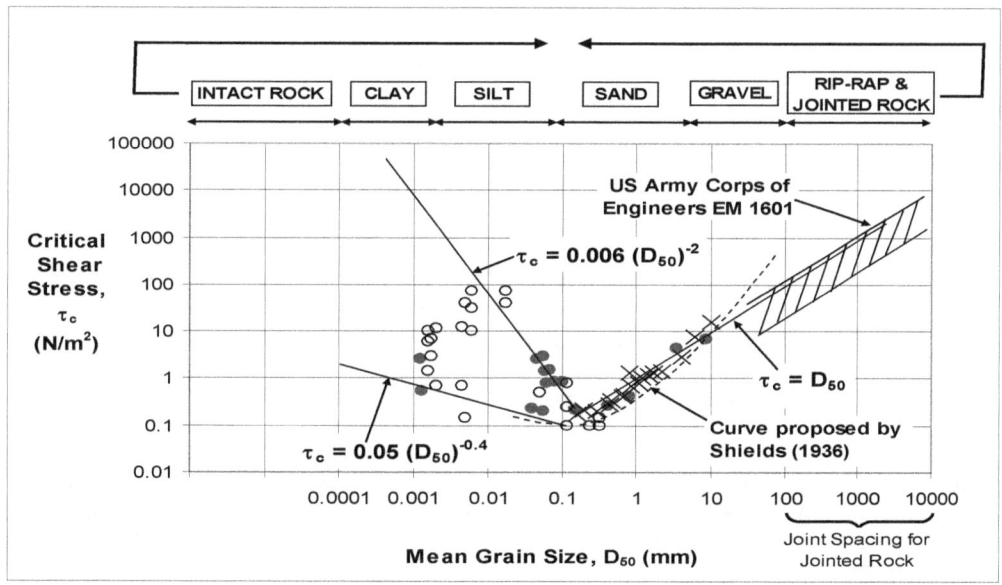

Figure 6.9. Critical shear stress versus particle size (Briaud et al. 2011).

The only reliable way of determining critical shear for silt and clay particles is to perform materials testing (see Chapter 4 and NCHRP 2004, Briaud et al. 2011). Figure 6.10 illustrates the results of material testing of a cohesive material. In addition to critical shear, the erosion rate ($\dot{z}$) versus excess shear ($\tau - \tau_c$) can also be determined. For a scour producing event, the initial erosion rate is determined by the initial, maximum shear. The rate will decrease as scour increases until the shear equals τ_c, at which point the scour ceases.

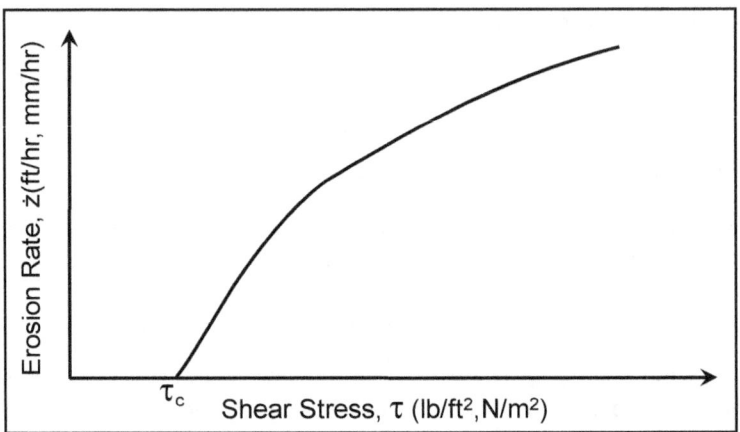

Figure 6.10. Example of critical shear and erosion rate from a material test (NCHRP 2004).

6.7.1 Ultimate Scour

As with the standard clear-water contraction scour equation, ultimate scour can be calculated for a particular hydraulic condition once the critical shear is known. The following equation can be used to compute ultimate scour for cohesive materials. Equation 6.6 is based on analysis on laboratory data (Briaud et al. 2011):

$$y_{s-ult} = 0.94 y_1 \left(\frac{1.83 V_2}{\sqrt{g y_1}} - \frac{K_u \sqrt{\frac{\tau_c}{\rho}}}{g n y_1^{1/3}} \right) \tag{6.6}$$

where:

y_1	=	Upstream average flow depth (ft, m)
V_2	=	Average flow depth in the contracted section (ft/s, m/s)
τ_c	=	Critical shear stress (lb/ft², N/m²)
n	=	Manning n
K_u	=	1.486 for U.S. Customary units and 1.0 for S.I.

It should be noted that although the upstream flow depth is used in the equation, the development of the equation assumes that the upstream flow depth is equal to the flow depth in the constriction. Briaud et al. (2011) indicate that this equation computes the centerline scour downstream of the entrance and that centerline scour in the vicinity of the entrance is 35 percent greater.

Including cohesion will typically reduce estimated ultimate scour in comparison to fine-sand in the clear-water contraction scour equation. In some cases, ultimate scour may not be reached during the life of a bridge if there is not sufficient duration of flooding. In order to estimate scour over the life of a bridge, much more information and additional calculations are required. The information includes the erosion rate versus excess shear curve, and flow magnitudes and durations for the life of the bridge. The calculations progress through the bridge-life flow hydrograph, where scour for a specific flood is added to scour from prior flows. Briaud (2011) provides generalized relationships for critical shear and erosion rates based on material types (Figure 6.11). For a specific flow, the initial shear stress can be computed from:

$$\tau = \gamma \left(\frac{V_2 n}{K_u}\right)^2 y_o^{-1/3} \qquad (6.7)$$

where variables are previously defined. If the shear stress does not exceed the critical value for that material (from test data or a figure such as 6.11), then no contraction scour will occur during that flow period. If the critical shear is exceeded, then ultimate scour for that flow condition is computed from Equation 6.6.

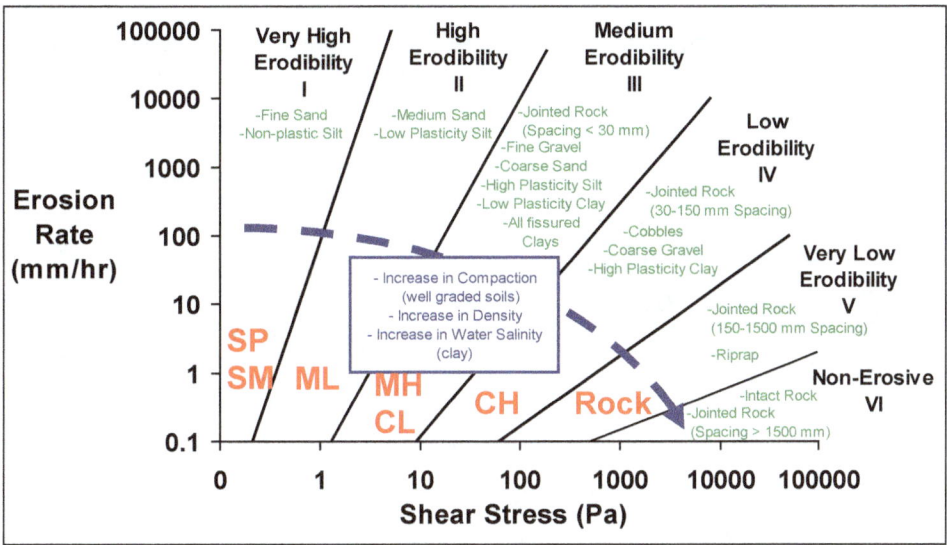

Figure 6.11. Generalized relationships for scour in cohesive materials (Briaud et al. 2011).

6.7.2 Time Rate of Scour

For the first event that produces contraction scour during the life of the bridge, the actual scour that will occur depends on the initial scour rate ($\dot{z}_i$), the ultimate scour for that flow, and the duration of the flow (t). The equation for the scour during the event is:

$$y_s(t) = \frac{t}{\dfrac{1}{\dot{z}_i} + \dfrac{t}{y_{s-ult}}} \qquad (6.8)$$

where:

$\dot{z}_i$ = Initial rate of scour (ft/hr, m/hr)
t = Duration of flow (hr)

It is important to note that the length scale for rate of scour be expressed in the same units as the ultimate scour. For subsequent scour events, scour will only occur when the ultimate scour for that event exceeds scour that has previously taken place. This will always occur when the shear exceeds any previously occurring shear, but may also occur for shear that is lower than previously occurring shear. Additional scour will occur for flow conditions when the ultimate scour for that flow condition is greater than the cumulative previous scour. Therefore, during the life of a bridge, scour in cohesive material is cumulative and the cumulative amount of scour can increase even when smaller events occur after larger events. Equation 6.8 is used to compute scour for subsequent scour events, but the time (t) used in the equation must be adjusted to account for the prior scour (t = $t_{event} + t_e$), where t_e is the equivalent time that that event would have required to reach the prior scour amount. Equivalent time (t_e) is illustrated in Figure 6.12 and is given by:

$$t_e = \frac{y_{s-ult} \, y_{s-prior}}{\dot{z}_1 (y_{s-ult} - y_{s-prior})} \qquad (6.9)$$

where:

$y_{s-prior}$ = Cumulative scour that has been reached in prior events (ft, m)

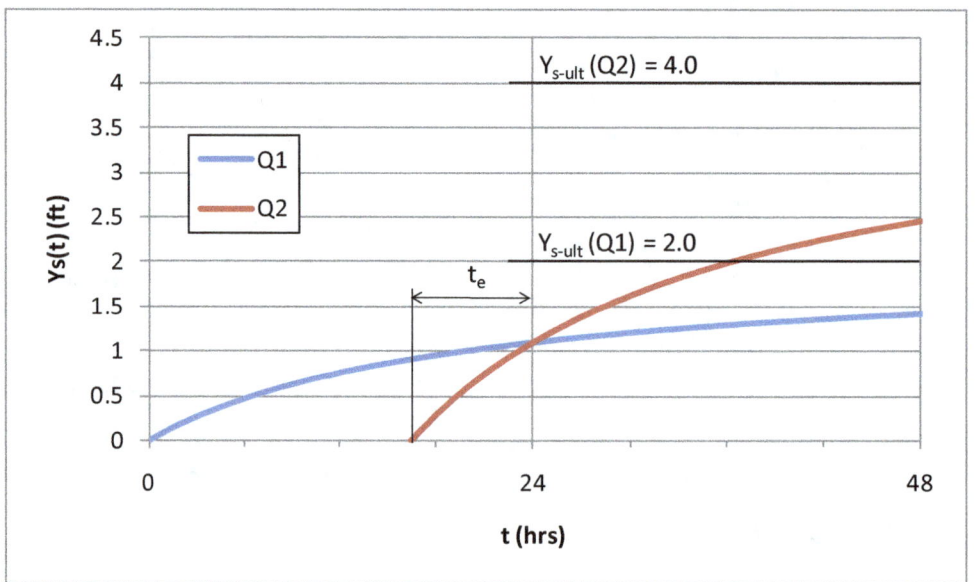

Figure 6.12. Illustration of time-dependent scour calculations.

The steps outlined above must be completed for all scouring events over the life of the bridge. Because the scour that a bridge will experience during one flood depends on the magnitude and duration of that flood and the amount of scour that has occurred in previous floods, the sequence of floods will also affect the total amount of scour over the life of a bridge. The greatest cumulative scour would occur if floods increase in magnitude over the life of a bridge and the smallest cumulative scour would occur if floods decrease in magnitude over the life of a bridge.

Although the information required to perform this level of calculation is significant and it must be estimated for the future life of a bridge, the effort may be warranted for extremely erosion resistant cohesive materials. Even when the cumulative scour is not estimated, ultimate scour should be computed for the scour design flood and scour design check flood (see Table 2.1), or controlling scour event using Equation 6.6. The hydraulic engineer must work closely with a geotechnical engineer to fully account for scour in cohesive materials.

6.8 CONTRACTION SCOUR IN ERODIBLE ROCK

Contraction scour can also occur in erodible rock. Concerns for some rock types include weathering and abrasion (see Chapter 4). In addition to hydraulic forces, channels in rock materials may degrade due to wetting and drying, freeze-thaw, abrasion, and chemical reactions. Some rock, such as weakly cemented sand stone and other friable rock, may be as erodible as sand while other rock may be extremely erosion resistant. The concepts from the previous section can also be applied to erodible rock; however, it is not only necessary to determine the critical shear and erosion rate information, but also account for these other potential factors. The hydraulic engineer must work closely with a geotechnical engineer and geologist to fully account for scour in rock.

6.9 SCOUR AT OPEN-BOTTOM CULVERTS

Open-bottom (bottomless or three-sided) culverts are structures that have natural channel materials as the bottom. Figure 6.13 shows a common type of open-bottom culvert that is over 10 feet (3 m) high and over 40 feet (12 m) wide. These cast-in-place, precast, or prefabricated structures may be rectangular in shape or have a more rounded top. They are typically founded on spread footings although pile foundations and pedestal walls are also used. Regardless of the foundation type, the structure may be highly susceptible to scour. Open-bottom culverts on spread footings are best suited for non-erodible rock but with caution and with scour protection can be used for other soils (see HEC-23 Design Guideline 18, FHWA 2009). Open-bottom culverts have several advantages over other crossing structures. The natural bottom material is more environmentally attractive than a traditional closed culvert, particularly where fish passage is a concern. They are also considered by many highway agencies to be economical alternatives to short bridges. They are more easily constructed than conventional bridges because they are commonly prefabricated.

Scour is greatest at the upstream corners of the culvert entrance. Pressure flow can greatly increase scour potential although pressure flow scour is not addressed in this section. The scour approach presented in this section accounts for combined contraction plus local scour at the upstream corners of the open-bottom culvert. Degradation is the only other scour component that may contribute to total scour. If dual open-bottom culverts (side-by-side) are used then the center foundation acts as a pier and must be designed to be stable for the total scour depth (degradation, contraction and pier scour) without a countermeasure.

6.9.1 Laboratory Investigations of Scour at Open-Bottom Culverts

FHWA sponsored two laboratory studies of scour at open-bottom culverts (FHWA 2003a, 2007, Kerenyi and Pagan-Ortiz 2007). The studies concluded that the scour is analogous to contraction scour caused by concentration of flow (primary flow) and to abutment scour caused by vortices and strong turbulence (secondary flow) (Figure 6.14). The studies included rectangular and arched shapes with and without wing walls (Figures 6.15 – 6.17). These figures show that scour is usually greatest at the upstream corners of the culvert entrance. The studies were performed for clear-water conditions (no sediment transport in the approach flow). Therefore, the scour equations apply only to clear-water conditions and should not be used for live-bed conditions. Future research will address the live-bed condition.

Figure 6.13. Open-bottom culvert on Whitehall Road over Euclid Creek in Cuyahoga County, OH.

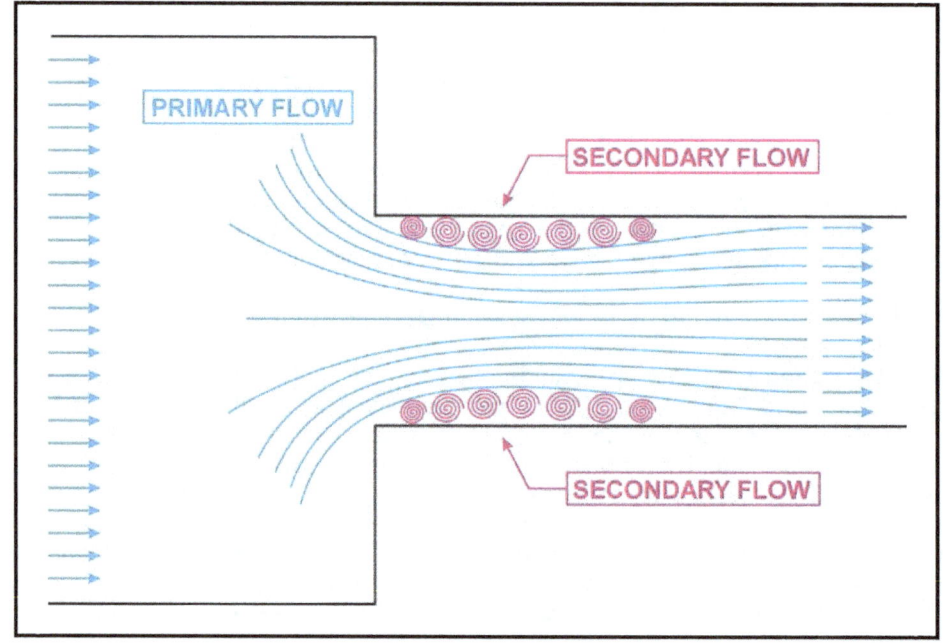

Figure 6.14. Flow concentration and separation zone (FHWA 2007).

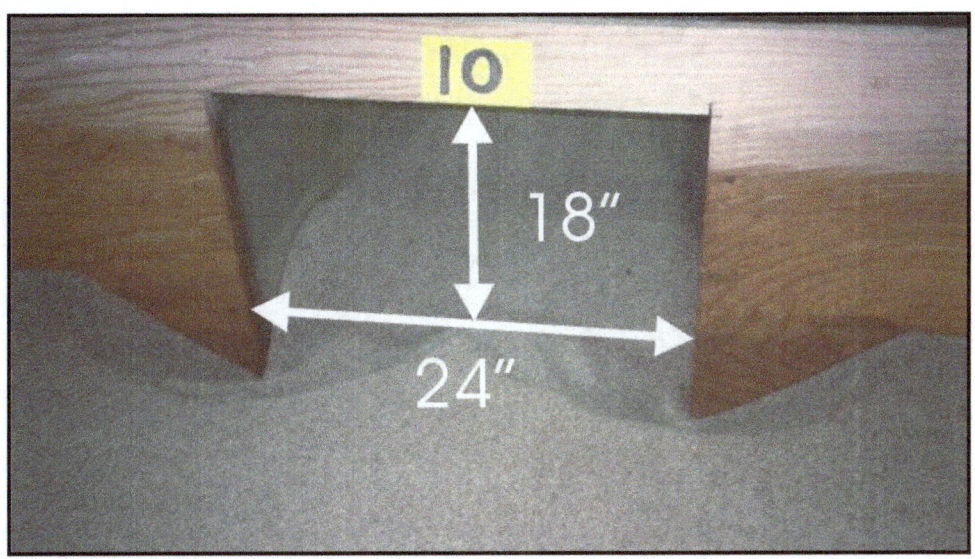

Figure 6.15. Rectangular model with vertical face (FHWA 2003a).

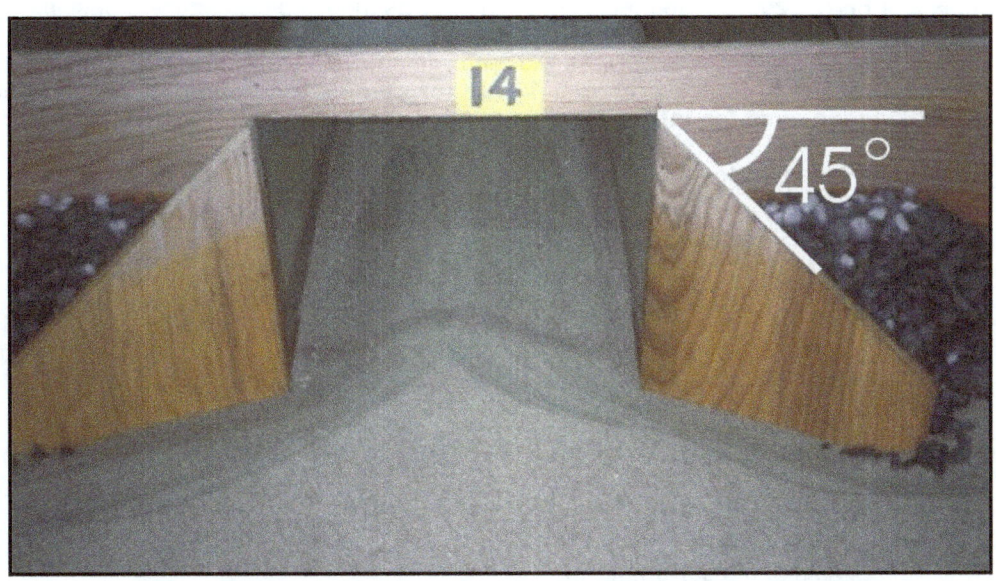

Figure 6.16. Rectangular model with wing walls (FHWA 2003a).

Figure 6.17. Arched model with wing walls (FHWA 2003a).

6.9.2 Clear-Water Scour Equation for Open-Bottom Culverts

The scour equations for open bottom culverts were developed from the second equation in Table 2 of FHWA 2007. The predictive equation was modified to produce a design equation with a reliability index (β) of 2.5 based on the laboratory data. There are separate scour relationships for open-bottom culverts with and without wing walls. **Each equation predicts the scour at the entrance corner but only for clear-water conditions. There are no available techniques for estimating live-bed scour at open-bottom culverts at this time.**

The clear-water equation for the with wing wall case is:

$$y_{max} = K_u Q_{BI}^{0.28} \left(\frac{Q}{W_c D_{50}^{1/3}} \right)^{0.26} \tag{6.10}$$

$$y_s = y_{max} - y_o \tag{6.11}$$

where:

y_{max}	=	Flow depth at culvert entrance corner including contraction and local scour, ft (m)
Q_{BI}	=	Discharge blocked by road embankment on one side of culvert, ft³/s (m³/s)
Q	=	Discharge through the culvert, ft³/s (m³/s)
W_c	=	Width of the culvert, ft (m)
D_{50}	=	Median diameter of the bed material, ft (m)
y_s	=	Scour at the culvert entrance corner, ft (m)
y_o	=	Flow depth prior to scour, ft (m)
K_u	=	0.84 English units
K_u	=	1.16 SI units

The equation for the without wing wall case is:

$$y_{max} = K_u Q_{BI}^{0.12} \left(\frac{Q}{W_c D_{50}^{1/3}} \right)^{0.60} \tag{6.12}$$

$$y_s = y_{max} - y_o \tag{6.13}$$

where:
K_u = 0.57 English units
K_u = 0.88 SI units

The total scour would also need to include estimated long-term degradation. For multiple barrel open-bottom culverts, the interior foundation would need to be designed for total scour including standard contraction scour, pier scour (for the interior foundation acting as a pier), and long-term degradation.

6.9.3 Example Problem

An open-bottom culvert is being installed in erodible materials. Median bed material size is 2.5 mm. Design discharge is 1000 ft³/s (28 m³/s), which is also flow through the culvert. Culvert width is 30 ft (9.1 m) and flow depth in the culvert (y_0) is 6.7 ft (2.0 m). Upstream flow depth is 7.3 ft (2.23 m) and upstream flow velocity is 2.8 ft/s (0.85 m). From Equation 6.1, critical velocity to move the 2.5 mm bed material in the upstream channel is 3.1 ft/s (0.96 m/s), confirming clear-water conditions. The flow blocked by each embankment is 195 ft³/s (5.52 m³/s). Computed clear-water contraction scour is 2.6 ft (0.79 m). Compute scour at the culvert entrance corner for with and without wing wall conditions.

With wing wall case (Equation 6.10):

$$y_{max} = K_u Q_{BI}^{0.28} \left(\frac{Q}{W_c D_{50}^{1/3}} \right)^{0.26} = 0.84 \times 195^{0.28} \left(\frac{1000}{30 \times 0.0082^{1/3}} \right)^{0.26} = 13.9 \text{ ft} (4.2 \text{ m})$$

$y_s = y_{max} - y_o = 13.9 - 6.7 = 7.2$ ft (2.2 m)

No wing wall case (Equation 6.12):

$$y_{max} = K_u Q_{BI}^{0.12} \left(\frac{Q}{W_c D_{50}^{1/3}} \right)^{0.60} = 0.57 \times 195^{0.12} \left(\frac{1000}{30 \times 0.0082^{1/3}} \right)^{0.26} = 23.0 \text{ ft} (7.0 \text{ m})$$

$y_s = y_{max} - y_o = 23.0 - 6.7 = 16.3$ ft (5.0 m)

These results confirm that wing walls significantly reduce the scour potential compared to the abrupt, square-corner inlet. Also, the scour at the corner of the culvert entrance is much greater than contraction scour alone. These scour estimates include both contraction and local scour, so only long-term degradation would need to be added to estimate total scour. Based on the laboratory data, a reliability index of 3.0 is achieved by multiplying y_{max} by 1.05. For the with wing wall case, y_{max} would be 13.9 x 1.05 = 14.6 ft (4.4 m) and the scour would increase to 14.6 – 6.7 = 7.9 ft (2.4 m). **These equations do not account for vertical contraction scour (pressure flow) conditions, which would be expected to produce much greater scour.**

6.10 PRESSURE FLOW SCOUR (VERTICAL CONTRACTION SCOUR)

6.10.1 Estimating Pressure Flow Scour

Prediction of pressure flow scour underneath an inundated deck in an extreme flood event is important for safe bridge design and for evaluation of scour at existing bridges. A formula calibrated with experimental data and Computational Fluid Dynamics (CFD) simulation was developed by FHWA (2012c) to calculate pressure flow scour depth under various bridge inundation conditions. The maximum scour depth is evaluated by using contraction scour equations combined with a correlation of separation zone thickness under the inundated bridge. Data from Arneson (1998), TRB (1998b), Umbrell et al. (1998), and the Turner-Fairbank Highway Research Center (FHWA 2012c) were used to develop the scour equations.

Figure 6.18 illustrates the flow characteristics at a fully submerged bridge superstructure. Note that the bridge "superstructure" mentioned in this section refers to a continuous cross section of the structural and non-structural elements that span the waterway and that can produce significant blockage when it is partially or fully inundated. Discharge under the superstructure can be conservatively assumed to be all approach flow below the top of the superstructure at height $h_b + T$, where h_b is the vertical size of the bridge opening prior to scour and T is the height of the obstruction including girders, deck, and parapet. For floods that do not create overtopping, all discharge upstream goes into the bridge opening. The depth at the location of maximum scour is comprised of three components: h_c, the vertically contracted flow height from the streamline bounding the separation zone under the superstructure at the maximum scour depth, y_s, the scour depth, and t, the maximum thickness of the flow separation zone. The separation zone does not convey any net mass from the upstream opening of the bridge to the downstream exit.

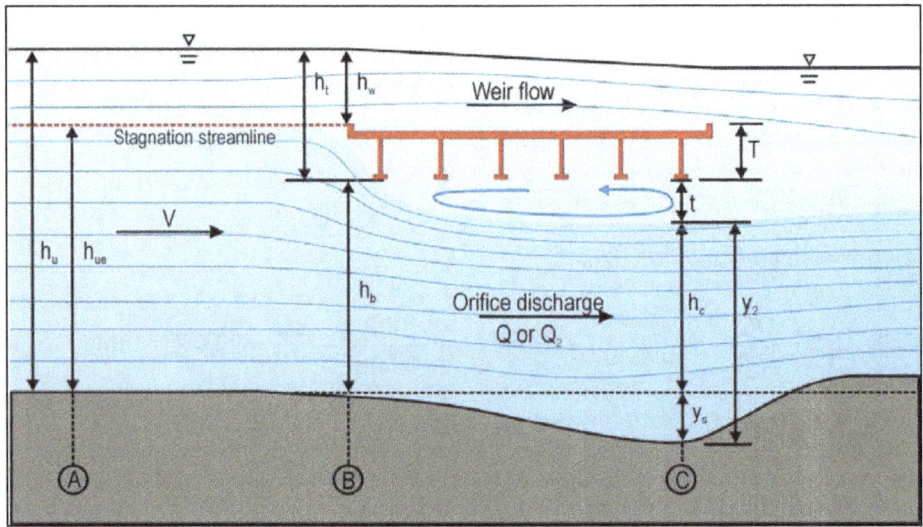

Figure 6.18. Vertical contraction and definition for geometric parameters.

The pressure scour depth y_s is determined by using the horizontal contraction scour equations to calculate the height, $y_s + h_c$, required to convey flow through the bridge opening at the critical velocity. This height is equivalent to y_2 (the average depth in the contracted section) in the clear-water contraction scour Equation 6.4 and the live-bed contraction scour Equation 6.2. Combining this relation with the definitions of t and h_b:

$$y_s = y_2 + t - h_b \tag{6.14}$$

Note that h_b in pressure flow scour is analogous to y_0 (existing depth in the contracted section before scour) in contraction scour. Comparing contraction scour Equations 6.3 and 6.5 with Equation 6.14, the scour depth of pressure flow can be significantly greater than that of non-pressure flow because depth available to convey flow through the opening under the bridge is reduced by the flow separation thickness, t.

Use Equation 6.4 to calculate y_2 for clear-water conditions and Equation 6.2 to calculate y_2 for live-bed conditions. For flow conditions that do not overtop the bridge or roadway approaches, all flow is through the bridge and the live-bed and clear-water equations can be applied directly. When flow overtops the bridge or approach roadway, the value of Q_2 (flow in the contracted channel) in the live-bed equation (Equation 6.2) or Q (discharge through the bridge) in the clear-water equation (Equation 6.4) should include only the flow through the bridge opening. This discharge is obtained from hydraulic models such as HEC-RAS or FST2DH.

For live-bed applications, the upstream channel discharge, Q_1 and channel flow depth, y_1, used in Equation 6.2 may also need to be adjusted. For non-overtopping flows Q_1 is not adjusted and $y_1 = h_{ue} = h_u$. For overtopping flows illustrated in Figure 6.18, Q_1 is adjusted and $y_1 = h_{ue} = h_b + T$, where T is the height of the obstruction including girders, deck, and parapet. If the bridge consists of railing with openings, the blockage height T extends up to the lower edge of the opening under the railing. The potential for debris blocking openings in the railing should be considered when determining T. For overtopping flows in live-bed conditions, Q_{ue} is used for Q_1 in Equation 6.2 and is calculated from the total channel discharge at the approach, Q_1, from:

$$Q_{ue} = Q_1 \left(\frac{h_{ue}}{h_u} \right)^{8/7} \tag{6.15}$$

where:

Q_{ue} = Effective channel discharge for live-bed conditions and bridge overtopping flow, ft³/s (m³/s)
Q_1 = Upstream channel discharge as defined for Equation 6.2, ft³/s (m³/s)
h_u = Upstream channel flow depth as defined for Equation 6.2, ft (m)
h_{ue} = Effective upstream channel flow depth for live-bed conditions and bridge overtopping, ft (m)

The separation zone thickness, t, is calculated using Equation 6.16:

$$\frac{t}{h_b} = 0.5 \left(\frac{h_b \cdot h_t}{h_u^2} \right)^{0.2} \left(1 - \frac{h_w}{h_t} \right)^{-0.1} \tag{6.16}$$

where:

h_b = Vertical size of the bridge opening prior to scour, ft (m)
h_t = Distance from the water surface to the lower face of the bridge girders, equals $h_u - h_b$, ft (m)
h_w = Weir flow height = $h_t - T$ for $h_t > T$, $h_w = 0$ for $h_t \leq T$

Note: Sufficient experimental data to determine the maximum thickness of the separated flow zone, t, beginning at the leading edge of a blunt entry slot cavity on a wall boundary are not available. Equation 6.16 was formulated based on dimensional analysis and CFD testing as a guide and then calibrated using the pressure flow scour data from Arneson (1998), TRB (1998b), Umbrell et al. (1998), and the Turner-Fairbank Highway Research Center (FHWA 2012c). A design safety factor was applied to the constant factor, and exponents were rounded to obtain a conservative estimate of separation zone thickness, t. Therefore t calculated using Equation 6.16 is expected to be larger than values that would be measured in a laboratory.

The use of Equations 6.2 or 6.4 in combination with Equation 6.14 incorporates the constriction of the channel and floodplain flows (lateral contraction) and pressure flow (vertical contraction). Pressure flow scour can occur even when there is no lateral contraction due to vertical contraction of the flow and the development of the flow separation zone.

6.10.2 Pressure Flow Scour Example Problems

Example Problem 1 - Clear-Water Application

Given:

All the flow is through the bridge with no overtopping.
There are no piers (clear span).
Upstream channel width and bridge opening width (W) = 40 ft (12.2 m)
Total discharge (Q) = 2800 ft³/s (79.3 m³/s)
Upstream channel discharge (Q_1) = 2000 ft³/s (56.6 m³/s)
Upstream floodplain discharge = 800 ft³/s (22.7 m³/s)
Upstream channel flow depth (h_u) = 10.0 ft (3.0 m)
Bridge opening height (h_b) = 8.0 ft (2.4 m)
Deck thickness (T) = 3 ft (0.91 m)
Bed material D_{50} = 15 mm (V_c = 6.0 ft/s, 1.8 m/s)
Upstream channel velocity (V=Q_1/(Wh_u)) = 2000/(40 x 10) = 5.0 ft/s (1.5 m/s)

Determine:

The magnitude of clear-water contraction scour for pressure flow conditions.

Solution:

1. Compute y_2 for flow through bridge using Eqn. 6.4.

$$y_2 = \left[\frac{K_u Q^2}{D_m^{2/3} W^2}\right]^{3/7} = \left[\frac{0.0077 \times 2800^2}{(1.25 \times 15.0 / 304.8)^{2/3} 40^2}\right]^{3/7} = 10.5 \text{ ft } (3.2 \text{m})$$

2. Compute separation zone thickness using Equation 6.16. For the non-overtopping case h_w = 0.

$$t = 0.5 \left(\frac{h_b \cdot h_t}{h_u^2}\right)^{0.2} \left(1 - \frac{h_w}{h_t}\right)^{-0.1} \cdot h_b = 0.5 \left(\frac{8 \times 2}{10^2}\right)^{0.2} \times 1 \times 8 = 2.77 \text{ ft } (0.85 \text{m})$$

3. Compute scour using Equation 6.14

   $y_s = y_2 + t - h_b = 10.5 + 2.77 - 8.0 = 5.27$ ft (1.6 m)

Note: This compares with only 0.5 ft (0.15 m) of clear-water contraction scour if the deck were not submerged.

Example Problem 2 - Live-Bed Application Without Overtopping

Given:

All the flow is through the bridge with no overtopping.
There are no piers (clear span).
Upstream channel width (W_1) and bridge opening width (W_2) = 40 ft (12.2 m)
Total discharge (Q_2) = 2800 ft³/s (79.3 m³/s)
Upstream channel discharge (Q_1) = 2000 ft³/s (56.6 m³/s)
Upstream floodplain discharge = 800 ft³/s (22.7 m³/s)
Upstream channel flow depth ($y_1 = h_u$) = 10.0 ft (3.0 m)
Bridge opening height (h_b) = 8.0 ft (2.4 m)
Deck thickness (T, not used in this example) = 3 ft (0.91 m)
Bed material D_{50} = 1.0 mm (V_c = 2.4 ft/s, 0.74 m/s)
Upstream channel velocity ($V = Q_1/(W_1 h_u)$) = 2000/(40 × 10) = 5.0 ft/s (1.5 m/s)

Determine:

The magnitude of live-bed contraction scour for pressure flow conditions.

Solution:

1. Calculate effective upstream channel flow depth (h_{ue})

   $h_{ue} = h_u$ (no overtopping) = 10.0 ft (3.0 m)

2. Compute y_2 for flow through bridge.

   $$y_2 = y_1 \left(\frac{Q_2}{Q_1}\right)^{6/7} \left(\frac{W_1}{W_2}\right)^{k_1} = 10.0 \left(\frac{2800}{2000}\right)^{6/7} \left(\frac{40}{40}\right)^{k_1} = 13.3 \text{ ft } (4.1\text{m})$$

3. Compute separation zone thickness using Equation 6.16. For the non-overtopping case $h_w = 0$.

   $$t = 0.5 \left(\frac{h_b \cdot h_t}{h_u^2}\right)^{0.2} \left(1 - \frac{h_w}{h_t}\right)^{-0.1} \cdot h_b = 0.5 \left(\frac{8 \times 2}{10^2}\right)^{0.2} \times 1 \times 8 = 2.77 \text{ ft } (0.85\text{m})$$

4. Compute scour using Equation 6.14

   $y_s = y_2 + t - h_b = 13.3 + 2.77 - 8.0 = 8.07$ ft (2.5 m)

Note: This compares with 3.3 ft (1.0 m) of live-bed contraction scour if the deck were not submerged.

Example Problem 3 - Live-Bed Application Including Overtopping

Given:

There are no piers (clear span).
Upstream channel width (W_1) and bridge opening width (W_2) = 40 ft (12.2 m)
Total discharge = 2800 ft³/s (79.3 m³/s)
Upstream channel discharge (Q_1) = 2000 ft³/s (56.6 m³/s)
Upstream floodplain discharge = 800 ft³/s (22.7 m³/s)
Discharge through the bridge (Q_2) = 2200 ft³/s (62.3 m³/s)
Upstream channel flow depth (h_u) = 12.0 ft (3.7 m)
Bridge opening height (h_b) = 8.0 ft (2.4 m)
Deck thickness (T) = 3 ft (0.91 m)
Bed material D_{50} = 1.0 mm (V_c = 2.5 ft/s, 0.77 m/s)
Upstream channel velocity ($V = Q_1/(W_1 h_u)$) = 2000/(40 × 12) = 4.2 ft/s (1.3 m/s)

Determine:

The magnitude of live-bed contraction scour for pressure flow conditions.

Solution:

1. Calculate effective upstream channel flow depth (h_{ue}) and discharge (Q_{ue}).

$$h_{ue} = h_b + T = 8.0 + 3.0 = 11 \text{ ft } (3.4 \text{ m})$$

$$Q_{ue} = Q_1 \left(\frac{h_{ue}}{h_u}\right)^{8/7} = 2000 \left(\frac{11}{12}\right)^{8/7} = 1811 \frac{\text{ft}^3}{\text{s}} \left(51.3 \frac{\text{m}^3}{\text{s}}\right)$$

2. Compute y_2 for flow through bridge.

$$y_2 = y_1 \left(\frac{Q_2}{Q_1}\right)^{6/7} \left(\frac{W_1}{W_2}\right)^{k_1} = 11 \left(\frac{2200}{1811}\right)^{6/7} \left(\frac{40}{40}\right)^{0.4} = 13.0 \text{ ft } (3.9 \text{ m})$$

3. Compute separation zone thickness for the overtopping case.

$$t = 0.5 \left(\frac{h_b \cdot h_t}{h_u^2}\right)^{0.2} \left(1 - \frac{h_w}{h_t}\right)^{-0.1} \cdot h_b = 0.5 \left(\frac{8 \times 4}{12^2}\right)^{0.2} \left(1 - \frac{4-3}{12-8}\right)^{-0.1} \times 8 = 3.05 \text{ ft } (0.93 \text{ m})$$

4. Compute scour using Equation 6.14

$$y_s = y_2 + t - h_b = 13 + 3.05 - 8.0 = 8.05 \text{ ft } (2.5 \text{ m})$$

Note: This compares with 4.0 ft (1.2 m) of live-bed contraction scour if the deck were not submerged and the total discharge was conveyed through the bridge opening.

Example Problem 4 - Clear-Water Application Including Overtopping

Given:

There are no piers (clear span).
Upstream channel width and bridge opening width (W) = 32 ft (9.7 m)
Total discharge = 2800 ft^3/s (79.3 m^3/s)
Upstream channel discharge (Q_1) = 2000 ft^3/s (56.6 m^3/s)
Upstream floodplain discharge = 800 ft^3/s (22.7 m^3/s)
Discharge through the bridge (Q) = 2200 ft^3/s (62.3 m^3/s)
Upstream channel flow depth (h_u) = 12.0 ft (3.7 m)
Bridge opening height (h_b) = 8.0 ft (2.4 m)
Deck thickness = 3 ft (0.91 m)
Bed material D_{50} = 15.0 mm (V_c = 6.2 ft/s, 1.9 m/s)
Upstream channel velocity ($V_u = Q_1/(Wh_u)$) = 2000/(32 x 12) = 5.2 ft/s (1.6 m/s)

Determine:

The magnitude of clear-water contraction scour for pressure flow conditions.

Solution:

1. Compute y_2 for flow through bridge.

$$y_2 = \left[\frac{K_u Q^2}{D_m^{2/3} W^2}\right]^{3/7} = \left[\frac{0.0077 \times 2200^2}{(1.25 \times 15.0/304.8)^{2/3} \, 32^2}\right]^{3/7} = 10.4 \text{ ft } (3.1\text{m})$$

2. Compute separation zone thickness for the overtopping case.

$$t = 0.5\left(\frac{h_b \cdot h_t}{h_u^2}\right)^{0.2}\left(1-\frac{h_w}{h_t}\right)^{-0.1} \cdot h_b = 0.5\left(\frac{8 \times 4}{12^2}\right)^{0.2}\left(1-\frac{4-3}{12-8}\right)^{-0.1} \times 8 = 3.05 \text{ ft } (0.93\text{m})$$

3. Compute scour using Equation 6.14

$$y_s = y_2 + t - h_b = 10.4 + 3.05 - 8.0 = 5.45 \text{ ft } (1.7 \text{ m})$$

Note: This compares with 2.3 ft (0.7 m) of clear water scour if the deck were not submerged and the total discharge was conveyed through the bridge opening.

(page intentionally left blank)

CHAPTER 7

PIER SCOUR

7.1 GENERAL

Local scour at piers is a function of bed material characteristics, bed configuration, flow characteristics, fluid properties, and the geometry of the pier and footing. The bed material characteristics are granular or non granular, cohesive or noncohesive, erodible or non erodible rock. Granular bed material ranges in size from silt to large boulders and is characterized by the D_{50} and a coarse size such as the D_{84} or D_{90} size. Cohesive bed material is composed of silt and clay, possibly with some sand which is bonded chemically. Rock may be solid, massive, or fractured. It may be sedimentary or igneous and erodible or non erodible (see discussion in Chapter 4).

Flow characteristics of interest for local pier scour are the velocity and depth just upstream of the pier, the angle the velocity vector makes to the pier (angle of attack), and free surface or pressure flow. Fluid properties are viscosity, and surface tension which for the field case can be ignored.

Pier geometry characteristics are its type, dimensions, and shape. Types of piers include single column, multiple columns, or rectangular; with or without friction or tip bearing piles; with or without a footing or pile cap; footing or pile cap in the bed, on the surface of the bed, in the flow or under the deck out of the flow. Important dimensions are the diameter for circular piers or columns, spacing for multiple columns, and width and length for solid piers. Shapes include round, square or sharp nose, circular cylinder, group of cylinders, or rectangular. In addition, piers may be simple or complex. A simple pier is a single shaft, column or multiple columns exposed to the flow. Whereas, a complex pier may have the pier, footing or pile cap, and piles exposed to the flow.

Local scour at piers has been studied extensively in the laboratory; however, there is limited field data. The laboratory studies have been mostly of simple piers, but there have been some laboratory studies of complex piers. Often the studies of complex piers are model studies of actual or proposed pier configurations. As a result of the many laboratory studies, there are numerous pier scour equations. In general, the equations are for live-bed scour in cohesionless sand-bed streams.

A graphical comparison by Jones of the more common equations is given in Figure 7.1 (TRB 1983). An equation given by Melville and Sutherland to calculate scour depths for live-bed scour in sand-bed streams has been added to the original figure. Some of the equations have velocity as a variable, normally in the form of a Froude Number. However, some equations, such as Laursen's (1960) do not include velocity. A Froude Number of 0.3 was used in Figure 7.1 for purposes of comparing commonly used scour equations. Jones also compared the equations with the available field data. His study showed that the Colorado State University (CSU) equation enveloped all the data, but gave lower values of scour than the Jain and Fischer, Laursen, Melville and Sutherland, and Neill equations (FHWA 2001 and 1979, Laursen 1980, Mellville and Sutherland 1988, TRB 1983). The CSU equation includes the velocity of the flow just upstream of the pier by including the Froude Number in the equation. On the basis of Jones' studies (TRB 1983) the CSU equation was recommended in the Interim Procedures that accompanied FHWA's Technical Advisory T5140.20 (FHWA 1988a and b). With modifications, the CSU equation was recommended in previous editions of HEC-18. The modifications were the addition of coefficients for the effect of bed forms, size of bed material, and wide piers.

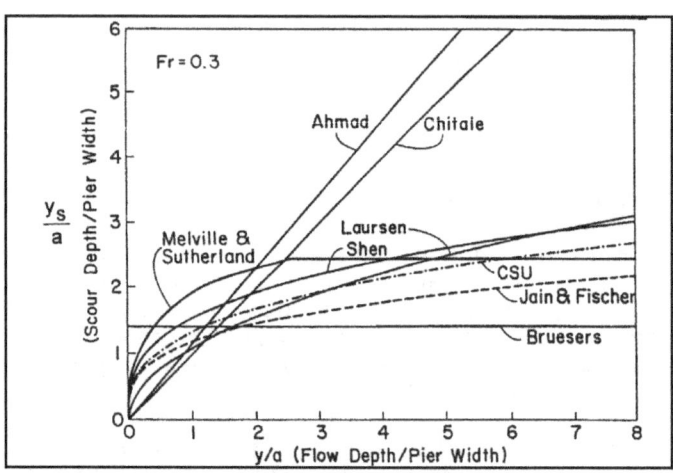

Figure 7.1. Comparison of scour equations for variable depth ratios (y/a) (after TRB 1983).

Mueller (1996) compared 22 scour equations using field data collected by the USGS (Landers et al. 1999). He concluded that the HEC-18 (CSU) equation was good for design because it rarely under predicted measured scour depth. However, it frequently over-predicted the observed scour. The data contained 384 field measurements of scour at 56 bridges. Figure 7.2 provides a definition sketch for pier scour variables where:

a = Pier diameter (or width) in the direction of flow
V = Approach velocity
y_1 = Flow depth
y_s = Depth of scour

From laboratory data, Melville and Sutherland (1988) reported 2.4 as an upper limit for the depth of scour to pier width ratio (y_s/a) for cylindrical piers. In these studies, the Froude Number was less than 1.0. Examination of laboratory data indicates the ratio of scour depth to pier width (y_s/a) rarely exceeds 2.3. However, values of y_s/a around 3.0 were obtained by Jain and Fischer for chute-and-pool flows with Froude Numbers as high as 1.5 (FHWA 1979). The largest value of y_s/a for antidune flow was 2.5 with a Froude Number of 1.2. These upper limits were derived for circular piers and were uncorrected for pier shape or for skew. Also, pressure flow, ice or debris can increase the ratio.

From the above discussion, the ratio of y_s/a can be as large as 3 at large Froude Numbers. Therefore, it is recommended that the maximum value of the ratio be taken as 2.4 for Froude Numbers less than or equal to 0.8 and 3.0 for larger Froude Numbers. These limiting ratio values apply only to round nose piers which are aligned with the flow.

7.2 HEC-18 PIER SCOUR EQUATION

The HEC-18 pier scour equation (based on the CSU equation) is recommended for both live-bed and clear-water pier scour. The equation predicts maximum pier scour depths. Basic applications include simple pier substructure configurations and riverine flow situations in alluvial sand-bed channels. The equation can be adapted for wide pier applications (Section 7.4), more complex (3-element) substructure configurations (Section 7.5), multiple columns skewed to the flow (Section 7.6), estimating scour from debris on piers (Section 7.7), and scour in tidal waterways (Chapter 9). An alternative approach that represents the complexity of the bridge pier scour flow field and the full range of pier geometries (narrow, transition and wide as described in Section 3.6.2) is presented in Section 7.3.

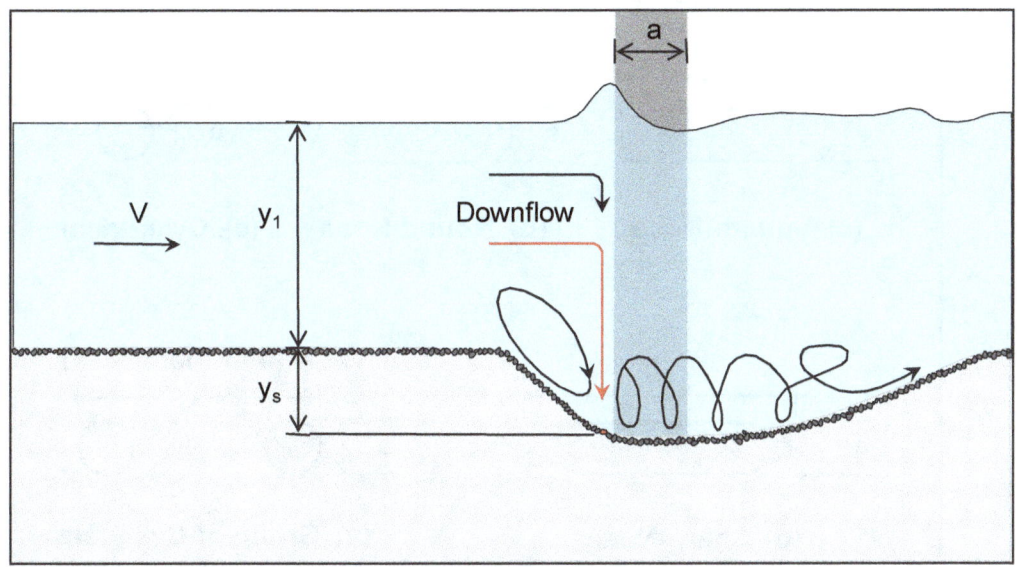

Figure 7.2. Definition sketch for pier scour.

The HEC-18 equation is:

$$\frac{y_s}{y_1} = 2.0 \, K_1 \, K_2 \, K_3 \left(\frac{a}{y_1}\right)^{0.65} Fr_1^{0.43} \tag{7.1}$$

As a Rule of Thumb, the maximum scour depth for round nose piers aligned with the flow is:

$y_s \leq 2.4$ times the pier width (a) for $Fr \leq 0.8$ (7.2)
$y_s \leq 3.0$ times the pier width (a) for $Fr > 0.8$

In terms of y_s/a, Equation 7.1 is:

$$\frac{y_s}{a} = 2.0 \, K_1 \, K_2 \, K_3 \left(\frac{y_1}{a}\right)^{0.35} Fr_1^{0.43} \tag{7.3}$$

where:

y_s = Scour depth, ft (m)
y_1 = Flow depth directly upstream of the pier, ft (m)
K_1 = Correction factor for pier nose shape from Figure 7.3 and Table 7.1
K_2 = Correction factor for angle of attack of flow from Table 7.2 or Equation 7.4
K_3 = Correction factor for bed condition from Table 7.3
a = Pier width, ft (m)
L = Length of pier, ft (m)
Fr_1 = Froude Number directly upstream of the pier = $V_1/(gy_1)^{1/2}$
V_1 = Mean velocity of flow directly upstream of the pier, ft/s (m/s)
g = Acceleration of gravity (32.2 ft/s²) (9.81 m/s²)

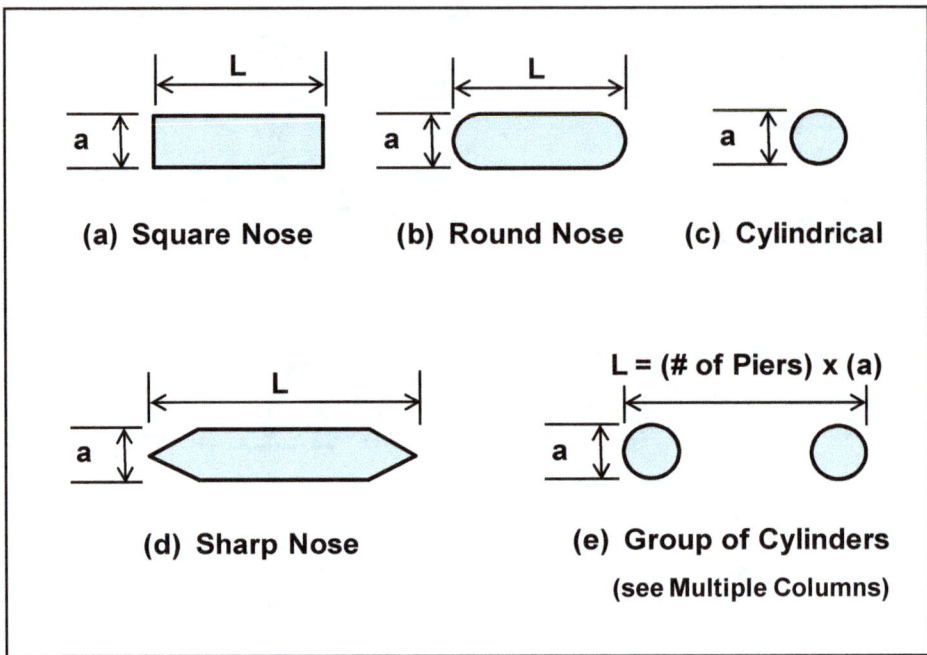

Figure 7.3. Common pier shapes.

The correction factor, K_2, for angle of attack of the flow, θ, is calculated using the following equation:

$$K_2 = (\cos\theta + \frac{L}{a}\sin\theta)^{0.65} \tag{7.4}$$

If L/a is larger than 12, use L/a = 12 as a maximum in Equation 7.4 and Table 7.2. Table 7.2 illustrates the magnitude of the effect of the angle of attack on local pier scour.

Table 7.1. Correction Factor, K_1, for Pier Nose Shape.

Shape of Pier Nose	K_1
(a) Square nose	1.1
(b) Round nose	1.0
(c) Circular cylinder	1.0
(d) Group of cylinders	1.0
(e) Sharp nose	0.9

Table 7.2. Correction Factor, K_2, for Angle of Attack, θ, of the Flow.

Angle	L/a=4	L/a=8	L/a=12
0	1.0	1.0	1.0
15	1.5	2.0	2.5
30	2.0	2.75	3.5
45	2.3	3.3	4.3
90	2.5	3.9	5.0

Angle = skew angle of flow
L = length of pier

Table 7.3. Increase in Equilibrium Pier Scour Depths, K_3, for Bed Condition.

Bed Condition	Dune Height ft	K_3
Clear-Water Scour	N/A	1.1
Plane bed and Antidune flow	N/A	1.1
Small Dunes	$10 > H \geq 2$	1.1
Medium Dunes	$30 > H \geq 10$	1.2 to 1.1
Large Dunes	$H \geq 30$	1.3

Notes:

1. The correction factor K_1 for pier nose shape should be determined using Table 7.1 for angles of attack up to 5 degrees. **For greater angles, K_2 dominates and K_1 should be considered as 1.0**. If L/a is larger than 12, use the values for L/a = 12 as a maximum in Table 7.2 and Equation 7.4.

2. The values of the correction factor K_2 should be applied only when the field conditions are such that the entire length of the pier is subjected to the angle of attack of the flow. Use of this factor will result in a significant over-prediction of scour if (1) a portion of the pier is shielded from the direct impingement of the flow by an abutment or another pier; or (2) an abutment or another pier redirects the flow in a direction parallel to the pier. For such cases, judgment must be exercised to reduce the value of the K_2 factor by selecting the effective length of the pier actually subjected to the angle of attack of the flow. **Equation 7.4 should be used for evaluation and design**. Table 7.2 is intended to illustrate the importance of angle of attack in pier scour computations and to establish a cutoff point for K_2 (i.e., a maximum value of 5.0).

3. The correction factor K_3 results from the fact that for plane-bed conditions, which is typical of most bridge sites for the flood frequencies employed in scour design, the maximum scour may be 10 percent greater than computed with Equation 7.1. In the **unusual** situation where a dune bed configuration **with large dunes** exists at a site during flood flow, the maximum pier scour may be 30 percent greater than the predicted equation value. This may occur on very large rivers, such as the Mississippi. For smaller streams that have a dune bed configuration at flood flow, the dunes will be smaller and the maximum scour may be only 10 to 20 percent larger than equilibrium scour. For antidune bed configuration the maximum scour depth may be 10 percent greater than the computed equilibrium pier scour depth.

4. Piers set close to abutments (for example at the toe of a spill through abutment) must be carefully evaluated for the angle of attack and velocity of the flow coming around the abutment.

7.3 FLORIDA DOT PIER SCOUR METHODOLOGY

Equation 7.1 has been included in all previous versions of HEC-18 and has been used for bridge scour evaluations and bridge design for countless bridges in the U.S. and worldwide. This equation, which was developed and modified over several decades, could be improved by including bed material size and a more detailed consideration of the bridge pier flow field (see Section 3.6.2). An NCHRP study (NCHRP 2011a) evaluated 22 pier scour equations and found that although the HEC-18 equation did well in comparison to the other equations, the Sheppard and Miller (2006) equation generally performed better for both laboratory and field data. A second NCHRP study (NCHRP 2011c) made slight modifications to the Sheppard and Miller equation to further improve its performance. As with the HEC-18 equation, the NCHRP equation includes flow velocity, depth and angle of attack, pier geometry and shape, but also includes particle size. The NCHRP equation combines pier geometry, shape, and angle of attack to compute an effective pier width, a*, and also distinguishes between clear-water and live-bed flow conditions.

The results of the NCHRP studies (NCHRP 2011a and c) were evaluated and expanded into a pier scour analysis methodology by Florida DOT (FDOT). FDOT has maintained this methodology in a Bridge Scour Manual (FDOT 2011) and developed supporting spread sheets for a wide range of pier scour applications. As noted, this methodology is based on a more complete dimensional analysis than the HEC-18 equation. Although the HEC-18 equation provides good results for most applications, the FDOT methodology should be considered as an alternative, particularly for wide piers (y/a < 0.2 as described in Section 3.6.2) in shallow flows with fine bed material. The FDOT methodology includes the following equations:

$$\frac{y_s}{a^*} = 2.5 f_1 f_2 f_3 \qquad \text{for } 0.4 \leq \frac{V_1}{V_c} < 1.0 \tag{7.5}$$

$$\frac{y_s}{a^*} = f_1 \left[2.2 \left(\frac{\frac{V_1}{V_c} - 1}{\frac{V_{lp}}{V_c} - 1} \right) + 2.5 f_3 \left(\frac{\frac{V_{lp}}{V_c} - \frac{V_1}{V_c}}{\frac{V_{lp}}{V_c} - 1} \right) \right] \qquad \text{for } 1.0 \leq \frac{V_1}{V_c} \leq \frac{V_{lp}}{V_c} \tag{7.6}$$

$$\frac{y_s}{a^*} = 2.2 f_1 \qquad \text{for } \frac{V_1}{V_c} > \frac{V_{lp}}{V_c} \tag{7.7}$$

$$f_1 = \tanh\left[\left(\frac{y_1}{a^*}\right)^{0.4}\right] \tag{7.8}$$

$$f_2 = \left\{ 1 - 1.2 \left[\ln\left(\frac{V_1}{V_c}\right) \right]^2 \right\} \tag{7.9}$$

$$f_3 = \left[\frac{\left(\frac{a^*}{D_{50}}\right)^{1.13}}{10.6 + 0.4\left(\frac{a^*}{D_{50}}\right)^{1.33}} \right] \tag{7.10}$$

where:

y_s = Pier scour depth, ft (m)
a^* = Effective pier width, ft (m)
V_1 = Mean velocity of flow directly upstream of the pier, ft/s (m/s)

V_{lp} = Velocity of the live-bed peak scour, ft/s (m/s)
V_c = Critical velocity for movement of D_{50} as defined below, ft/s (m/s)
D_{50} = Median particle size of bed material, ft (m)

$$V_{lp} = 5V_c \text{ or } 0.6\sqrt{gy_1} \quad \text{(whichever is greater)} \tag{7.11}$$

$$V_c = 5.75 u_c^* \log\left(5.53 \frac{y_1}{D_{50}}\right) \tag{7.12}$$

where:

D_{50} = Median bed material, ft (m)

and:

$$u_c^* = K_u\left(0.0377 + 0.041 D_{50}^{1.4}\right) \quad \text{for} \quad 0.1 \text{ mm} < D_{50} < 1 \text{ mm} \tag{7.13}$$

$$u_c^* = K_u\left(0.1 D_{50}^{0.5} - 0.0213/D_{50}\right) \quad \text{for} \quad 1 \text{ mm} < D_{50} < 100 \text{ mm} \tag{7.14}$$

(Note: u^*_c equation requires D_{50} in mm, not ft or m.)

where:

K_u = 1.0 English units
K_u = 0.3048 SI
D_{50} = Median bed material (mm)

The effective pier width, a^*, is the projected width of the pier times the shape factor, K_{sf}.

$$a^* = K_{sf} a_{proj} \tag{7.15}$$

The shape factor for a circular or round nosed pier is 1.0 and for a square end pier the shape factor depends on the angle of attack.

$$K_{sf} = 1.0 \quad \text{for circular or round nosed piers} \tag{7.16}$$

$$K_{sf} = 0.86 + 0.97\left(\left|\frac{\pi\theta}{180} - \frac{\pi}{4}\right|\right)^4 \quad \text{for square nosed piers} \tag{7.17}$$

where:

θ = flow angle of attack in degrees.

The projected width of the pier is:

$$a_{proj} = a\cos\theta + L\sin\theta \tag{7.18}$$

where:

a_{proj} = Projected pier width in direction of flow, ft (m)
a = Pier width, ft (m)
L = Pier length, ft (m)

The methodology can be accessed through a spreadsheet available at the Florida Department of Transportation website. It can also be computed from the equations presented above or by following the following steps.

1. Calculate V_c using Equation 7.12
2. Calculate V_{lp} using Equation 7.11
3. Calculate a^* using Equation 7.15
4. Calculate f_1 using Equation 7.8 (note: values of hyperbolic tangent are provided in Table 7.4)
5. Calculate f_3 using Equation 7.10
6. Calculate $\dfrac{y_{s-c}}{a^*}$ and y_{s-c} (defined below)
7. Calculate $\dfrac{y_{s-lp}}{a^*}$ and y_{s-lp} (defined below)
8. If $V_1 < 0.4V_c$, then $y_s = 0.0$
9. If $0.4V_c < V_1 \leq V_c$, then calculate f_2 and $y_s = f_2 y_{s-c}$
10. If $V_1 \geq V_{lp}$, then $y_s = y_{s-lp}$
11. If $V_c < V_1 < V_{lp}$, then calculate y_s from:

$$y_s = y_{s-c} + \frac{(V_1 - V_c)}{(V_{lp} - V_c)}(y_{s-lp} - y_{s-c}) \qquad (7.19)$$

Note that Equation 7.19 is an equivalent, but simplified version of Equation 7.6. y_{s-c} is the scour at critical velocity for bed material movement (V_c) and is equal to $2.5f_1f_3a^*$. y_{s-lp} is the scour at live-bed peak velocity (V_{lp}) and is equal to $2.2f_1a^*$. The Florida DOT spreadsheet uses y_{s-c} as the design scour value when it is greater than y_{s-lp}.

The FDOT methodology for pier scour includes four regions as shown in Figure 7.4.

- Scour Region I (Step 8, above) is for clear-water conditions with velocity too low to produce scour, which occurs for velocities less than $0.4V_c$. However, field data in the NCHRP (2011c) report include observed scour for this condition, although it was only observed on one occasion for laboratory data.

- Scour Region II is for clear-water conditions with flow velocity large enough to produce pier scour ($V_c > V_1 > 0.4V_c$) as defined by Step 9, above.

- Scour Region IV is defined by the live-bed peak velocity (V_{lp}), where the maximum live-bed scour occurs at $5V_c$ or greater. Any velocity greater than V_{lp} is assigned the scour, y_{s-lp}, computed for V_{lp} (Step 10).

- Live-bed scour that occurs for flow velocities between critical velocity and the live-bed peak velocity ($V_c < V_1 < V_{lp}$) occurs in scour Region III as defined by Step 11 and Equation 7.19.

Table 7.4. Hyperbolic Tangent of X.										
X	0.00	0.01	0.02	0.03	0.04	0.05	0.06	0.07	0.08	0.09
0.0	0.00	0.01	0.02	0.03	0.04	0.05	0.06	0.07	0.08	0.09
0.1	0.10	0.11	0.12	0.13	0.14	0.15	0.16	0.17	0.18	0.19
0.2	0.20	0.21	0.22	0.23	0.24	0.24	0.25	0.26	0.27	0.28
0.3	0.29	0.30	0.31	0.32	0.33	0.34	0.35	0.35	0.36	0.37
0.4	0.38	0.39	0.40	0.41	0.41	0.42	0.43	0.44	0.45	0.45
0.5	0.46	0.47	0.48	0.49	0.49	0.50	0.51	0.52	0.52	0.53
0.6	0.54	0.54	0.55	0.56	0.56	0.57	0.58	0.58	0.59	0.60
0.7	0.60	0.61	0.62	0.62	0.63	0.64	0.64	0.65	0.65	0.66
0.8	0.66	0.67	0.68	0.68	0.69	0.69	0.70	0.70	0.71	0.71
0.9	0.72	0.72	0.73	0.73	0.74	0.74	0.74	0.75	0.75	0.76
1.0	0.76	0.77	0.77	0.77	0.78	0.78	0.79	0.79	0.79	0.80
1.1	0.80	0.80	0.81	0.81	0.81	0.82	0.82	0.82	0.83	0.83
1.2	0.83	0.84	0.84	0.84	0.85	0.85	0.85	0.85	0.86	0.86
1.3	0.86	0.86	0.87	0.87	0.87	0.87	0.88	0.88	0.88	0.88
1.4	0.89	0.89	0.89	0.89	0.89	0.90	0.90	0.90	0.90	0.90
1.5	0.91	0.91	0.91	0.91	0.91	0.91	0.92	0.92	0.92	0.92
1.6	0.92	0.92	0.92	0.93	0.93	0.93	0.93	0.93	0.93	0.93
1.7	0.94	0.94	0.94	0.94	0.94	0.94	0.94	0.94	0.94	0.95
1.8	0.95	0.95	0.95	0.95	0.95	0.95	0.95	0.95	0.95	0.96
1.9	0.96	0.96	0.96	0.96	0.96	0.96	0.96	0.96	0.96	0.96
2.0	0.96	0.96	0.97	0.97	0.97	0.97	0.97	0.97	0.97	0.97
2.1	0.97	0.97	0.97	0.97	0.97	0.97	0.97	0.97	0.97	0.98
2.2	0.98	0.98	0.98	0.98	0.98	0.98	0.98	0.98	0.98	0.98
2.3	0.98	0.98	0.98	0.98	0.98	0.98	0.98	0.98	0.98	0.98
2.4	0.98	0.98	0.98	0.98	0.98	0.99	0.99	0.99	0.99	0.99
2.5	0.99	0.99	0.99	0.99	0.99	0.99	0.99	0.99	0.99	0.99
2.6	0.99	0.99	0.99	0.99	0.99	0.99	0.99	0.99	0.99	0.99
2.7	0.99	0.99	0.99	0.99	0.99	0.99	0.99	0.99	0.99	0.99
2.8	0.99	0.99	0.99	0.99	0.99	0.99	0.99	0.99	0.99	0.99
2.9	0.99	0.99	0.99	0.99	0.99	0.99	0.99	0.99	0.99	0.99
>=3.0	1.00	1.00	1.00	1.00	1.00	1.00	1.00	1.00	1.00	1.00

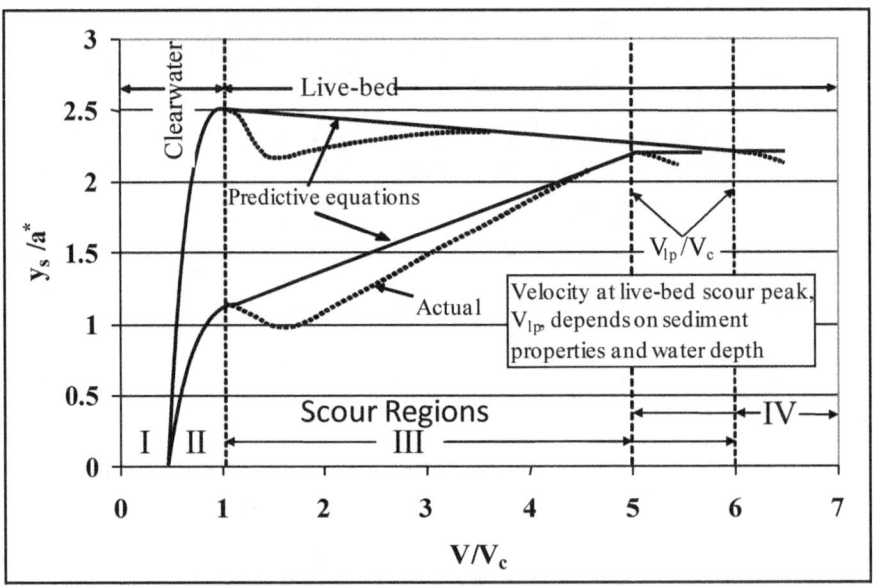

Figure 7.4. Scour for FDOT methodology (after Florida DOT 2011).

An example of an application of the FDOT methodology is presented in Section 7.10.6.

7.4 PIER SCOUR AT WIDE PIERS

Flume studies on scour depths at wide piers in shallow flows and field observations of scour depths at bascule piers in shallow flows indicate that existing equations, including the CSU equation, overestimate scour depths. Johnson and Torrico (TRB 1994) suggest the following equations for a K_w factor to be used to correct Equation 7.1 or 7.3 for wide piers in shallow flow (see Section 7.2). **The correction factor should be applied when the ratio of depth of flow (y) to pier width (a) is less than 0.8 (y/a < 0.8); the ratio of pier width (a) to the median diameter of the bed material (D_{50}) is greater than 50 (a/D_{50} > 50); and the Froude Number of the flow is subcritical.**

$$K_w = 2.58 \left(\frac{y}{a}\right)^{0.34} Fr_1^{0.65} \quad \text{for } V/V_c < 1 \tag{7.20}$$

$$K_w = 1.0 \left(\frac{y}{a}\right)^{0.13} Fr_1^{0.25} \quad \text{for } V/V_c \geq 1 \tag{7.21}$$

where:

K_w = Correction factor to Equation 7.1 or 7.3 for wide piers in shallow flow. The other variables as previously defined.

Engineering judgment should be used in applying K_w because it is based on limited data from flume experiments. Engineering judgment should take into consideration the volume of traffic, the importance of the highway, cost of a failure (potential loss of lives and dollars) and the change in cost that would occur if the K_w factor is used.

7.5 SCOUR FOR COMPLEX PIER FOUNDATIONS

7.5.1 Introduction

As Salim and Jones (1995, 1996, 1999) point out most pier scour research has focused on **solid piers** with limited attention to the determining scour depths for (1) pile groups, (2) pile groups and pile caps, or (3) pile groups, pile caps and solid piers exposed to the flow. The three types of exposure to the flow may be by design or by scour (long-term degradation, general (contraction) scour, and local scour, in addition to stream migration). In the general case, the flow could be obstructed by three substructural elements, herein referred to as the scour-producing components, which include the pier stem, the pile cap or footing, and the pile group. Nevertheless, research has provided methods and equations to determine scour depths for complex pier foundations as an extension of the pier scour equations for routine applications (Section 7.2). The results of this research are recommended for use and are given in the following sections. These procedures can be expected to produce conservative results. Physical Model studies are still recommended for complex piers with unusual features such as staggered or unevenly spaced piles or for major bridges where conservative scour estimates are not economically acceptable. However, the methods presented in this section provide a good estimate of scour for a variety of complex pier situations.

The steps listed below are recommended for determining the depth of scour for any combination of the three substructural elements exposed to the flow (Jones and Sheppard 2000), but engineering judgment is an essential element in applying the design graphs and equations presented in this section as well as in deciding when a more rigorous level of evaluation is warranted. Engineering judgment should take into consideration the volume of traffic, type of traffic (school bus, ambulance, fire trucks, local road, interstate, etc.), the importance of the highway, cost of a failure (potential loss of life and dollars) and the increase in cost that would occur if the most conservative scour depth is used. The stability of the foundation should be checked for:

- The scour depths should be determined for the scour design flood or smaller discharge if it causes deeper scour and the scour design check flood, as recommended in this manual (see Table 2.1).
- If needed use computer programs HEC-RAS (USACE 2010a), FST2DH (FHWA 2003b), etc.) to compute the hydraulic variables.
- Total scour depth is determined by separating the scour producing components, determining the scour depth for each component and adding the results. The method is called **"Superposition of the Scour Components."**
- Analyze the complex pile configuration to determine the components of the pier that are exposed to the flow or will be exposed to the flow which will cause scour.
- Determine the scour depths for each component exposed to the flow using the equations and methods presented in the following sections.
- Add the components to determine the total scour depths.
- Plot the scour depths and analyze the results using an interdisciplinary team to determine their reliability and adequacy for the bridge, flow and site conditions, safety and costs.
- Conduct a physical model study (Section 7.9) if engineering judgment determines it will reduce uncertainly, increase the safety of the design and/or reduce cost.

7.5.2 Superposition of Scour Components Method of Analysis

The components of a complex pier are illustrated in Figure 7.5 (Jones and Sheppard 2000). This is followed by a definition of the variables. Note that the pile cap can be above the water surface, at the water surface, in the water or on the bed. The location of the pile cap may result from design or from long-term degradation and/or contraction scour. The pile group, as illustrated, is in uniform (lined up) rows and columns. This may not always be the case. The support for the bridge in many flow fields and designs may require a more complex arrangement of the pile group. In more complex pile group arrangements, the methods of analysis given in this manual may give smaller or larger scour depths.

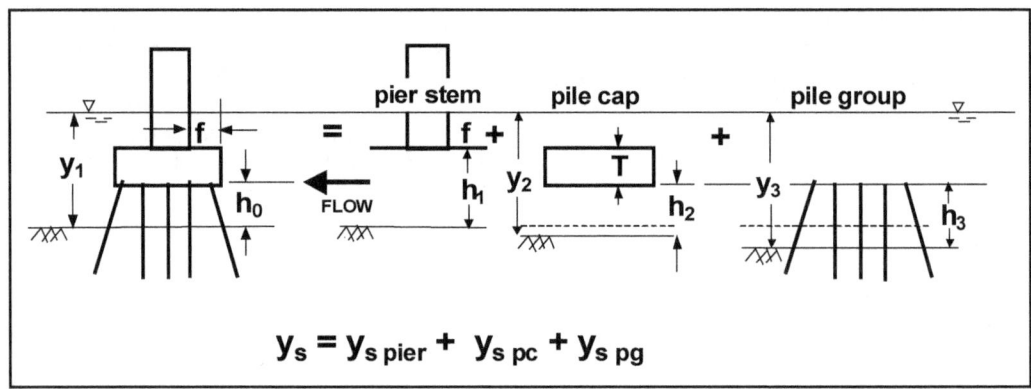

Figure 7.5. Definition sketch for scour components for a complex pier.

The variables illustrated in Figure 7.5 and others used in computations are as follows:

- f = Distance between front edge of pile cap or footing and pier, ft (m)
- h_o = Height of the pile cap above bed at beginning of computation, ft (m)
- h_1 = $h_o + T$ = height of the pier stem above the bed before scour, ft (m)
- h_2 = $h_o + y_{s\,pier}/2$ = height of pile cap after pier stem scour component has been computed, ft (m)
- h_3 = $h_o + y_{s\,pier}/2 + y_{s\,pc}/2$ = height of pile group after the pier stem and pile cap scour components have been computed, ft (m)
- S = Spacing between columns of piles, pile center to pile center, ft (m)
- T = Thickness of pile cap or footing, ft (m)
- y_1 = Approach flow depth at the beginning of computations, ft (m)
- y_2 = $y_1 + y_{s\,pier}/2$ = adjusted flow depth for pile cap computations ft (m)
- y_3 = $y_1 + y_{s\,pier}/2 + y_{s\,pc}/2$ = adjusted flow depth for pile group computations, ft (m)
- V_1 = Approach velocity used at the beginning of computations, ft/sec (m/sec)
- V_2 = $V_1(y_1/y_2)$ = adjusted velocity for pile cap computations, ft/sec (m/sec)
- V_3 = $V_1(y_1/y_3)$ = adjusted velocity for pile group computations, ft/sec (m/sec)

Total scour from superposition of components is given by:

$$y_s = y_{s\,pier} + y_{s\,pc} + y_{s\,pg} \qquad (7.22)$$

where:

- y_s = Total scour depth, ft (m)

$y_{s\ pier}$ = Scour component for the pier stem in the flow, ft (m)
$y_{s\ pc}$ = Scour component for the pier cap or footing in the flow, ft (m)
$y_{s\ pg}$ = Scour component for the piles exposed to the flow, ft (m)

Each of the scour components is computed from the basic pier scour Equation 7.1 using an equivalent sized pier to represent the irregular pier components, adjusted flow depths and velocities as described in the list of variables for Figure 7.5, and height adjustments for the pier stem and pile group. The height adjustment is included in the equivalent pier size for the pile cap. In the following sections guidance for calculating each of the components is given.

7.5.3 Determination of the Pier Stem Scour Depth Component

The need to compute the pier stem scour depth component occurs when the pier cap or the footing is in the flow and the pier stem is subjected to sufficient flow depth and velocity as to cause scour. The first computation is the scour estimate, $y_{s\ pier}$, for a full depth pier that has the width and length of the pier stem using the basic pier equation (Equation 7.1). In Equation 7.1, a_{pier} is the pier width and other variables in the equation are as defined previously. This base scour estimate is multiplied by $K_{h\ pier}$, given in Figure 7.6 as a function of h_1/a_{pier} and f/a_{pier}, to yield the pier stem scour component as follows:

$$\frac{y_{s\,pier}}{y_1} = K_{h\,pier}\left[2.0 K_1 K_2 K_3 \left(\frac{a_{pier}}{y_1}\right)^{0.65} \left(\frac{V_1}{\sqrt{gy_1}}\right)^{0.43}\right] \quad (7.23)$$

where:

$K_{h\ pier}$ = Coefficient to account for height of pier stem above bed and shielding effect by pile cap overhang distance "f" in front of pier stem (from Figure 7.6)

The quantity in the square brackets in Equation 7.23 is the basic pier scour ratio as if the pier stem were full depth and extended below the scour.

7.5.4 Determination of the Pile Cap (Footing) Scour Depth Component

The need to compute the pile cap or footing scour depth component occurs when the pile cap is in the flow by design, or as the result of long-term degradation, contraction scour, and/or by local scour attributed to the pier stem above it. As described below, there are two cases to consider in estimating the scour caused by the pile cap (or footing). Equation 7.1 is used to estimate the scour component in both cases, but the conceptual strategy for determining the variables to be used in the equation is different (partly due to limitations in the research that has been done to date). In both cases the wide pier factor, K_w, in Section 7.4 may be applicable for this computation.

Case 1: The bottom of the pile cap is above the bed and in the flow either by design or after the bed has been lowered by scour caused by the pier stem component. The strategy is to reduce the pile cap width, a_{pc}, to an equivalent full depth solid pier width, a^*_{pc}, using Figure 7.7. The equivalent pier width, an adjusted flow depth, y_2, and an adjusted flow velocity, V_2, are then used in Equation 7.1 to estimate the scour component.

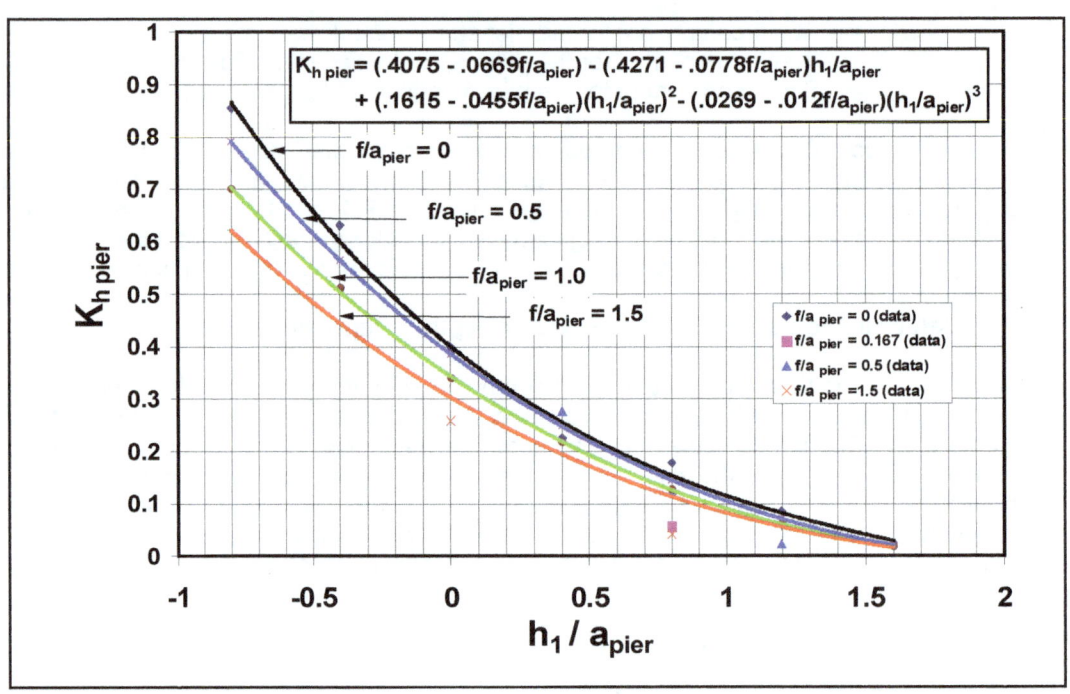

Figure 7.6. Suspended pier scour ratio (Jones and Sheppard 2000).

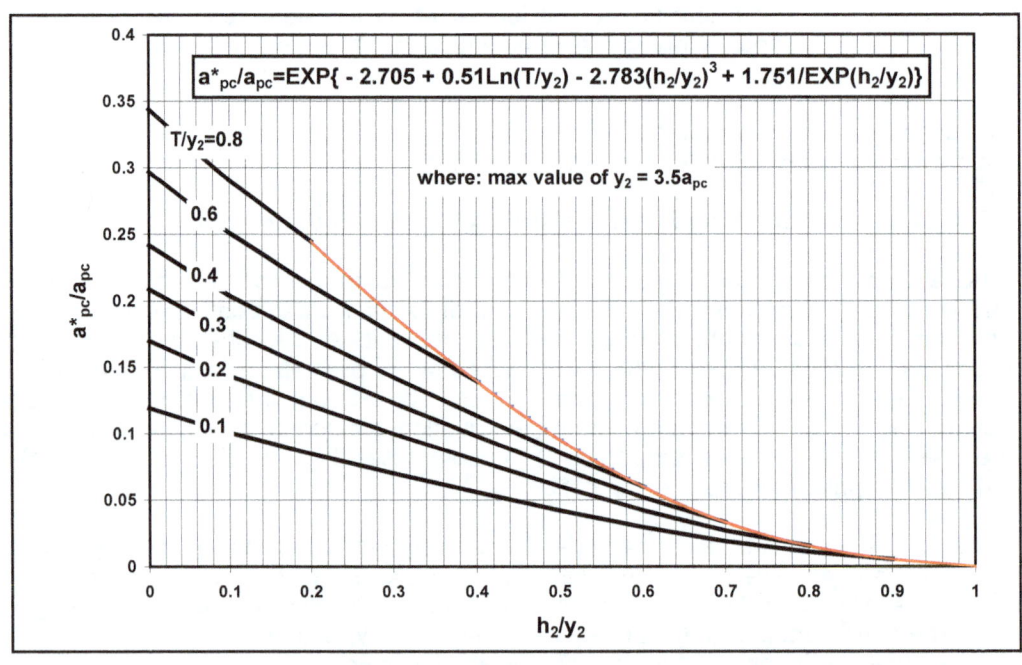

Figure 7.7. Pile cap (footing) equivalent width (Jones and Sheppard 2000).

Case 2: The bottom of the pile cap or footing is on or below the bed. The strategy is to treat the pile cap or exposed footing like a short pier in a shallow stream of depth equal to the height to the top of the footing above bed. The portion of the flow that goes over the top of the pile cap or footing is ignored. Then, the full pile cap width, a_{pc}, is used in the computations, but the exposed footing height, y_f, (in lieu of the flow depth), and the average velocity, V_f, in the portion of the profile approaching the footing are used in Equation 7.1 to estimate the scour component.

An inherent assumption in this second case is that the footing is deeper than the expected scour depth so it is <u>not necessary</u> to add the pile group scour as a third component in this case. If the bottom of the pile cap happens to be right on the bed, either the Case 1 or Case 2 method could be applied, but they won't necessarily give the same answers. If both methods are tried, then engineering judgment should dictate which one to accept.

Details for determining the pile cap or footing scour component for these two cases are described in the following paragraphs.

Case 1. Bottom of the Pile Cap (Footing) in the Flow above the Bed

T = Thickness of the pile cap exposed to the flow, ft (m)
h_2 = $h_o + y_{s\,pier}/2$, ft (m)
y_2 = $y_1 + y_{s\,pier}/2$, = adjusted flow depth, ft (m)
V_2 = $V_1(y_1/y_2)$ = adjusted flow velocity, ft/s (m/s)

where:

h_o = Original height of the pile cap above the bed, ft (m)
y_1 = Original flow depth at the beginning of the computations before scour, ft (m)
$y_{s\,pier}$ = Pier stem scour depth component, ft (m)
V_1 = Original approach velocity at the beginning of the computations, ft/s (m/s)

Determine a^*_{pc}/a_{pc} from Figure 7.7 as a function of h_2/y_2 and T/y_2 (note that the maximum value of $y_2 = 3.5\,a_{pc}$).

Compute $a^*_{pc} = (a^*_{pc}/a_{pc})\,a_{pc}$; where a^*_{pc} is the width of the equivalent pier to be used in Equation 7.1 and a_{pc} is the width of the original pile cap. Compute the pile cap scour component, $y_{s\,pc}$ from Equation 7.1 using a^*_{pc}, y_2, and V_2 as the pier width, flow depth, and velocity parameters, respectively. The rationale for using the adjusted velocity for this computation is that the near bottom velocities are the primary currents that produce scour and they tend to be reduced in the local scour hole from the overlying component. **For skewed flow use the L/a for the original pile cap as the L/a for the equivalent pier to determine K_2.** Apply the wide pier correction factor, K_w, if (1) the total depth, $y_2 < 0.8\,a^*_{pc}$, (2) the Froude Number $V_2/(g\,y_2)^{1/2} < 1$, and (3) $a^*_{pc} > 50\,D_{50}$. The scour component equation for the Case 1 pile cap can then be written:

$$\frac{y_{spc}}{y_2} = 2.0\,K_1 K_2 K_3 K_w \left(\frac{a^*_{pc}}{y_2}\right)^{0.65} \left(\frac{V_2}{\sqrt{g y_2}}\right)^{0.43} \qquad (7.24)$$

Next, the pile group scour component should be computed. This is discussed in Section 7.5.5.

Case 2. Bottom of the Pile Cap (Footing) Located On or Below the Bed.

One limitation of the procedure described above is that the design chart in Figure 7.7 has not been developed for the case of the bottom of the pile cap or footing being below the bed (i.e., negative values of h_2). In this case, use a modification of the exposed footing procedure that has been described in previous editions of HEC-18. The previous procedure was developed from experiments in which the footing was never undermined by scour and tended to be an over predictor if the footing is undermined.

As for Case 1:

$$y_2 = y_1 + y_{s\ pier}/2, \text{ ft (m)}$$
$$V_2 = V_1(y_1/y_2), \text{ ft/s (m/s)}$$

The average velocity of flow at the exposed footing (V_f) is determined using the following equation:

$$\frac{V_f}{V_2} = \frac{\ln\left(10.93\frac{y_f}{k_s} + 1\right)}{\ln\left(10.93\frac{y_2}{k_s} + 1\right)} \tag{7.25}$$

where:

V_f = Average velocity in the flow zone below the top of the footing, ft/s (m/s)
V_2 = Average adjusted velocity in vertical of flow approaching the pier, ft/s (m/s)
ln = Natural log to the base e
y_f = $h_1 + y_{s\ pier}/2$ = distance from the bed (after degradation, contraction scour, and pier stem scour) to the top of the footing, ft (m)
k_s = Grain roughness of the bed (normally taken as the D_{84} for sand size bed material and 3.5 D_{84} for gravel and coarser bed material), ft (m)
y_2 = Adjusted depth of flow upstream of the pier, including degradation, contraction scour and half the pier stem scour, ft (m)

See Figure 7.8 for an illustration of variables.

Compute the pile cap scour depth component, $y_{s\ pc}$ from Equation 7.1 using the full pile cap width, a_{pc}, y_f, V_f as the width, flow depth, and velocity parameters, respectively. The wide pier factor K_w in Section 7.4 should be used in this computation if (1) the total depth $y_2 < 0.8\ a_{pc}$, (2) the Froude Number $V_2/(gy_2)^{1/2} < 1$, and (3) $a_{pc} > 50\ D_{50}$. Use y_2/a_{pc} to compute the K_w factor if it is applicable. The scour component equation for the case 2 pile cap or footing can then be written:

$$\frac{y_{spc}}{y_f} = 2.0 K_1 K_2 K_3 K_w \left(\frac{a_{pc}}{y_f}\right)^{0.65} \left(\frac{V_f}{\sqrt{gy_f}}\right)^{0.43} \tag{7.26}$$

In this case assume the pile cap scour component includes the pile group scour and compute the total scour depth as:

$$y_s = y_{s\ pier} + y_{s\ pc} \quad \text{(For Case 2 only)} \tag{7.27}$$

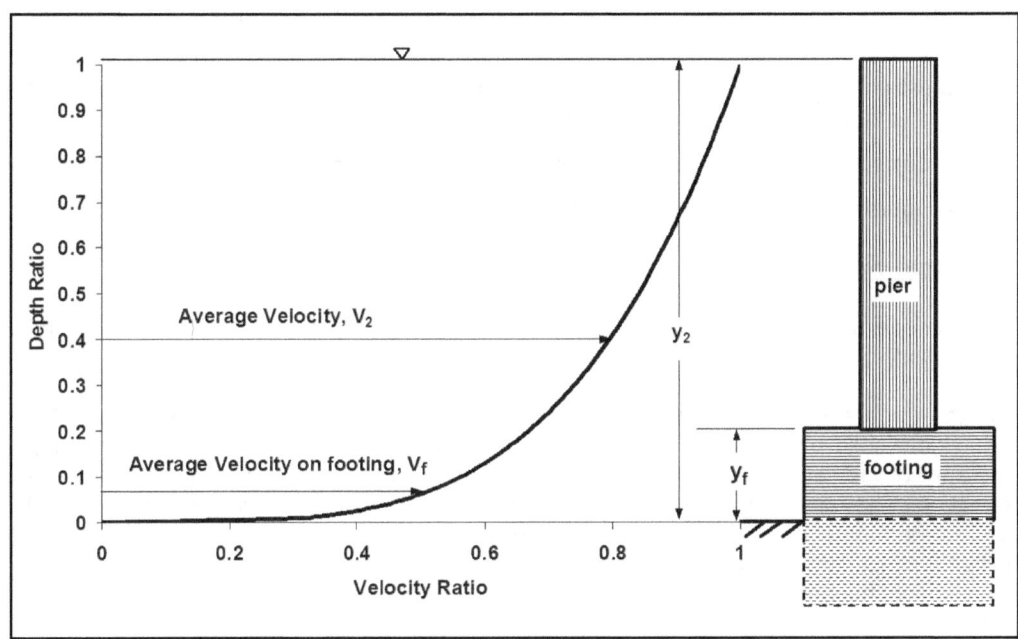

Figure 7.8. Definition sketch for velocity and depth on exposed footing.

In earlier editions of HEC-18, the recommendation was to use the larger of the exposed footing scour estimate or the pier stem scour estimate, treating the pier stem portion as a full depth pier that extended below the scour depth. **Now the recommendation is to add the components using a more realistic estimate of the pier stem component and using an adjusted approach velocity, V_2, to calculate V_f and the wide pier correction in the computations for the exposed footing component.**

7.5.5 Determination of the Pile Group Scour Depth Component

Research by Jones (USDOT 1989), Salim and Jones (1995, 1996, 1999), and by Smith (1999), has provided a basis for determining pile group scour depth by taking into consideration the spacing between piles, the number of pile rows and a height factor to account for the pile length exposed to the flow. Guidelines are given for analyzing the following typical cases:

- Special case of piles aligned with each other and with the flow. No angle of attack.
- General case of the pile group skewed to the flow, with an angle of attack, or pile groups with staggered rows of piles.

The strategy for estimating the pile group scour component is the same for both cases, but the technique for determining the projected width of piles is simpler for the special case of aligned piles. The strategy is as follows:

- Project the width of the piles onto a plane normal to the flow.

- Determine the effective width of an equivalent pier that would produce the same scour if the pile group penetrated the water surface.
- Adjust the flow depth, velocity and exposed height of the pile group to account for the pier stem and pile cap scour components previously calculated.
- Determine pile group height factor based on exposed height of pile group above the bed.
- Compute the pile group scour component using a modified version of Equation 7.1.

Projected width of piles

For the special case of aligned piles, the projected width, a_{proj}, onto a plane normal to the flow is simply the width of the collapsed pile group as illustrated in Figure 7.9.

For the general case, Smith (1999) determined that a pile group could be represented by an equivalent solid pier that has an effective width, a^*_{pg}, equal to a spacing factor multiplied by the sum of the non-overlapping projected widths of the piles onto a plane normal to the flow direction. The aligned pile group is a special case in which the sum of the non-overlapping projected widths happens to be the same as the width of the collapsed pile group. The procedure for the general case is the same as the procedure for the aligned pile groups except for the determination of the width of the equivalent solid which is a more tedious process for the general case. The sum of the projected widths can be determined by sketching the pile group to scale and projecting the outside edges of each pile onto the projection plane as illustrated in Figure 7.10 or by systematically calculating coordinates of the edges of each pile along the projection plane. The coordinates are sorted in ascending order to facilitate inspection to eliminate double counting of overlapping areas.

Smith attempted to derive weighting factors to adjust the impact of piles according to their distance from the projection plane, but concluded that there was not enough data and the procedure would become very cumbersome with weighting factors. **A reasonable alternative to using weighting factors is to exclude piles other than the two rows and one column closest to the plane of projection as illustrated by the bold outlines in Figure 7.10**.

Effective width of an equivalent full depth pier

The effective width of an equivalent full depth pier is the product of the projected width of piles multiplied by a spacing factor and a number of aligned rows factor (used for the special case of aligned piles only).

$$a^*_{pg} = a_{proj} K_{sp} K_m \qquad (7.28)$$

where:

a_{proj} = Sum of non-overlapping projected widths of piles (see Figures 7.9 and 7.10)
K_{sp} = Coefficient for pile spacing (Figure 7.11)
K_m = Coefficient for number of aligned rows, m, (Figure 7.12 - note that K_m is constant for all S/a values when there are more than 6 rows of piles)
K_m = 1.0 for skewed or staggered pile groups

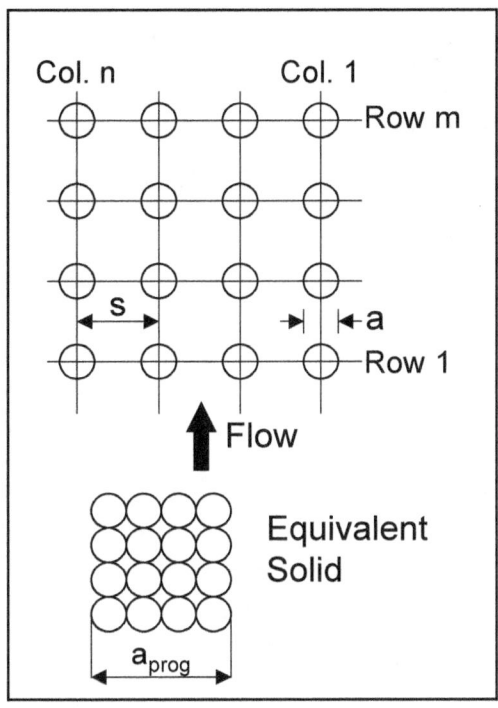

Figure 7.9. Projected width of piles for the special case of aligned flow.

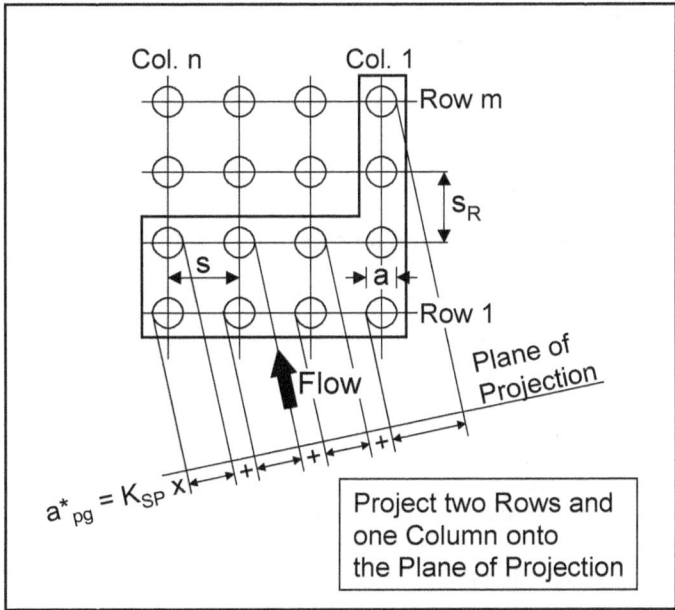

Figure 7.10. Projected width of piles for the general case of skewed flow.

7.19

The number of rows factor, K_m, is 1.0 for the general case of skewed or staggered rows of piles because the projection technique for skewed flow accounts for the number of rows and is already conservative for staggered rows.

Adjusted flow depth and velocity

The adjusted flow depth and velocity to be used in the pier scour equation are as follows:

$$y_3 = y_1 + y_{s\,pier}/2 + y_{s\,pc}/2, \text{ ft (m)} \tag{7.29}$$

$$V_3 = V_1 (y_1/y_3), \text{ ft/s (m/s)} \tag{7.30}$$

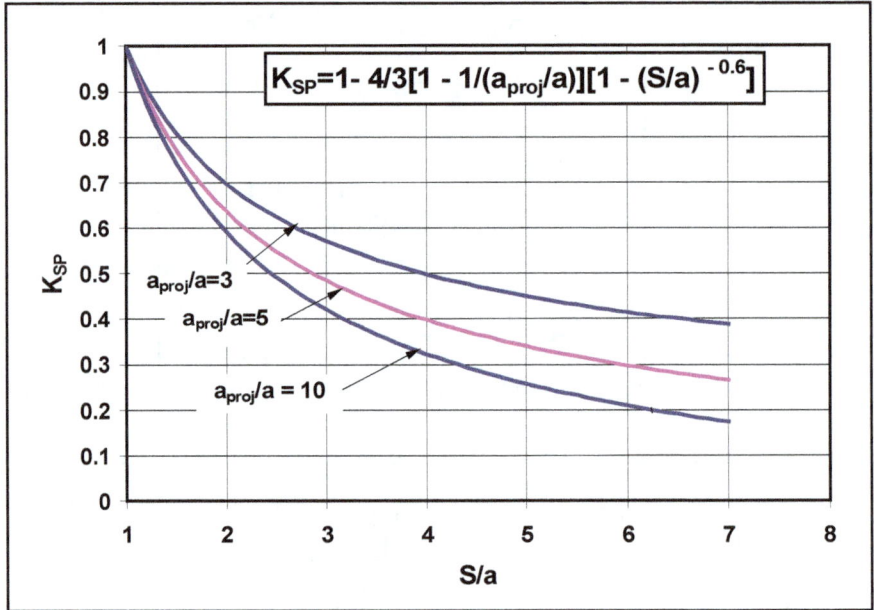

Figure 7.11. Pile spacing factor (refer to Sheppard 2001).

The scour equation for a pile group can then be written as follows:

$$\frac{Y_{spg}}{y_3} = K_{hpg}\left[2.0 K_1 K_3 \left(\frac{a^*_{pg}}{y_3}\right)^{0.65}\left(\frac{V_3}{\sqrt{gy_3}}\right)^{0.43}\right] \tag{7.31}$$

where:

$K_{h\,pg}$ = Pile group height factor given in Figure 7.13 as a function of h_3/y_3 (note that the maximum value of $y_3 = 3.5\, a^*_{pg}$)

h_3 = $h_0 + y_{s\,pier}/2 + y_{s\,pc}/2$ = height of pile group above the lowered stream bed after pier and pile cap scour components have been computed, ft (m)

K_2 from Equation 7.1 has been omitted because pile widths are projected onto a plane that is normal to the flow. The quantity in the square brackets is the scour ratio for a solid pier of width, a^*_{pg}, if it extended to the water surface. This is the scour ratio for a full depth pile group.

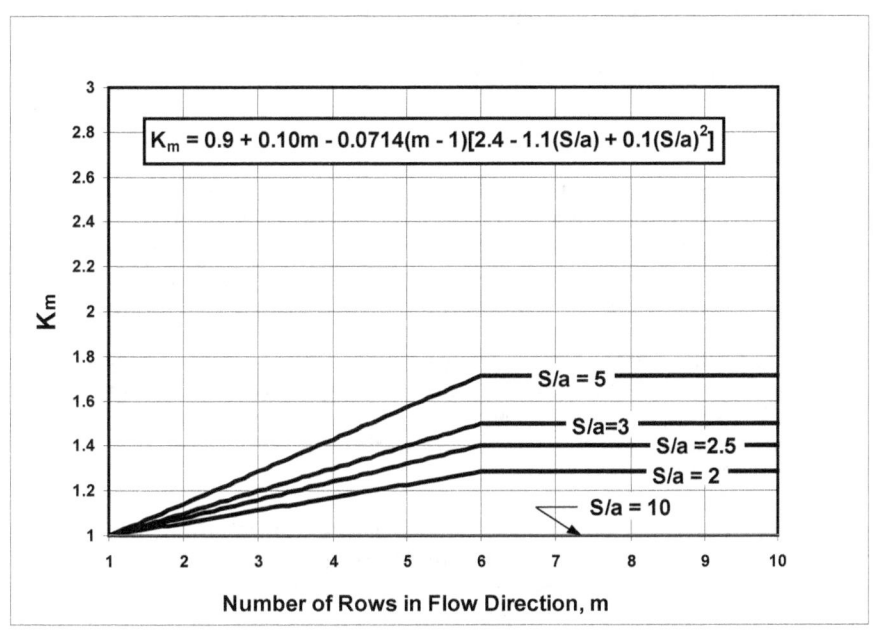

Figure 7.12. Adjustment factor for number of aligned rows of piles (refer to Sheppard 2001).

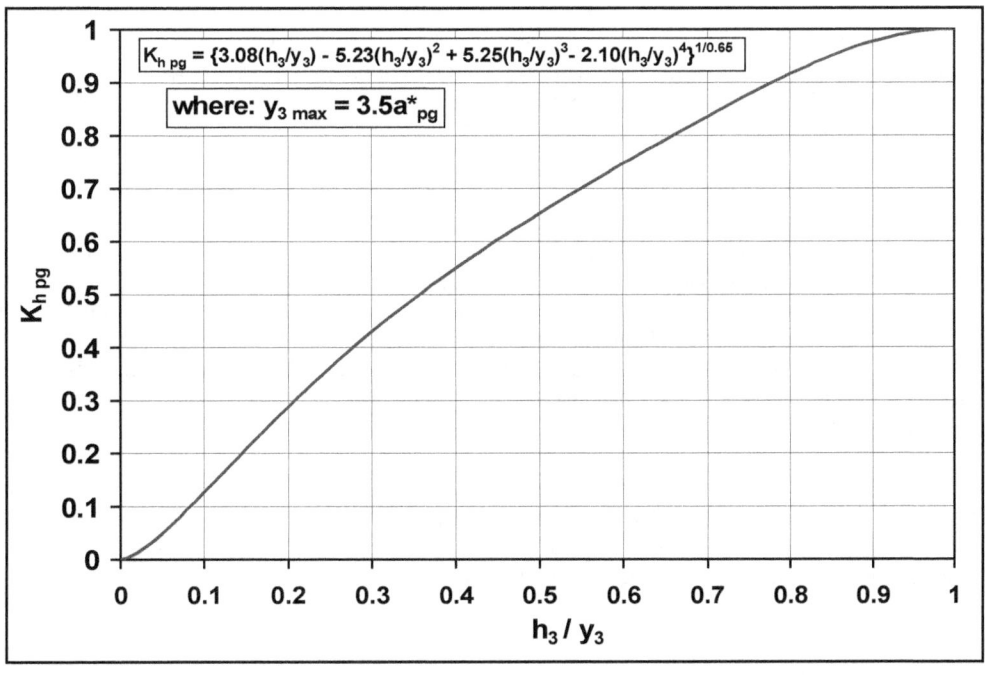

Figure 7.13. Pile group height adjustment factor (refer to Sheppard 2001).

7.5.6 Determination of Total Scour Depth for the Complex Pier

The total scour for the complex pier from Equation (7.22) is:

$y_s = y_{s\,pier} + y_{s\,pc} + y_{s\,pg}$

The guidelines described in this section can be used to compute scour for a simple full depth pile group in which case the first two components will be zero and the pile group height factor will be 1.0. Engineering judgment must be used if debris is considered a factor in which case it would be logical to treat the pile group and debris as a vertical extension of the pile cap and to compute scour using the Case 2 pile cap procedure described previously.

In cases of complex pile configurations where costs are a major concern, where significant savings are anticipated, and/or for major bridge crossings, physical model studies are still the best guide. Nevertheless, the guidelines described in this section provide a first estimate and a good indication of what can be anticipated from a physical model study.

In many complex piers, the pile groups have a different number of piles in a row or column, the spacing between piles is not uniform, and the widths of the piles may not all be the same. An estimate of the scour depth can be obtained using the methods and equations in this section. However, again it is recommended that a physical model study be conducted to arrive at the final design and to determine the scour depths.

7.6 MULTIPLE COLUMNS SKEWED TO THE FLOW

For multiple columns (illustrated as a group of cylinders in Figure 7.14) skewed to the flow, the scour depth depends on the spacing between the columns. The correction factor for angle of attack would be smaller than for a solid pier. Raudkivi (1986) in discussing effects of alignment states "...the use of cylindrical columns would produce a shallower scour; for example, with five-diameter spacing the local scour can be limited to about 1.2 times the local scour at a single cylinder."

In application of Equation 7.1 with multiple columns spaced less than 5 pier diameters apart, the pier width 'a' is the total projected width of all the columns in a single bent, normal to the flow angle of attack (Figure 7.14). For example, three 6.6 ft (2.0 m) cylindrical columns spaced at 10.0 m (33 ft) would have an 'a' value ranging between 6.6 and 33 ft (2.0 and 6.0 m), depending upon the flow angle of attack. **This composite pier width would be used in Equation 7.1 to determine depth of pier scour.** The correction factor K_1 in Equation 6.1 for the multiple column would be 1.0 regardless of column shape. The coefficient K_2 would also be equal to 1.0 since the effect of skew would be accounted for by the projected area of the piers normal to the flow.

The scour depth for multiple columns skewed to the flow can also be determined by determining the K_2 factor using Equation 7.4 and using it in Equation 7.1. The width "a" in Equation 7.1 would be the width of a single column. An example problem illustrates two methods of obtaining the scour depth for multiple columns (Section 7.10.5).

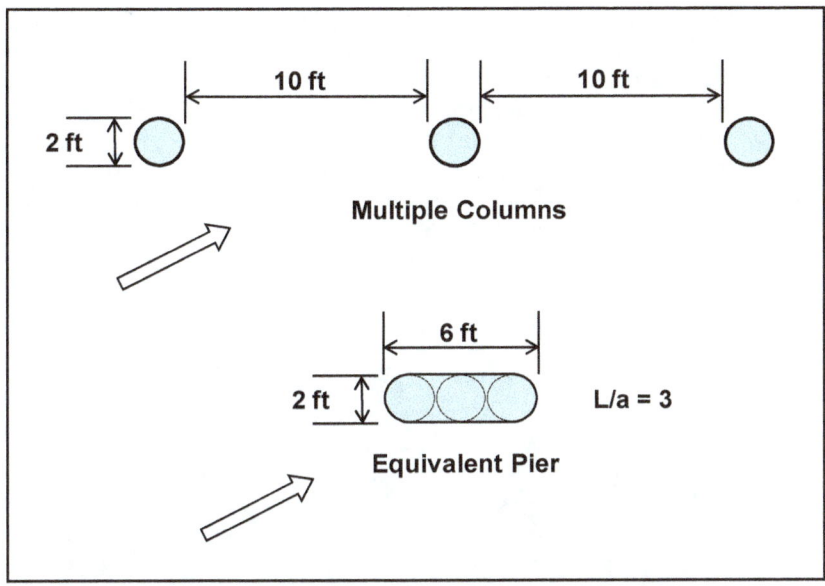

Figure 7.14. Multiple columns skewed to the flow.

If the multiple columns are spaced 5 diameter or greater apart; and debris is not a problem, limit the scour depths to a maximum of 1.2 times the local scour of a single column.

The depth of scour for a multiple column bent will be analyzed in this manner except when addressing the effect of debris lodged between columns. If debris is evaluated, it would be logical to consider the multiple columns and debris as a solid elongated pier. The appropriate L/a value and flow angle of attack would then be used to determine K_2 in Equation 7.4.

Additional laboratory studies are necessary to provide guidance on the limiting flow angles of attack for given distance between multiple columns beyond which multiple columns can be expected to function as solitary members with minimal influence from adjacent columns.

7.7 SCOUR FROM DEBRIS ON PIERS

Floating woody debris (drift) that lodges and accumulates at bridge piers creates additional obstruction to flow, and transforms the pier geometry into one that is effectively wider than if debris were not present. Equations have been developed to estimate the effective width based on the original (debris-free) pier geometry, and size and shape of the debris accumulation. This section provides current guidance on estimating the effect of debris on pier scour.

7.7.1 Debris Accumulation on Piers

Obviously, for design and scour assessment estimating the size and shape of debris on bridge piers is largely a matter of experience and judgment. Most woody debris is derived from bank failures on the main channel and major tributaries upstream of the bridge, but watershed condition and land management practices can influence the recruitment of debris as well. Also, maintenance practices vary from district to district and from state to state. Some agencies allow debris accumulations to grow to very large dimensions before they are removed, whereas other agencies aggressively remove debris even when only a few logs are present to prevent the potential snagging of additional material.

The best approach to estimating the size, shape, and dimensions of debris accumulations is to review the history and maintenance practices of highway agencies in the region. Experience with debris on similar river systems having similar watershed characteristics will provide valuable insight on potential problems in the future. NCHRP Report 653, "Effects of Debris on Pier Scour" (NCHRP 2010a) and HEC-20 (FHWA 2012b) provide guidance on estimating the delivery of floating debris at a bridge, and the processes by which it accumulates on the structure. Field data collection sheets and a flowchart-based approach for estimating debris at piers are presented in the NCHRP report. Figure 7.15 provides a photo of woody debris accumulation at a bridge pier.

Figure 7.15. Woody debris at a bridge pier. Note how debris can be much wider than the pier itself (also note debris on lower chord of bridge deck). Source: NCHRP Web-Only Document 48 (see NCHRP 2010a).

7.7.2 Debris Size and Shape

NCHRP Report 653, "Effects of Debris on Pier Scour," (NCHRP 2010a) provides extensive insight into scour processes at piers when debris loading is present. The primary variables involve the shape of the debris blockage, and the dimensions of the debris mass compared to the pier width.

The shape of debris accumulations can be generally idealized as either rectangular or triangular. Rectangular shapes represent a more extreme blockage of flow, and therefore they create more scour. Triangular debris shapes are somewhat streamlined, and create a flow pattern that is not as severe at the base of the pier. However, both shapes result in more blockage compared to a pier without debris, and therefore both result in additional scour. Figures 7.16 and 7.17 illustrate the idealized dimensions as described in NCHRP Report 653.

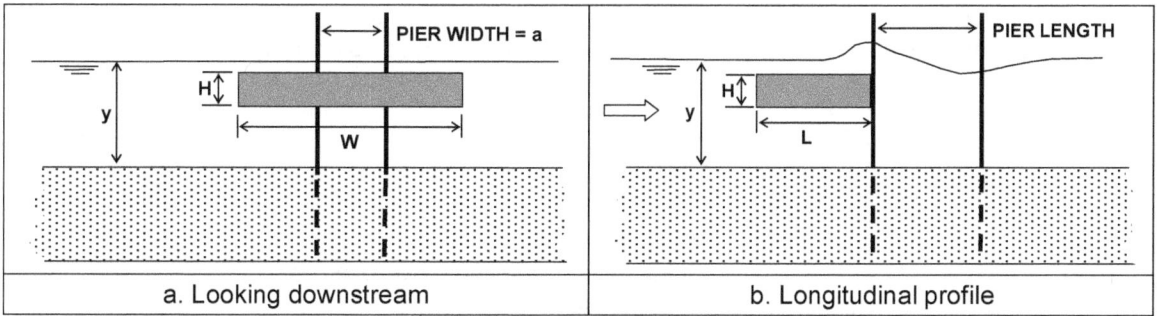

Figure 7.16. Idealized dimensions of rectangular debris accumulations (modified from NCHRP Report 653).

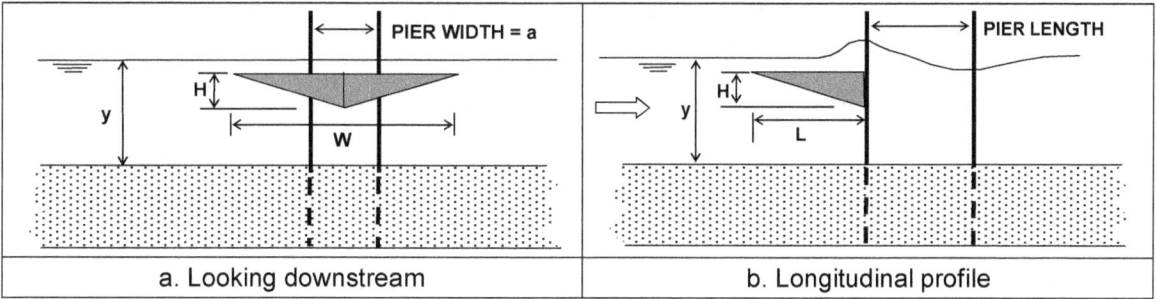

Figure 7.17. Idealized dimensions of triangular debris accumulations (modified from NCHRP Report 653).

Perhaps the most important result from the NCHRP research is the fact that the greatest amount of debris-induced scour at the pier occurs when the length of the debris in the upstream direction (dimension "L" in the above figures) is equal to the approach flow depth. When the debris accumulation has grown to this dimension, the plunging flow created by the debris is focused at the base of the pier, reinforcing the horseshoe vortex.

During the NCHRP study, it was also found that when debris floats at the water surface during a flood event, all of the approach flow is forced to plunge beneath the debris, with no flow going over the top. Therefore, the guidance developed for estimating pier scour with debris assumes that the debris is floating at the surface during a flood, which is a very likely condition during the peak of a flood event.

7.7.3 Effective Pier Width with Debris

Based on the results of the NCHRP study, a relatively simple equation was developed that can be used to estimate the equivalent pier width, denoted a^*_d, when debris is present. The equation considers the shape, width and height of the debris in addition to the unobstructed pier width and the depth of the approach flow. The equation yields an equivalent pier width that can be used in the HEC-18 pier scour equation (Section 7.2) to estimate the local scour depth at the pier. As previously noted, the most severe scour at the pier occurs when the length of the debris (in the upstream direction) is equal to the flow depth. **This formulation of the equivalent pier width has not been validated for the FDOT methodology (Section 7.3).**

Based on the shape of the debris (rectangular vs. triangular), the effective width of a pier a^*_d with debris loading when the length of the debris is equal to the flow depth, and the debris is floating at the water surface (i.e., the most conservative case) is estimated as:

$$a^*_d = \frac{K_1(HW) + (y - K_1H)a}{y} \tag{7.32}$$

where:

a^*_d	=	Effective width of pier when debris is present, ft (m)
a	=	Width of pier perpendicular to flow, ft (m)
K_1	=	0.79 for rectangular debris, 0.21 for triangular debris
H	=	Height (thickness) of the debris, ft (m)
W	=	Width of debris perpendicular to the flow direction, ft (m)
y	=	Depth of approach flow, ft (m)

An example problem using this approach is presented in Section 7.10.7.

Debris lodged on a pier can increase local scour at a pier. The debris may increase pier width and deflect a component of flow downward. This increases the transport of sediment out of the scour hole. When floating debris is lodged on the pier, the scour depth can be estimated by assuming that the pier width is larger than the actual width. The problem is in determining the increase in pier width to use in the pier scour equation. Furthermore, at large depths, the effect of the debris on scour depth should diminish (for additional discussion, see HEC-20 (FHWA 2012b).

7.8 TOPWIDTH OF SCOUR HOLES

The topwidth of a scour hole in cohesionless bed material from one side of a pier or footing can be estimated from the following equation (Richardson and Abed 1993):

$$W = y_s (K + \cot \theta) \tag{7.33}$$

where:

W	=	Topwidth of the scour hole from each side of the pier or footing, ft (m)
y_s	=	Scour depth, ft (m)
K	=	Bottom width of the scour hole related to the depth of scour
θ	=	Angle of repose of the bed material ranging from about 30° to 44°

The angle of repose of cohesionless material in air ranges from about 30° to 44°. Therefore, if the bottom width of the scour hole is equal to the depth of scour y_s (K = 1), the topwidth in cohesionless sand would vary from 2.07 to 2.80 y_s. At the other extreme, if K = 0, the topwidth would vary from 1.07 to 1.8 y_s. Thus, the topwidth could range from 1.0 to 2.8 y_s and depends on the bottom width of the scour hole and composition of the bed material. In general, the deeper the scour hole, the smaller the bottom width. In water, the angle of repose of cohesionless material is less than the values given for air; therefore, a topwidth of 2.0 y_s is suggested for practical applications (Figure 7.18).

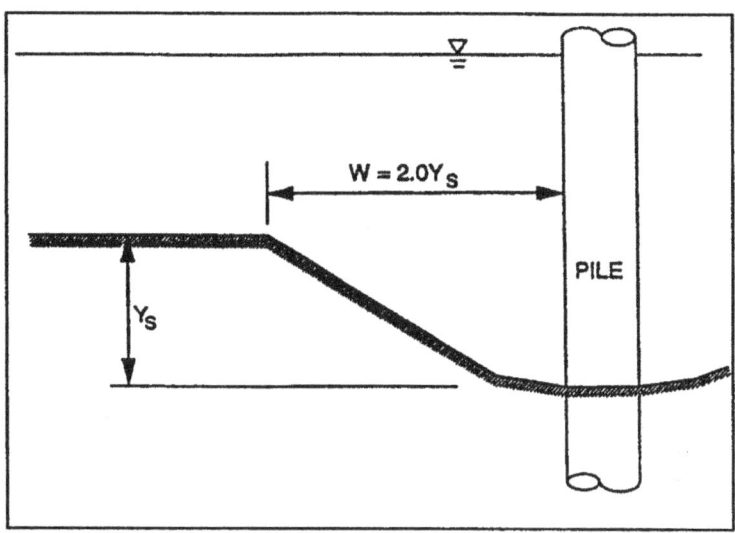

Figure 7.18. Topwidth of scour hole.

7.9 PHYSICAL MODEL STUDIES

For unusual or complex pier foundation configurations a physical model study should be made. The scale between model and prototype is based on the Froude criteria, that is, the Froude number for the model should be the same as for the prototype. In general it is not possible to scale the bed material size. Also, at flood flows in sand bed streams the sediment transport conditions will be live-bed and the bed configuration will be plane bed. Whereas, in the model live-bed transport conditions will be ripples or dunes. These are incomparable pier scour conditions. Therefore, it is recommended that a bed material be used that has a critical velocity just below the model velocity (i.e., clear-water scour conditions). This will usually give the maximum scour depth; but a careful study of the results needs to be made by persons with field and model scour experience. For additional discussion of the use of physical modeling in hydraulic design, see HEC-23 (FHWA 2009).

7.10 PIER SCOUR EXAMPLE PROBLEMS

7.10.1 Example Problem 1 - Scour at a Simple Solid Pier

Given:

>Pier geometry: a = 4.0 ft (1.22 m), L = 59 ft (18 m), round nose
>Flow variables: y_1 = 10.2 ft (3.12 m), V_1 = 11.02 ft/s (3.36 m/s)
>Angle of attack = 0 degrees, g = 32.2 ft/s^2 (9.81 m/s^2)
>Froude No. = $11.02/(32.2 \times 10.2)^{0.5}$ = 0.61
>Bed material: D_{50} = 7.3 mm (0.024 ft), D_{95} 0.32 mm (0.0011 ft)
>Bed Configuration: Plane bed

Determine:

>The magnitude of pier scour depth.

Solution:

Use Equation 7.1.

$$\frac{y_s}{y_1} = 2.0 K_1 K_2 K_3 \left(\frac{a}{y_1}\right)^{0.65} Fr_1^{0.43}$$

$y_s / 10.2 = 2.0 \times 1.0 \times 1.0 \times 1.1 \times (4.0 / 10.2)^{0.65} \times 0.61^{0.43} = 0.97$

$y_s = 0.97 \times 10.2 = 9.9$ ft (3 m)

Check:

$y_{s\,max} = 2.4a$
$y_{s\,max} = 2.4 (4.0) = 9.6$ ft

7.10.2 Example Problem 2 - Angle of Attack

Given:

Same as Problem 1 but angle of attack is 20 degrees

Solution:

Use Equation 7.4 to compute K_2

$K_2 = (\cos \theta + L / a \sin \theta)^{0.65}$

If L/a is larger than 12, use L/a = 12 as a maximum in Equation 7.4 (see Table 7.2).
L/a = 59/4.0 = 14.8 > 12 use 12
$K_2 = (\cos 20 + 12 \sin 20)^{0.65} = 2.86$
$y_s = 9.9 \times 2.86 = 28.3$ ft (8.6 m)

7.10.3 Example Problem 3 - Scour at Complex Piers (Solid Pier on an Exposed Footing)

Given:

The pier in Problem 1 (Section 7.10.1) is on a 8.0 ft (2.44 m) wide by 5.25 ft (1.6 m) high by 65 ft (19.81 m) long rectangular footing. Footing extends 2.5 ft (0.76 m) upstream from the pier. The footing is on an unspecified pile foundation. The footing is exposed 4.92 ft (1.5 m) by long-term degradation. Determine local pier scour.

Data:

Pier geometry; a_{pier} = 4.0 ft (1.22 m), L = 59 ft (18 m), round nose
Pile cap or footing geometry, a_{pc} (or a_f) = 8 ft (2.44 m), L = 65 ft (19.81 m), T = 5.12 ft (1.60 m), f = 2.5 ft (0.76 m)
Approach flow: y_1 = 10.2 ft (3.12 m), V_1 = 11.02 ft/s (3.36 m/s)
Angle of attack = 0 degrees
Froude No. = $11.02/(32.2 \times 10.2)^{0.5} = 0.61$
Bed material: D_{50} = 0.32 mm, D_{84} = 7.3 mm, Plane bed
h_0 = 4.92 - 5.25 = -0.33 ft (1.5 – 1.6 = -0.10 m)
See sketch below:

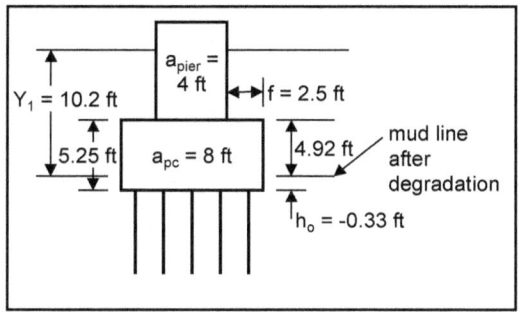
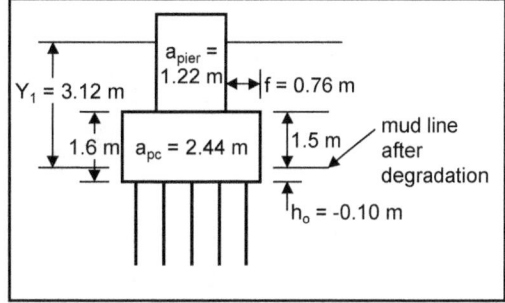

Local Scour from Pier Stem

$f = 2.5$ ft (0.76 m)

$h_1 = h_o + T = -0.33 + 5.25 = 4.92$ ft (1.50 m)

$K_{h\,pier}$ = function (h_1/a_{pier}, f/a_{pier}) (from Figure 7.6)

$h_1/a_{pier} = 4.92/4.0 = 1.23$

$f/a_{pier} = 2.5/4 = 0.62$

$K_{h\,pier} = 0.06$

$$\frac{y_{s\,pier}}{y_1} = K_{hpier}\left[2.0 K_1 K_2 K_3 \left(\frac{a_{pier}}{y_1}\right)^{0.65}\left(\frac{V_1}{\sqrt{gy_1}}\right)^{0.43}\right]$$

$$\frac{y_{s\,pier}}{y_1} = 0.06\left[2.0(1.0)\ (1.0)\ (1.1)\left(\frac{4.0}{10.2}\right)^{0.65}\left(\frac{11.02}{\sqrt{32.2 \times 10.2}}\right)^{0.43}\right]$$

$y_{s\,pier} = 0.06 \times [0.97] \times 10.2 = 0.6$ ft (0.18 m)

Note: the quantity in the square brackets is the scour ratio for a full depth pier.

Local Scour from the Pile Cap or Footing

Assume the average bed elevation in the vicinity of the pier lowers by ½ the pier stem scour.

$y_2 = y_1 + y_{s\,pier}/2 = 10.2 + 0.3 = 10.5$ ft (3.21 m)

$V_2 = V_1(y_1/y_2) = 11.02\,(10.2/10.5) = 10.7$ ft/s (3.26 m/s)

$h_2 = h_o + y_{s\,pier}/2 = -0.33 + 0.3 = -0.03$ ft (-0.01 m)

The bottom of the pile cap is below the adjusted mud line; use Case 2 computations for an exposed footing.

$y_f = h_1 + y_{s\,pier}/2 = 4.92 + 0.3 = 5.22$ ft (1.59 m)

The velocity on the footing is:

$$\frac{V_f}{V_2} = \frac{\ln\left(10.93\frac{y_f}{k_s}+1\right)}{\ln\left(10.93\frac{y_2}{k_s}+1\right)} = \frac{\ln\left(10.93\frac{5.22}{.024}+1\right)}{\ln\left(10.93\frac{10.5}{0.024}+1\right)} = 0.92$$

Note: assume $k_s = D_{84} = 7.3$ mm = 0.024 ft

$V_f = 0.92 \times V_2 = 0.92 \times 10.7 = 9.84$ ft/s (2.99 m/s)

$$\frac{y_{s\,footing}}{y_f} = 2.0 K_1 K_2 K_3 K_w \left(\frac{a_f}{y_f}\right)^{0.65}\left(\frac{V_f}{\sqrt{gy_f}}\right)^{0.43}$$

$$\frac{y_{s\,footing}}{y_f} = 2.0\,(1.1)\ (1.0)\ (1.1)\ (1.0)\left(\frac{8.0}{5.22}\right)^{0.65}\left(\frac{9.84}{\sqrt{32.2\times 5.22}}\right)^{0.43} = 2.83$$

Note that $y_2/a_f = 1.31$ (>0.8); use $K_W = 1.0$

$y_{s\,footing} = 2.83 y_f = 2.83 \times 5.22 = 14.8$ ft (4.5 m)

Total Local Pier Scour Depth

$y_s = y_{s\,pier} + y_{s\,footing} = 14.8 + 0.6 = 15.4$ ft (4.68 m)

7.10.4 Example Problem 4 - Scour at a Complex Pier with Pile Cap in the Flow

During the design of the new Woodrow Wilson Bridge over the Potomac River, several complex pier configurations were tested in physical model studies. The purpose of this problem is to analyze local scour for the possible condition that the main channel migrated to the pier configured as shown in the figure below. It was determined that the water surface elevations would be +7.3 ft (2.23 m) and + 9.7 ft (2.96 m) for the Q_{100} and the Q_{500} events respectively and the velocities in the main channel would be 11.2 ft/sec (3.41 m/s) and 14 ft/sec (4.27 m/s) for the Q_{100} and Q_{500} events respectively. The following computations are for the Q_{100} event:

Initial parameters:

$y_1 = 51.8$ ft (15.79 m)
$V_1 = 11.2$ ft/sec (3.41 m/s)
$a_{pier} = 32$ ft (9.754 m)
$a_{pc} = 53.25$ ft (16.23 m)
$h_0 = 25.5$ ft (7.77 m)
$h_1 = h_0 + T = 41.5$ ft (12.65 m) (resolution of the pile cap thickness below)
S = 13.75 ft (4.19 m) (center to center spacing of piles)
T = 16 ft (4.88 m) (assign half of tapered portion of cap to pile cap and half to pier)
f = 8.62 ft (2.627 m) (see figure)
zero angle of attack

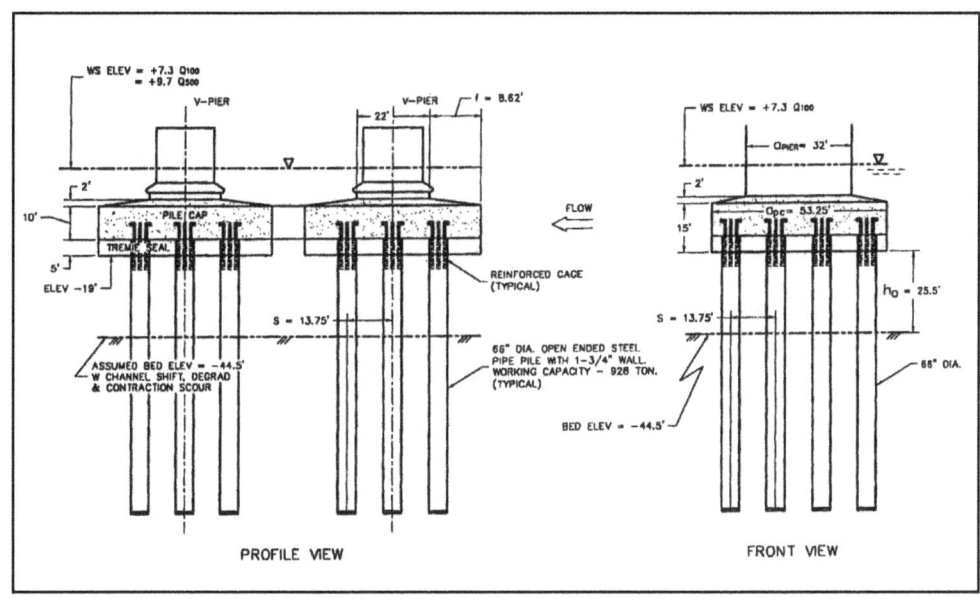

Model of complex pier geometry for the Woodrow Wilson Bridge.

Pier Stem Component

$f/a_{pier} = 8.62/32 = 0.27$

$h_1/a_{pier} = 41.5/32 = 1.30$

$K_{h\ pier} = 0.062$ (from Figure 7.6)

$$\frac{y_{s\,pier}}{y_1} = K_{hpier}\left[2.0 K_1 K_2 K_3 \left(\frac{a_{pier}}{y_1}\right)^{0.65}\left(\frac{V_1}{\sqrt{gy_1}}\right)\right]^{0.43}$$

$$\frac{y_{s\,pier}}{51.8} = 0.062\left[2.0(1.1)\ (1.0)\ (1.1)\left(\frac{32}{51.8}\right)^{0.65}\left(\frac{11.2}{\sqrt{32.2)51.8}}\right)^{0.43}\right] = 0.0629$$

The quantity in the brackets is the scour ratio for a full depth pier that extends below the scour hole.

$y_{s\ pier} = 0.0629 \times 51.8$ ft $= 3.2$ ft $(0.99$ m$)$

Pile Cap Component

$h_2 = h_0 + y_{s\ pier}/2 = 25.5 + 1.6 = 27.1$ ft $(8.27$ m$)$

$y_2 = y_1 + y_{s\ pier}/2 = 51.8 + 1.6 = 53.4$ ft $(16.28$ m$)$

$V_2 = V_1 \times (y_1/y_2) = 11.2 \times (51.8/53.4) = 10.9$ ft/s $(3.31$ m/s$)$

Note: For Figure 7.7, $y_{2max} = 3.5\, a_{pc} = 186.38 > 53.4$; use $y_2 = 53.4$ ft (16.28 m)

$h_2/y_2 = 0.51$

$T/y_2 = 16/53.4 = 0.30$

$$\frac{a^*_{pc}}{a_{pc}} = 0.07 \qquad \text{(from Figure 7.7)}$$

$a^*_{pc} = 0.07 \times 53.25 = 3.7$ ft (1.10 m)

This is the width of a full depth pier that would produce the same scour depth as the isolated pile cap will produce.

$$\frac{y_{spc}}{y_2} = 2.0 K_1 K_2 K_3 K_w \left(\frac{a^*_{pc}}{y_2}\right)^{0.65} \left(\frac{V_2}{\sqrt{gy_2}}\right)^{0.43}$$

$$\frac{y_{spc}}{53.4} = 2.0(1.1)(1.0)(1.1)(1.0)\left(\frac{3.7}{53.4}\right)^{0.65}\left(\frac{10.9}{\sqrt{32.2}\,(53.4)}\right)^{0.43} = 0.24$$

Note that $y_2/a^*_{pc} = 14.4$ (>0.8); use $K_w = 1.0$

$y_{s\,pc} = 0.24 \times 53.4 = 12.8$ ft (3.84 m)

Pile Group Component

$h_3 = h_0 + (y_{s\,pier} + y_{s\,pc})/2 = 25.5 + (3.2 + 12.8)/2 = 33.5$ ft (10.19 m)

$y_3 = y_1 + (y_{s\,pier} + y_{s\,pc})/2 = 51.8 + (3.2 + 12.8)/2 = 59.8$ ft (18.20 m)

$V_3 = V_1 \times (y_1/y_3) = 11.2 \times (51.8/59.8) = 9.7$ ft/s (2.95 m/s)

$a_{proj} = 4 \times 5.5 = 22.0$ ft (6.71 m) (from Figure 7.9)

$a_{proj}/a = 22.0/5.5 = 4.0$

$S/a = 13.75/5.5 = 2.5$ (relative center to center spacing of piles)

$K_{sp} = 0.58$ (from Figure 7.11)

$K_m = 1.16$ (From Figure 7.12 for three rows per foundation; foundations separated)

$a^*_{pg} = K_{sp} \times K_m \times a_{proj} = 0.58 \times 1.16 \times 22.0 = 14.8$ ft (4.51 m)

Note: in Figure 7.13, $y_{3\,max} = 3.5 \times a^*_{pg} = 51.8 < 59.8$; use $y_3 = 51.8$ ft (15.79 m)

$h_3/y_3 = 33.5/51.8 = 0.65$

$K_{h\,pg} = 0.79$ (from Figure 7.13)

$$\frac{y_{spg}}{y_3} = K_{hpg}\left[2.0 K_1 K_2 K_3 \left(\frac{a^*_{pg}}{y_3}\right)^{0.65}\left(\frac{V_3}{\sqrt{gy_3}}\right)^{0.43}\right]$$

$$\frac{y_{spg}}{51.8} = 0.79\left[2.0\ 1.0(1.0)\ 1.1\left(\frac{14.8}{51.8}\right)^{0.65}\left(\frac{9.7}{\sqrt{32.2}\ (51.8)}\right)^{0.43}\right] = 0.41$$

$y_{s\,pg} = 0.41 \times 51.8 = 21.24$ ft (6.47 m)

Total Estimated Scour

$y_s = y_{s\,pier} + y_{s\,pc} + y_{s\,pg} = 3.7 + 12.8 + 21.24 = 37.74$ ft (11.3 m)

7.10.5 Example Problem 5 - Scour at Multiple Columns

Calculate the scour depth for a pier that consists of six 16-inch (0.406 m) columns spaced at 7.5 ft (2.29 m) with a flow angle of attack of 26 degrees. Debris is not a problem and there is no armoring at this site.

Data:

Columns: 6 columns 1.33 ft (0.406 m), spaced 7.5 ft (2.29 m)
Velocity: $V_1 = 11.16$ ft/s (3.4 m/s); Depth: $y_1 = 20.0$ ft (6.1 m)
Angle of attack: 26 degrees
Assume $K_3 = 1.1$ for plane bed condition

Determine the depth of local scour:

Two methods of calculating the scour depth will be illustrated.

a. Scour depth according to Raudkivi (1986) is 1.2 times the local scour of a single column, if S/a > 5.0.

Spacing coefficient = S/a = 7.5/1.33 = 5.6; S/a > 5.0

$$\frac{y_s}{20} = 2.0 \times 1.0 \times 1.0 \times 1.1 \left(\frac{1.33}{20}\right)^{0.65}\left(\frac{11.16}{\sqrt{32.2 \times 20}^{0.5}}\right)^{0.43} = 0.266$$

$y_s = 20 \times 0.266 \times 1.2 = 6.4$ ft (1.95 m)

b. Compare this value with that computed for columns spaced 6.0 ft (1.8 m) apart.

Spacing coefficient = S/a = 6.0/1.33 = 4.51; S/a < 5.0

$K_2 = (\cos 26 + 8.0/1.33 \sin 26)^{0.65} = 2.27$

$$\frac{y_s}{20} = 2.0(1.0)(2.27)(1.1)(1.0)\left(\frac{1.33}{20}\right)^{0.65}\left(\frac{11.16}{(32.2 \times 20)^{0.5}}\right)^{0.43} = 0.603$$

$y_s = 20 \times 0.603 = 12.1$ ft (3.68 m)

7.10.6 Example Problem 6 – Florida DOT Pier Scour Methodology

Given:

Same as Problem 2.

Step 1. Calculate V_c for D_{50} = 7.3 mm (0.24 ft) and y_1 = 10.2 ft

$$u_c^* = K_u \left(0.1 D_{50}^{0.5} - 0.0213/D_{50}\right) = 1.0 \left(0.1 \times 7.3^{0.5} - 0.0213/7.3\right) = 0.27 \text{ ft/s} \ (0.081 \text{ m/s})$$

$$V_c = 5.75 u_c^* \log\left(5.53 \frac{y_1}{D_{50}}\right) = 5.75 \times 0.27 \log\left(5.53 \frac{10.2}{0.024}\right) = 5.2 \text{ ft/s} \ (1.6 \text{ m/s})$$

Step 2. Calculate V_{lp} for D_{50} = 7.3 mm and y_1 = 10.2 ft

5 x V_c = 5 x 5.2 = 26.0 ft/s (7.9 m/s)

$0.6\sqrt{gy_1} = 0.6\sqrt{32.2 \times 10.2} = 10.9 \text{ ft/s} (3.3 \text{ m/s})$

V_{lp} = 26.0 ft/s (7.9 m/s)

Step 3. Calculate a* for round nose pier with a = 4.0 ft and L = 48 ft.

Note: This example assumes the maximum L/a of 12.0 applies to the FDOT methodology.

K_{sf} = 1.0

$$a_{proj} = a\text{Cos}\theta + L\text{Sin}\theta = 4.0\text{Cos}(20) + 48\text{Sin}(20) = 20.2 \text{ ft} (6.16 \text{ m})$$

a* = K_{sf} x a_{proj} = 1.0 x 20.2 = 20.2 ft (6.16 m)

Step 4. Calculate f_1

$$f_1 = \tanh\left[\left(\frac{y_1}{a^*}\right)^{0.4}\right] = \tanh\left[\left(\frac{10.2}{20.2}\right)^{0.4}\right] = \tanh(0.76) = 0.64$$

Step 5. Calculate f_3

D_{50} = 7.3/304.8 = 0.024 ft (0.0073 m)

a*/D_{50} = 20.2/0.024 = 842

$$f_3 = \left[\frac{\left(\frac{a^*}{D_{50}}\right)^{1.13}}{10.6 + 0.4\left(\frac{a^*}{D_{50}}\right)^{1.33}}\right] = \left[\frac{842^{1.13}}{10.6 + 0.4(842)^{1.33}}\right] = 0.65$$

Step 6. Calculate $\dfrac{y_{s-c}}{a^*}$ and y_{s-c}

$$\dfrac{y_{s-c}}{a^*} = 2.5 f_1 f_3 = 2.5 \times 0.64 \times 0.65 = 1.04$$

$y_{s-c} = 1.04 a^* = 1.04 \times 20.2 = 21.0$ ft (6.4 m)

Step 7. Calculate $\dfrac{y_{s-lp}}{a^*}$ and y_{s-lp}

$$\dfrac{y_{s-lp}}{a^*} = 2.2 f_1 = 2.2 \times 0.64 = 1.41$$

$y_{s-lp} = 1.41 a^* = 1.41 \times 20.2 = 28.5$ ft (8.68 m)

Skip to Step 11 because V_1 is greater than V_c and less than V_{lp}

Step 11. Calculate y_s using Equation 7.19

$$y_s = y_{s-c} + \dfrac{(V_1 - V_c)}{(V_{lp} - V_c)}(y_{s-lp} - y_{s-c}) = 21.0 + \dfrac{(11.02 - 5.2)}{(26.0 - 5.2)}(28.5 - 21.0) = 23.1 \text{ft} (7.04 \text{m})$$

This scour result compares with 28.3 ft (8.6 m) for Example Problem 2, a reduction of 5.2 ft (1.58 m) or 19 percent. It is also possible for the FDOT methodology to produce greater scour for specific flow, pier, and sediment conditions. The Florida DOT spreadsheet could also be used to solve this problem, though with slightly different results. The spreadsheet uses a different method than the NCHRP (2011c) report for computing V_c, resulting in 4.98 ft/s (1.52 m/s). Therefore, the live-bed peak velocity is also changed to 24.9 ft/s (7.69 m/s). The spreadsheet only includes circular and rectangular piers, but not round nose piers. If a circular pier with a width of a_{proj} = 20.2 feet (6.16 m) is entered into the FDOT spreadsheet, the resulting scour is 23.3 feet (7.1 m), which is very close to the result of this problem. If a rectangular pier is entered, the scour reduces to 21.8 feet (6.6 m). This reduction occurs because the corner of the rectangular pier is oriented into the flow and the shape factor, K_{sf}, becomes 0.9.

7.10.7 Example Problem 7 - Pier Scour with Debris

The application of the effective width approach for estimating scour at a pier with debris accumulation is illustrated by the following example:

Debris characteristics: A new bridge is planned to replace an older bridge that has had a history of debris problems. Past inspection reports and maintenance records have documented debris accumulations that have extended up to 30 ft (9 m) upstream of the pier before the debris was removed. Debris masses have tended to be triangular in shape, with a top width at the pier of about 20 ft (6 m), and it typically accumulates to a thickness of about 4 ft (1.2 m) before it is removed by maintenance forces, as is typical practice in this district.

Design hydraulic conditions: The new pier is to be round nose with a width of 2.5 ft (0.76 m) perpendicular to the flow. The design flood depth is 11.5 ft (3.5 m) and the approach velocity upstream of the bridge is 9 ft/s (2.7 m/s).

Step 1: Calculate the effective width of the pier a^*_d for the design conditions, assuming that a debris mass has accumulated on the pier, either before a flood event, or during the event itself. Using Equation 7.32:

$$a^*_d = \frac{K_1(HW) + (y - K_1H)a}{y}$$

For triangular shaped debris, $K_1 = 0.21$ from Section 7.7. Using the given data, the thickness of the debris at the pier is H = 4 ft (1.2 m) and the width of the debris at the pier is W = 20 ft (6 m). With a design flow depth of 11.5 ft (3.5 m), the effective width of the pier including debris is:

$$a^*_d = \frac{K_1(HW) + (y - K_1H)a}{y} = \frac{0.21(4 \times 20) + (11.5 - (0.21 \times 4)) \times 2.5}{11.5} = 3.8 \text{ ft } (1.2 \text{ m})$$

Step 2: The effective width of the pier with a triangular-shaped debris mass, under these conditions, is 3.8 ft (1.2 m) compared to an unobstructed (i.e., debris-free) width of 2.5 ft (0.76 m). The effective width is used in the HEC-18 equation (Equation 7.3) for pier scour, considering that the shape of the pier itself has no effect once debris has accumulated:

$$\frac{y_s}{a^*_d} = 2.0 \, K_1 \, K_2 \, K_3 \left(\frac{y_1}{a^*_d}\right)^{0.35} Fr_1^{0.43}$$

The Froude number under these flow conditions is $Fr_1 = \frac{V}{\sqrt{gy}} = \frac{9}{\sqrt{(32.2)(11.5)}} = 0.47$

The calculated pier scour with triangular-shaped debris using the effective width a^*_d is:

$$y_s = (3.8) \, 2.0 \, (1.0) \, (1.0) \, (1.1) \left(\frac{11.5}{3.8}\right)^{0.35} (0.47)^{0.43} = 8.9 \text{ ft } (2.7 \text{ m})$$

By comparison, the pier scour <u>without</u> debris is calculated using the unblocked pier width:

$$y_s = (2.5) \, 2.0 \, (1.0) \, (1.0) \, (1.1) \left(\frac{11.5}{2.5}\right)^{0.35} (0.47)^{0.43} = 6.8 \text{ ft } (2.1 \text{ m})$$

In this example, the triangular-shaped debris increased the pier scour by an additional 2.1 ft (0.6 m) compared to the same pier without debris.

By way of further comparison, if the debris had been considered to be *rectangular* in shape, the K_1 factor in the effective width equation would be 0.79, and the effective pier width would be:

$$a*_d = \frac{K_1(HW)+(y-K_1H)a}{y} = \frac{0.79(4 \times 20)+(11.5-(0.79 \times 4)) \times 2.5}{11.5} = 7.3 \text{ ft } (2.2 \text{ m})$$

Using the effective width calculated with a *rectangular*-shaped debris mass, the pier scour would be:

$$y_s = (7.3)\, 2.0\, (1.0)\, (1.0)\, (1.1) \left(\frac{11.5}{7.3}\right)^{0.35} (0.47)^{0.43} = 13.6 \text{ ft } (4.1 \text{ m})$$

Obviously, the shape of the debris mass (i.e., triangular vs. rectangular) has a considerable effect on the estimated pier scour when debris is expected. This is a result of the more streamlined shape of triangular debris masses compared to the blunt shape presented by rectangular-shaped debris accumulations and the plunging flow that reinforces the horseshoe vortex at the base of the pier.

Note: This example emphasizes that the choice of the debris shape factor K_1 is a matter of experience and judgment on the part of the hydraulic engineer in consultation with inspection and maintenance personnel regarding typical debris management history and practice in his/her jurisdiction.

7.10.8 Comprehensive Example

Additional pier scour problems are included in the Comprehensive Example in Appendix D.

7.11 PIER SCOUR IN COARSE BED MATERIALS

Prior editions of HEC-18 included a factor (K_4) for coarse-bed armoring that reduced scour computed with Equation 7.1. A USGS study (USGS 2011) concluded that the K_4 factor in the fourth edition of HEC-18 (2001) performed well with data collected in Montana. Additional research by FHWA (2012d) has developed a coarse-bed pier scour equation using data collected at the FHWA J. Sterling Jones Hydraulics Research Laboratory and field data collected by the USGS. The coarse-bed pier scour equation is only for clear-water conditions where the approach flow velocity is less than the critical velocity (V_c) for initiation of bed-material motion. The coarse-bed pier scour equation is:

$$y_s = 1.1 K_1 K_2 a^{0.62} y_1^{0.38} \tanh\left(\frac{H^2}{1.97\sigma^{1.5}}\right) \tag{7.34}$$

where y_s, K_1, K_2, a, y_1, and V_1 are in either US Customary or SI units defined as in Equation 7.1 and:

H	=	Densimetric particle Froude Number = $\dfrac{V_1}{\sqrt{g(S_g-1)D_{50}}}$
S_g	=	Sediment specific gravity
g	=	Acceleration due to gravity (32.2 ft/s², 9.81 m/s²)
D_{50}	=	Median bed material size ft (m)
σ	=	Sediment gradation coefficient = D_{84}/D_{50}

The equation is only applicable to clear-water flow conditions and to coarse bed materials with $D_{50} \geq 20$ mm and $\sigma \geq 1.5$. Hyperbolic tangent values (tanh) are provided in Table 7.4.

Example Application:

Pier geometry: a = 4.0 ft (1.22 m), circular (K_1 and K_2 =1.0)
Flow variables: y_1 = 10.2 ft (3.12 m), V_1 = 11.02 ft/s (3.36 m/s)
Bed Material: D_{50} = 150 mm, D_{84} = 255 mm, σ = 255/150 = 1.70, S_g = 2.65
Critical Velocity (Equation 6.1), V_c = 13.0 ft/s (4.0 m/s)

$$H = \frac{V_1}{\sqrt{g(S_g - 1)D_{50}}} = \frac{11.02}{\sqrt{32.2(2.65 - 1)(150/304.8)}} = 2.16$$

$$y_s = 1.1 K_1 K_2 a^{0.62} y_1^{0.38} \tanh\left(\frac{H^2}{1.97\sigma^{1.5}}\right) = 1.1 \times 1.0 \times 1.0 \times 4.0^{0.62} 10.2^{0.38} \tanh\left(\frac{2.16^2}{1.97 \times 1.7^{1.5}}\right)$$

y_s = 5.0 ft (1.51 m)

This result compares with 9.9 ft (3.0 m) for pier scour using Equation 7.1 applied to smaller size sediments.

7.12 PIER SCOUR IN COHESIVE MATERIALS

Pier scour in cohesive materials generally progresses more slowly and is more dependent on soil properties than for non-cohesive sediments. The properties include critical velocity, critical shear stress, and the erosion rate for hydraulic conditions that exceed the critical value. Briaud et al. (2011) present a pier scour equation for cohesive material that incorporates the critical velocity for initiation of erosion. The equation simplifies to:

$$y_s = 2.2 K_1 K_2 a^{0.65} \left(\frac{2.6V_1 - V_c}{\sqrt{g}}\right)^{0.7} \tag{7.35}$$

where y_s, K_1, K_2, a, and V_1 are in either US Customary or SI units defined as in Equation 7.1 and:

V_c = Critical velocity for initiation of erosion of the cohesive material, ft/s (m/s)
g = Acceleration due to gravity (32.2 ft/s^2, 9.81 m/s^2)

The critical velocity can be determined through material testing (see Chapter 4, NCHRP 2004, or Briaud et al. 2011) or it can be estimated for various types of materials using an erosion rate of 0.1 mm/hr in Figure 4.7. The computed scour is the maximum potential scour for the hydraulic condition sustained for sufficient time. Because scour in cohesive materials progresses more slowly than in non-cohesive materials, the maximum scour may not be reached during a flood or even over the life of the bridge. Therefore, the scour expected over the life of a bridge may need to account for time dependency. The method for computing time-dependent contraction scour in cohesive materials as discussed in Section 6.7.2 also applies to pier scour. Equations 6.8 and 6.9 are applied to a time series of flows that are expected for the life of the bridge. These flows should include extreme design events as discussed in Chapter 2 (Table 2.1). To calculate incremental scour using Equations 6.8 and 6.9, the initial rate of scour and the ultimate scour must be determined for each flow condition in the time series of flows. Ultimate scour is determined using Equation 7.35. The initial rate of scour is determined from material testing, estimated from Figure 6.11 from shear stress, or from Figure 4.7 from velocity. The shear stress or velocity at the pier must; however, be increased to account for flow acceleration and increased turbulence.

Briaud (2011) provides an equation for estimating maximum shear stress at a pier. Because the shear stress is used only to calculate the initial rate of scour and not the amount of scour, a more simplified approach presented in HEC-23 (FHWA 2009) can be used to compute shear stress at the pier. This approach is uses the equation:

$$\tau_{pier} = \frac{\gamma}{y_1^{1/3}} \left(\frac{nKV_1}{K_u} \right)^2 \tag{7.36}$$

where:

τ_{pier} = Shear stress at the pier, lb/ft^2, Pa
γ = unit weight of water, 62.4 lb/ft^3, 9810 N/m^3
y_1 = Depth of flow at the pier, ft (m)
n = Manning n of the channel bed
K = Velocity coefficient = 2.0 for circular piers and 2.5 for square piers
K_u = 1.486 for English units and 1.0 for SI

Example Application:

Pier geometry: a = 4.0 ft (1.22 m), circular (K_1 and K_2 =1.0)
Flow variables: y_1 = 10.2 ft (3.12 m), V_1 = 11.02 ft/s (3.36 m/s), duration is 2 days.
Bed Material: low plasticity clay, V_c = 10.5 ft/s (3.2 m/s)

$$y_s = 2.2 K_1 K_2 a^{0.65} \left(\frac{2.6 V_1 - V_c}{\sqrt{g}} \right)^{0.7} = 2.2 \times 1.0 \times 1.0 \times 4^{0.65} \left(\frac{2.6 \times 11.02 - 10.5}{\sqrt{32.2}} \right)^{0.7}$$

y_s = 12.2 ft (3.72 m)

This result compares with 9.9 ft (3.0 m) for pier scour from Equation 7.1 applied to non-cohesive sediments. The channel bed Manning n is 0.022. Therefore, the shear stress can be estimated as:

$$\tau_{pier} = \frac{\gamma}{y_1^{1/3}} \left(\frac{nKV_1}{K_u} \right)^2 = \frac{62.4}{10.2^{1/3}} \left(\frac{0.022 \times 2 \times 11.02}{1.486} \right)^2 = 3.1 \text{lb/ft}^2 (147 \text{Pa})$$

For this shear stress and material determine from material testing that the initial erosion rate is 0.3 ft/hr (0.09 m/hr). From Equation 6.8,

$$y_s(t = 48 \text{hr}) = \frac{t}{\frac{1}{\dot{z}_1} + \frac{t}{y_s}} = \frac{48}{\frac{1}{0.3} + \frac{48}{12.2}} = 6.6 \text{ft} (2.0 \text{m})$$

For each additional flow period, only flows that can produce scour greater than prior achieved scour, in this case 6.6 feet (2.0 m), would need to be considered. For these additional flows, the equivalent time to reach the prior scour is computed from Equation 6.9 and the cumulative scour is computed from Equation 6.8.

7.13 PIER SCOUR IN ERODIBLE ROCK

Two modes of scour and erosion in rock are presented in this section. The first mode, referred to as quarrying and plucking, involves the removal of relatively intact rock blocks in fissured and jointed rock masses which are otherwise relatively hard and durable compared to soils. The second mode is a more gradual and progressive process caused by bedload abrasion of the rock surface over relatively long periods of time, such that the pier is exposed to many flood events over the life of the bridge.

7.13.1 Quarrying and Plucking

The Erodibility Index classification described in Chapter 4 is used to estimate the depth of pier scour when quarrying and plucking of intact rock blocks is the dominant mode of erosion. Rock discontinuities have the greatest influence on scour processes in this mode. A schematic diagram of this process is presented in Figure 7.19.

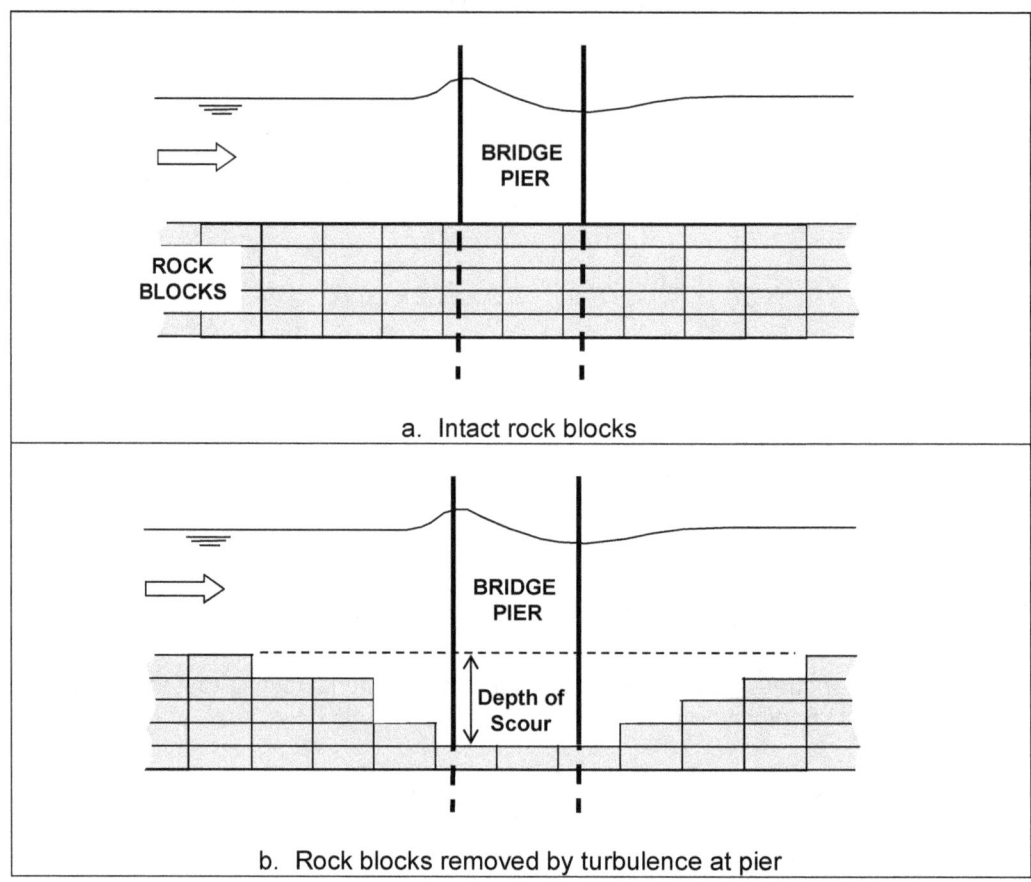

Figure 7.19. Conceptual model of quarrying and plucking at a bridge pier.

As discussed in Chapter 4 (Section 4.7.2), the Erodibility Index, K, is calculated as:

$$K = (M_s)(K_b)(K_d)(J_s) \tag{7.37}$$

where:

K = Erodibility Index
M_s = Intact rock mass strength parameter
K_b = Block size parameter
K_d = Shear strength parameter
J_s = Relative orientation parameter

As a general rule, rock masses on which bridge piers are founded typically exhibit Erodibility Index K values ranging from 0.1 (very poor rock) up to 10,000 or greater (very good rock). For K values greater than 0.1, the critical stream power P_c in SI units (kilowatts per square meter) for initiating quarrying and plucking is related to K as given by Annandale (2006):

$$P_c = K^{0.75} \tag{7.38}$$

where:

K = Erodibility Index
P_c = Critical stream power necessary to initiate scour, KW/m²

As developed by Annandale, the stream power of the approach flow upstream of a bridge pier (i.e., at a location not affected by the additional turbulence caused by the pier itself) is calculated by considering the turbulence production near the bed of the stream:

$$P_a = 7.853 \rho \left(\frac{\tau}{\rho}\right)^{3/2} \tag{7.39}$$

where:

P_a = Stream power of approach flow, W/m²
ρ = Mass density of water, 1000 kg/m³
τ = Bed shear stress of approach flow, N/m² or Pa

Note that P_a in Equation 7.39 is expressed in units of W/m², whereas the critical shear stress P_c given by Equation 7.38 is expressed in units of KW/m². To convert P_a to KW/m², the value resulting from Equation 7.39 must be divided by 1000.

In the vicinity of a bridge pier, the downward flow at the upstream face of the pier creates additional local turbulence in the form of the horseshoe vortex. As scour occurs, the stream power P at the bottom of the scour hole decreases as the scour hole becomes deeper. Scour will continue until the stream power at the bottom of the scour hole becomes less than the critical stream power P_c, at which point the scouring process can no longer be sustained. The relationship relating the relative depth of the scour hole to the stream power at the bottom of the hole for a variety of pier shapes (round, square, and rectangular) can be expressed as:

$$\frac{P}{P_a} = 8.42 \, e^{-0.712 \left(\frac{y_s}{b}\right)} \tag{7.40}$$

where:

- P = Stream power at the bottom of the scour hole, W/m²
- P_a = Stream power of the approach flow near the stream bed, W/m²
- y_s = Depth of scour hole, m
- b = Pier width perpendicular to the flow direction, m

7.13.2 Example Problem - Erodibility Index Method for Rock

The application of the Erodibility Index method to estimate scour at a pier founded on erodible rock is illustrated by the following example:

Rock mass characteristics: A relatively soft sandstone is characterized by an unconfined compressive strength of about 5 MPa. The rock quality designation RQD of core samples obtained from this material is 20. The sandstone exhibits three joint/fissure sets with relatively rough, planar joints greater than 5 mm wide that are filled with noncohesive, slightly clayey material. The dip angle of the closer spaced joint set is 30 degrees in the direction of stream flow, and the ratio of joint spacing is approximately 1:4.

Design hydraulic conditions: The pier is rectangular with a width of 2.5 ft (0.76 m) perpendicular to the flow. The design flood depth is 11.5 ft (3.5 m) and the approach velocity upstream of the bridge is 9 ft/s (2.7 m/s). The slope of the energy grade line is 0.0035 ft/ft.

Step 1: Calculate the Erodibility Index K for the sandstone (Equation 7.37):

$$K = (M_s)(K_b)(K_d)(J_s)$$

From the information presented in Chapter 4 the following parameter values are determined:

$M_s = 3.95$ (from Table 4.21)

$K_b = \dfrac{RQD}{J_n} = \dfrac{20}{2.73} = 7.33$ (from Table 4.22)

$K_d = \dfrac{J_r}{J_a} = \dfrac{1.5}{10.0} = 0.15$ (from Tables 4.23 and 4.24)

$J_s = 0.55$ (from Table 4.25)

$K = (3.95)(7.33)(0.15)(0.55) = 2.4$

Step 2: Calculate the critical stream power required to initiate scouring (quarrying and plucking) of the sandstone blocks in SI units (Equation 7.38):

$P_c = K^{0.75} = (2.4)^{0.75} = 1.93 \, KW/m^2$

Step 3: Calculate the approach shear stress τ in SI units, using the local approach depth upstream of the pier instead of the hydraulic radius and assuming the stream is relatively straight in the bridge reach (Equation 4.3):

$$\tau = \gamma y S_f = (9800)(3.5)(0.0035) = 120.1 \, N/m^2$$

Step 4: Calculate the approach stream power P_a in W/m² and convert to KW/m² (Equation 7.39):

$$P_a = 7.853\rho\left(\frac{\tau}{\rho}\right)^{3/2} = 7.853(1000)\left(\frac{120.1}{1000}\right)^{3/2} = 326.9 \, W/m^2$$

$$P_a = \frac{326.9 \, W/m^2}{1000} = 0.327 \, KW/m^2$$

Step 5: Calculate the local stream power at the pier as a function of pier width and scour hole depth to find the maximum depth of scour at the pier, using a critical stream power of 1.93 KW/m², approach stream power $P_a = 0.327$ KW/m² and pier width b = 0.76 m (Equation 7.40). The calculation results are shown in Table 7.5.

$$\frac{P}{P_a} = 8.42 \, e^{-0.712\left(\frac{y_s}{b}\right)}$$

Table 7.5. Calculation Results for Example Problem.

y_s/b	P/P_a	P (KW/m²)	P > P_c ?	y_s (m)
0.10	7.84	2.56	yes	0.08
0.20	7.30	2.39	yes	0.15
0.30	6.80	2.22	yes	0.23
0.40	6.33	2.07	yes	0.30
0.50	5.90	1.93	no	0.38

The calculation results presented in Table 7.5 indicate that for this example problem, the pier scour during the design flood is expected to be 0.38 meters, or about 1.25 feet. Note that in this example, the critical stream power for the sandstone is greater than the stream power of the approach flow; in other words, the streambed would not be expected to erode were it not for the presence of a bridge pier.

7.13.3 Abrasion

Abrasion is the gradual wearing away of rock surfaces by the more-or-less continuous movement of water and bedload sediment over the bed of a channel (NCHRP 2011e). At bridge piers founded on rock, the local flow acceleration, horseshoe vortex and increased turbulence in the immediate vicinity of the pier result in more vigorous particle movement compared to the unobstructed channel. Therefore, abrasion scour at piers is much more pronounced than the general scouring of the streambed by this process.

Abrasion Resistance of Rock:

The resistance of rock to abrasion from bedload movement is related to the physical properties of the rock mass. Dickenson and Baillie (1999) used a modification of the conventional slake durability test (ASTM D4644) to characterize the abrasion resistance of rock materials found in streambeds of Western Oregon. They eliminated oven drying from the ASTM slake durability procedure because complete drying was not representative of the streambeds they were studying, and used the modified test to determine an "abrasion number" based on the slope of the percent weight loss vs. time data from the test. They disregarded the first few readings of the test data because of rounding of sharp sample edges which is not representative of longer-term sample behavior.

Keaton et al. (NCHRP 2011e) further modified the test and developed a method to relate the sample weight loss data from the test to an equivalent scour depth vs. equivalent stream power relationship based on the dimensions of the slake durability test drum and its rate of rotation. Figure 7.20 presents data from this test using a bulk sample of siltstone from the Sacramento River near Redding, California.

The slope of the linear regression line through the test data is defined by Keaton et al. (NCHRP 2011e) as the "Geotechnical Scour Number" and is used to estimate abrasion scour at piers. In Figure 7.20, for example, the Geotechnical Scour Number for the siltstone sample is 0.00018; tests on other rock types compared with actual scour measured at four field sites with bridge piers founded in erodible rock resulted in Figure 7.21, which relates pier scour to the total amount of work done by the flow on the stream bed over the life of the bridge.

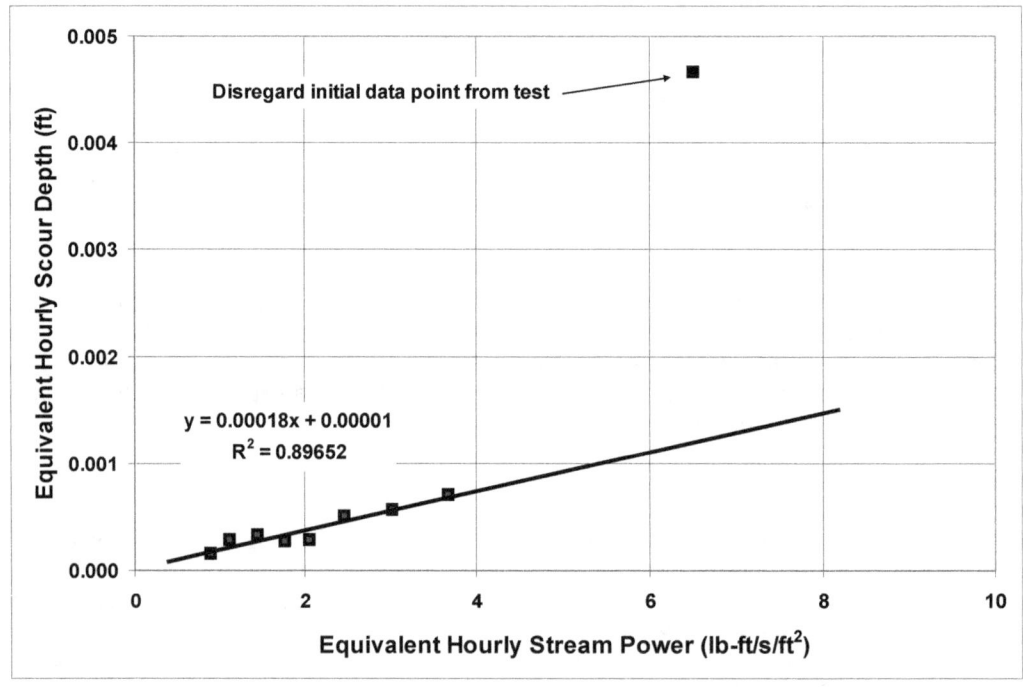

Figure 7.20. Example data from modified slake durability test (NCHRP 2011e).

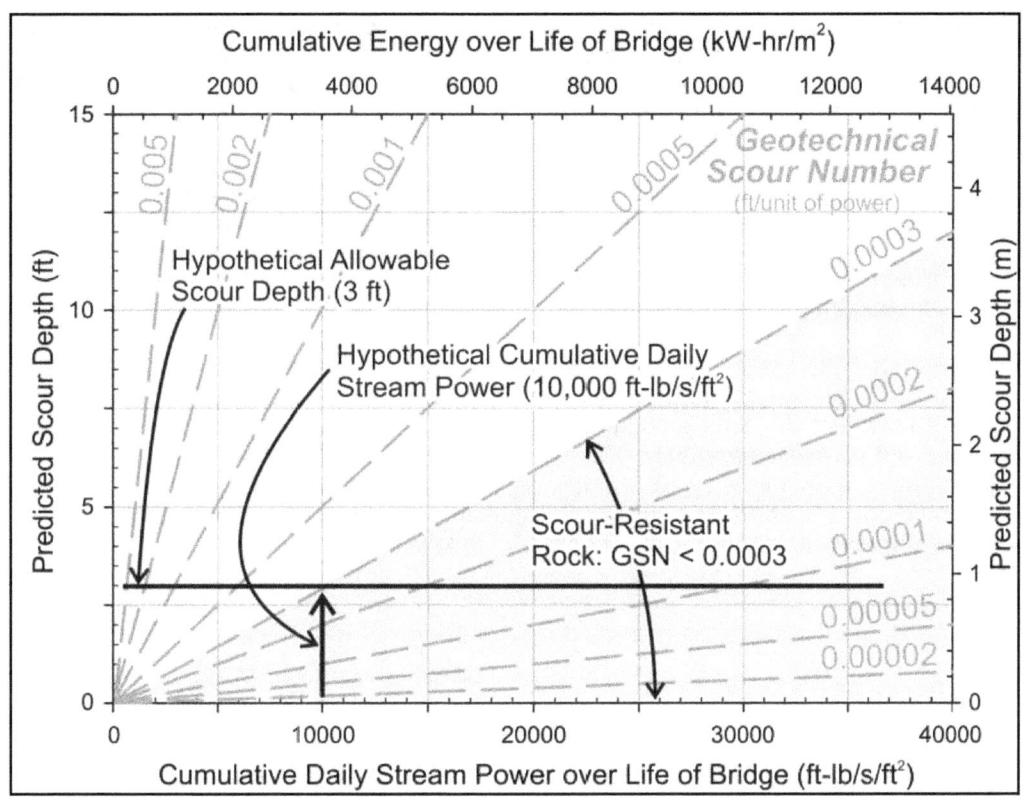

Figure 7.21. Pier scour in rock as a function of stream power and Geotechnical Scour Number (NCHRP 2011e).

Stream Power and Duration:

Because the process of abrasion is gradual, in quantifying pier scour the effect of duration is more important than the hydraulic conditions that exist temporarily at the peak of a flood event. The product of power and time is work; therefore, integrating a time series of stream power over many years and many floods provides a meaningful measure of the total amount of work done by the flow on the streambed for the time period being considered. Such integration has been performed and described as "Cumulative Daily Stream Power" by Keaton et al. (NCHRP 2011e). The integral, denoted Ω, is the area under the curve of stream power versus time for any particular duration, and is expressed in units of work per unit area (e.g., lb-ft/ft^2, KW-hr/m^2).

$$y_s = (GSN)(\Omega) \tag{7.41}$$

where:

y_s = Pier scour depth due to abrasion, ft (m)
GSN = Geotechnical Scour Number from modified slake durability test, ft per unit of stream power (m per unit of stream power)
Ω = Cumulative stream power, lb-ft-day/s/ft^2 (KW-hr/m^2)

The concept of accumulating stream power to quantify work per unit area is illustrated by the use of a time series of average daily flows typically obtained from USGS gaging station records. Knowing the velocity V, depth of flow y, and energy slope S_f as functions of discharge, a time series of average daily stream power can be constructed as shown in Figure 7.22, which illustrates typical data from one water year.

In Figure 7.22(a), it is assumed that there is a threshold discharge that initiates bedload motion, therefore a threshold discharge must be exceeded before abrasion processes can occur. The <u>effective</u> stream power causing abrasion is illustrated in Figure 7.22(b). This figure illustrates the concept that over the course of a typical year, only a few flood events may contribute to the abrasive work that creates scour at a bridge pier.

In the Figure 7.22, the total effective amount of work Ω done by the stream during the water year is found by integrating the daily stream power curve, as shown in Figure 7.23. This figure shows that the total amount of work done in this particular water year is 266 lb-ft-day/s/ft^2 in U.S. customary units. Converting to SI units and using the convention of KW-hr as the unit of work, this is 93 KW-hr/m^2 over the course of the water year. Over many years, the total amount of effective work done by the stream on its bed can be quantified as the sum of the daily work done by individual events.

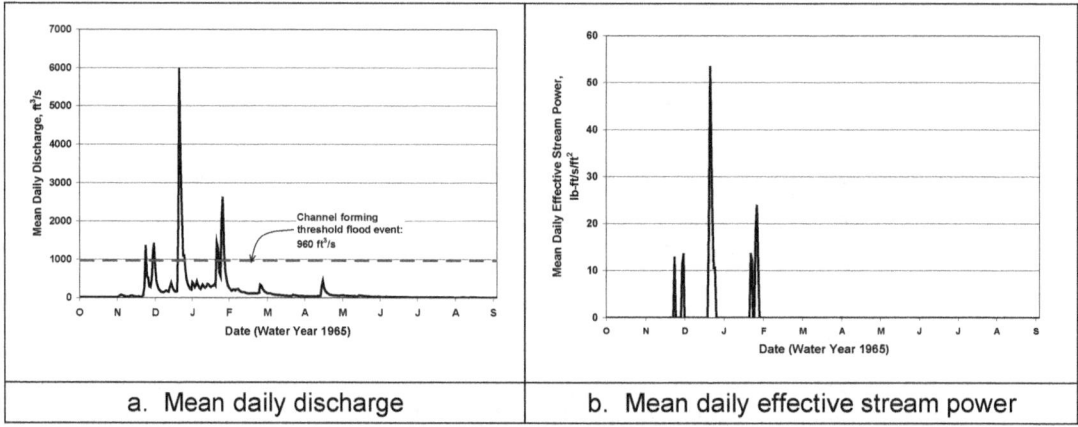

| a. Mean daily discharge | b. Mean daily effective stream power |

Figure 7.22. Transforming a mean daily flow series to mean daily effective stream power.

Given a future cumulative hydraulic loading Ω_{fut}, the Geotechnical Scour Number can be used to estimate the future scour associated with that loading, for the particular rock formation. Estimates of future scour may then be made for a variety of purposes:

- Estimating scour over the remaining life of the structure
- Estimating scour at other existing structures with foundations in the same (or similar) rock formation
- Estimating scour at proposed structures on the same (or similar) rock formation

The difficulty with the above approach is quantifying the cumulative effective hydraulic load in the future. Considering that this method is concerned with long-term abrasion processes, a threshold concept will typically apply, and the practitioner must exercise judgment in determining an appropriate threshold considering local hydraulic conditions at the pier. Once a reasonable threshold is established, only the effects of larger, relatively infrequent events that exceed this threshold over the life of the structure need be considered.

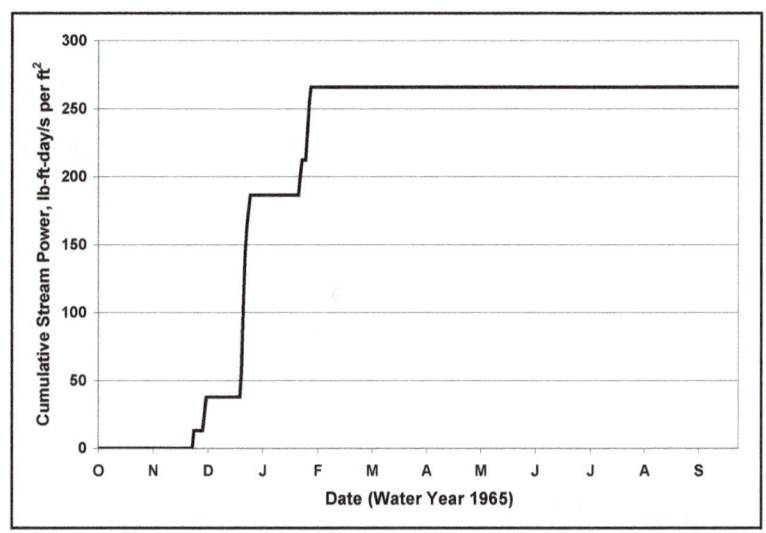

Figure 7.23. Cumulative stream power for the water year in Figure 7.22.

7.13.4 Example Problem - Long-Term Abrasion of Rock

Step 1: Develop long-term flow series: Figure 7.24 shows 71 years of daily stream flow from USGS gaging station on the Sacramento River near Redding, California. A threshold discharge of 30,300 ft^3/s (860 m^3/s) represents the initiation of bedload movement based on particle size and stream geometry.

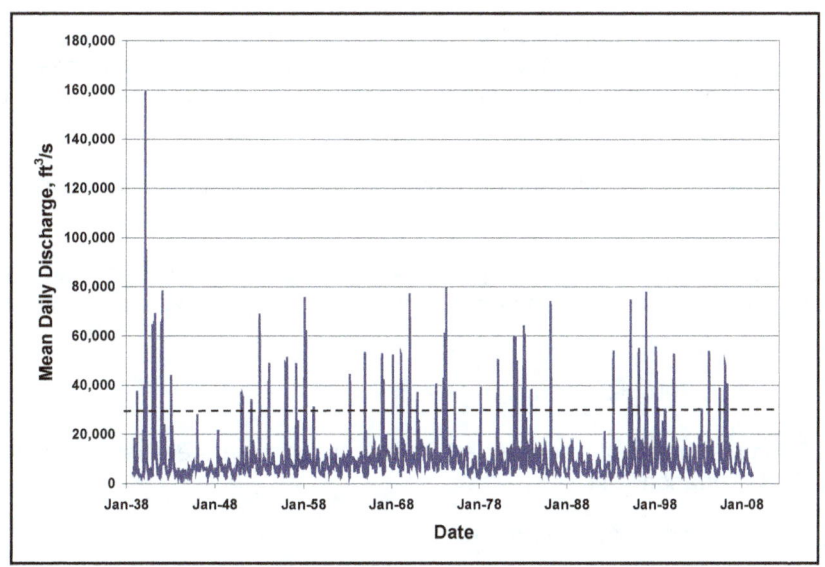

Figure 7.24. Mean daily flow, Sacramento River near Redding, CA 1938 - 2009.

Step 2: Develop long-term cumulative stream power: The daily flow series is converted to a daily stream power and the cumulative effect of many floods over the entire period of record is plotted in Figure 7.25.

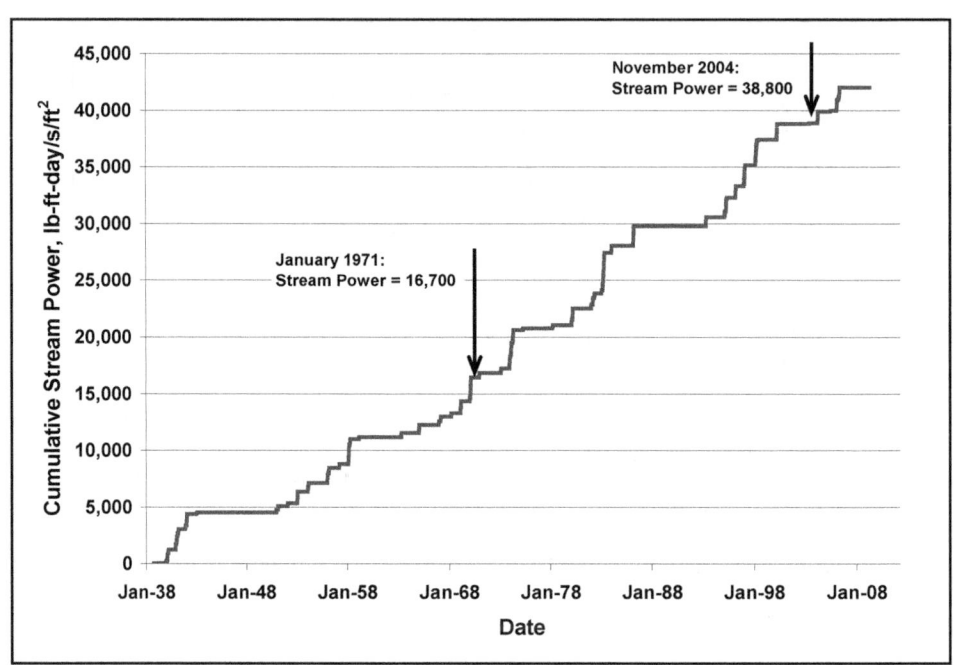

Figure 7.25. Cumulative daily stream power, Sacramento River near Redding, CA 1938 - 2009.

Step 3: Determine the Geotechnical Scour Number for the rock of the streambed:

From the modified slake durability test described in Section 7.13.3, a siltstone sample from this bridge site exhibits a Geotechnical Scour Number of 0.00018 based on the data presented in Figure 7.20. The units are feet of scour per hour per unit of hourly stream power.

Step 4: Calculate the total amount of work done by the stream on its bed for the time period of interest.

In this example, surveys of the bridge piers were conducted in early 1971 and late 2004 (nearly 34 years). During this time period, the total amount of work is calculated from Figure 7.25:

$\Omega = (38,800) - (16,700) = 22,100$ lb-ft-day/s/ft^2

Step 5: Estimate the amount of pier scour that occurred in the siltstone between 1971 and 2004 (Equation 7.41):

$y_s = (0.00018)(22,100) = 4.0$ ft (1.2 m)

This value of pier scour is compared to the observed pier scour of approximately 5 feet (1.5 m) at Piers 4, 5, and 6 at the State Route 273 bridge over the Sacramento River that occurred from 1971 through 2004.

CHAPTER 8

EVALUATING LOCAL SCOUR AT ABUTMENTS

8.1 GENERAL

Scour occurs at abutments when the abutment and roadway embankment obstruct the flow. Several causes of abutment failures during post-flood field inspections of bridge sites have been documented (TRB 1998a):

- Overtopping of abutments or approach embankments
- Lateral channel migration or stream widening processes
- Contraction scour
- Local scour at one or both abutments

Abutment damage is often caused by a combination of these factors. As a general rule, the abutments most vulnerable to damage are those located at or near the channel banks. Where abutments are set back from the channel banks, especially on wide floodplains, large local scour holes have been observed with scour depths of as much as four times the approach flow depth on the floodplain.

The flow obstructed by the abutment and approach highway embankment accelerates and often forms a vortex starting at the upstream end of the abutment and running along the toe of the abutment. Generally a wake vortex forms at the downstream end of the abutment (Figure 8.1).

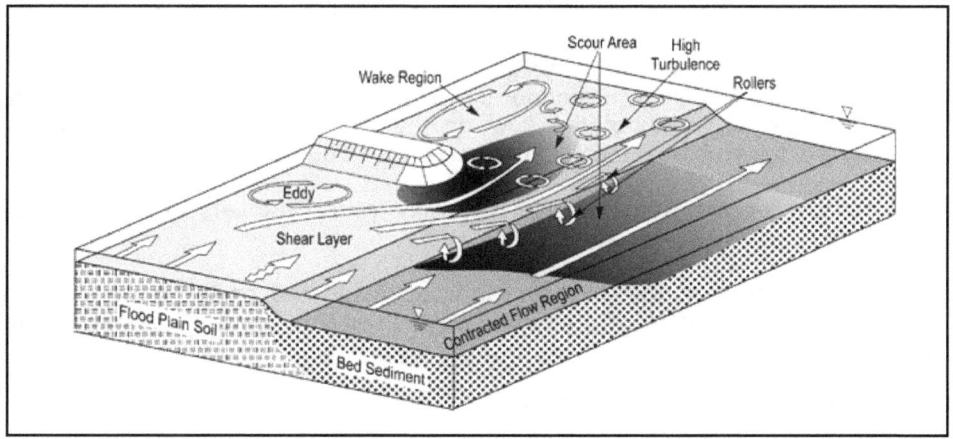

Figure 8.1. Schematic representation of abutment scour in a compound channel (NCHRP 2011b).

The vortex that forms at the downstream end of the abutment is similar to the wake vortex that forms downstream of a pier. Research has been conducted to determine the depth and location of the scour hole that develops from the vortex that occurs at the abutment, and numerous abutment scour equations have been developed to predict this scour depth.

Abutment failures and erosion of the roadway embankment fill also occur from the action of the downstream wake vortex. An example of abutment and approach erosion of a bridge due to the action of the various vortices is shown in Figure 8.2.

Figure 8.2. Scour of bridge abutment and approach embankment.

The types of failures described above are initiated as a result of the obstruction to the flow caused by the abutment and highway embankment and subsequent contraction and turbulence of the flow at the abutments. There are other conditions that develop during major floods, particularly on wide floodplains, that are more difficult to foresee but that need to be considered in the hydraulic analysis and design of the substructure (TRB 1998a):

- Gravel pits on the floodplain upstream of a structure can capture the flow and divert the main channel flow out of its normal banks into the gravel pit. This can result in an adverse angle of attack of the flow on the downstream highway with subsequent breaching of the embankment and/ or failure of the abutment.

- Levees can become weakened and fail with resultant adverse flow conditions at the bridge abutment.

- Debris can become lodged at piers and abutments and on the bridge superstructure, modifying flow conditions and creating adverse angles of attack of the flow on bridge piers and abutments.

8.2 ABUTMENT SCOUR EQUATIONS

8.2.1 Overview

Equations for predicting abutment scour depths such as Liu et al. (1961), Laursen (1980), Froehlich (TRB 1989), and Melville (1992) are based entirely on laboratory data. The problem is that little field data on abutment scour exist. Liu et al.'s equations were developed by dimensional analysis of the variables with a best-fit line drawn through the laboratory data. Laursen's equations are based on inductive reasoning of the change in transport relations due to the acceleration of the flow caused by the abutment. Froehlich's equations were derived from dimensional analysis and regression analysis of the available laboratory data.

Melville's equations were derived from dimensional analysis and development of relations between dimensionless parameters using best-fit lines through laboratory data.

Until recently, the equations in the literature were developed using the abutment and roadway approach length as one of the variables. This approach may result in excessively conservative estimates of scour depth since the discharge in the laboratory flume intercepted by the abutment is directly related to the abutment length; whereas, in the field, this is rarely the case.

Figure 8.3. illustrates the difference. Thus, equations for predicting abutment scour would be more applicable to field conditions if they included the discharge intercepted by the embankment rather than embankment length. Sturm (1999, FHWA 1999a) concluded that a discharge distribution factor is the appropriate variable to use on local scour depth rather than abutment length.

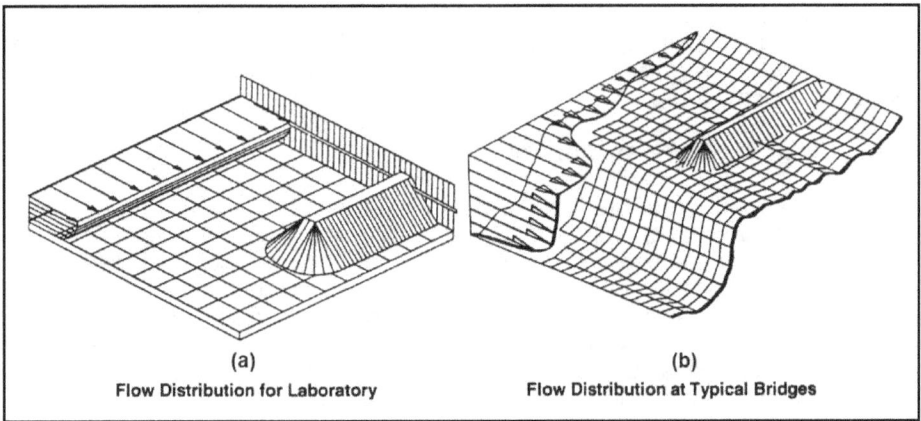

Figure 8.3. Comparison of (a) laboratory flow characteristics to (b) field flow conditions.

Abutment scour depends on the interaction of the flow obstructed by the abutment and roadway approach and the flow in the main channel at the abutment. The discharge returned to the main channel at the abutment is not simply a function of the abutment and roadway length in the field case. Abutment scour depth depends on abutment shape, discharge in the main channel at the abutment, discharge intercepted by the abutment and returned to the main channel at the abutment, sediment characteristics, cross-sectional shape of the main channel at the abutment (especially the depth of flow in the main channel and depth of the overbank flow at the abutment), and alignment. In addition, field conditions may have tree-lined or vegetated banks, low velocities, and shallow depths upstream of the abutment. Most of the early laboratory research failed to replicate these field conditions.

Research sponsored by the National Cooperative Highway Research Program of the Transportation Research Board developed an approach to determining abutment scour that includes the discharge intercepted by an abutment and its approach rather than abutment and approach length (Sturm and Chrisochoides 1998). In addition, Maryland State Highway Administration developed a method to determine scour depths at abutments (Chang and Davis 1999a and b).

NCHRP (2011b) conducted an evaluation of abutment scour processes and prediction methods. The conclusions and recommendations that pertain to abutment scour evaluation and abutment design include:

- Contraction scour should be viewed as the reference scour depth for calculating abutment scour. Abutment scour should be taken as the product of the contraction scour caused by flow acceleration through the constricted opening multiplied by a factor accounting for large-scale turbulence. This approach would replace the current approach for adding contraction scour to a separately computed abutment scour.
- Abutments should be designed to have a minimum setback distance from the channel bank of the main channel with riprap protection of the embankment and a riprap apron to protect against scour. The setback distance should accommodate the apron width recommended in HEC-23 (FHWA 2009).
- Two-dimensional models should be used on all but the simplest bridge crossings as a matter of course.

Abutment foundations should be designed to be safe from long-term degradation, lateral migration, and contraction scour; and protected from local scour with countermeasures such as riprap, guide banks, or dikes. The equations provided in this chapter should be used as guides in the design.

8.2.2 Abutment Scour Parameter Determination

Many of the abutment scour prediction equations presented in the literature use the length of an abutment (embankment) projected normal to flow as an independent variable. In practice, the length of embankment projected normal to flow that is used in these relationships is determined from the results of 1-dimensional hydraulic models such as HEC-RAS (USACE 2010a) which assume an average velocity over the entire cross section (Figure 8.3a). In reality, conveyance and associated velocity and flow depth at the outer extremes of a floodplain are much less, particularly in wide and shallow heavily vegetated floodplains (Figure 8.3b). This flow is typically referred to as "ineffective" flow. When applying abutment scour equations that use the length of embankment projected normal to flow, it is imperative that the length used be the length of embankment blocking "live" flow.

The length of embankment blocking "live" flow can be determined from a graph of conveyance versus distance across a representative cross-section upstream of the bridge (Figure 8.4). If a relatively large portion of a cross-section is required to convey a known amount of discharge in the floodplain, then the length of embankment blocking this flow should probably not be included when determining the length of embankment for use in the abutment scour prediction relationship. Alternately, if the flow in a significant portion of the cross-section has low velocity and/or is shallow, then the length of embankment blocking this flow should probably not be used either. HEC-RAS (USACE 2010a) can easily compute conveyance versus distance across a cross section.

For example, Figure 8.4 shows the plan view of an embankment blocking three equal conveyance tubes on the right floodplain at a bridge. Since the right conveyance tube occupies the majority of floodplain but conveys only one-third of the floodplain flow, it should not be included in the "live" flow area for determining L'. In this case the length of embankment, L', blocking the "live" flow is approximately the length of the two inner conveyance tubes. In the event that the conveyance versus distance graph does not show a conclusive break point between "live" flow and ineffective flow, an alternative procedure is to estimate L' as the width of the conveyance tube directly upstream of the abutment times the total number of conveyance tubes (including fractional portions) obstructed by the embankment. This length is more representative of the uniform flow conditions in the laboratory experiments used to develop abutment scour equations.

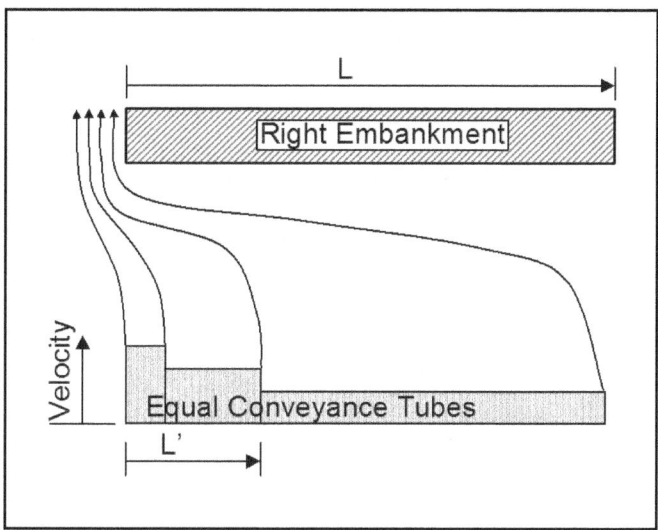

Figure 8.4. Determination of length of embankment blocking live flow for abutment scour estimation.

8.3 ABUTMENT SITE CONDITIONS

Abutments can be set back from the natural stream bank, placed at the bankline or, in some cases, actually set into the channel itself. Common designs include stub abutments placed on spill-through slopes, and vertical wall abutments, with or without wingwalls. Scour at abutments can be live-bed or clear-water scour. The bridge and approach road can cross the stream and floodplain at a skew angle and this will have an effect on flow conditions at the abutment. Finally, there can be varying amounts of overbank flow intercepted by the approaches to the bridge and returned to the stream at the abutment. More severe abutment scour will occur when the majority of overbank flow returns to the bridge opening directly upstream of the bridge crossing. Less severe abutment scour will occur when overbank flows gradually return to the main channel upstream of the bridge crossing.

8.4 ABUTMENT SKEW

The skew angle for an abutment (embankment) is depicted in Figure 8.5. For an abutment angled downstream, the scour depth is decreased, whereas the scour depth is increased for an abutment angled upstream. An equation and guidance for adjusting abutment scour depth for embankment skew are given in Section 8.6.1.

8.5 ABUTMENT SHAPE

There are three general shapes of abutments: (1) spill-through abutments, (2) vertical walls without wing walls, and (3) vertical-wall abutments with wing walls (Figure 8.6). These shapes have varying angles to the flow. As shown in Table 8.1, depth of scour is approximately double for vertical-wall abutments as compared with spill-through abutments. Similarly, scour at vertical wall abutments with wingwalls is reduced to 82 percent of the scour of vertical wall abutments without wingwalls.

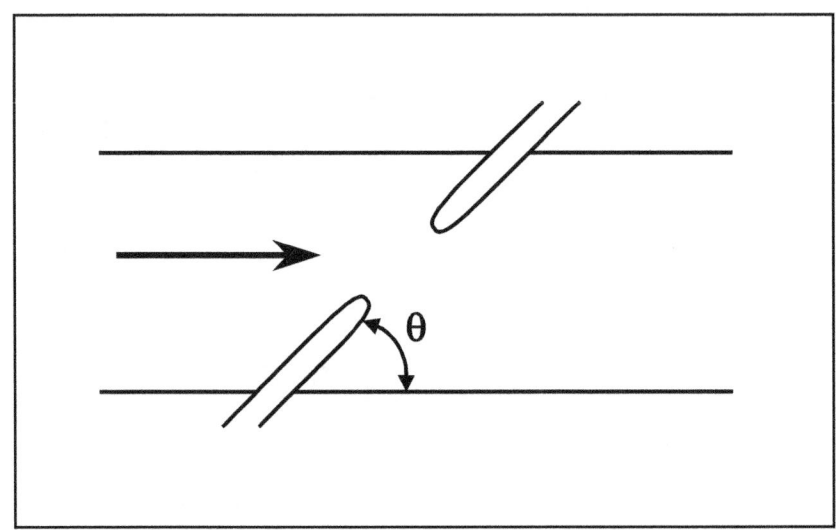

Figure 8.5. Orientation of embankment angle, θ, to the flow.

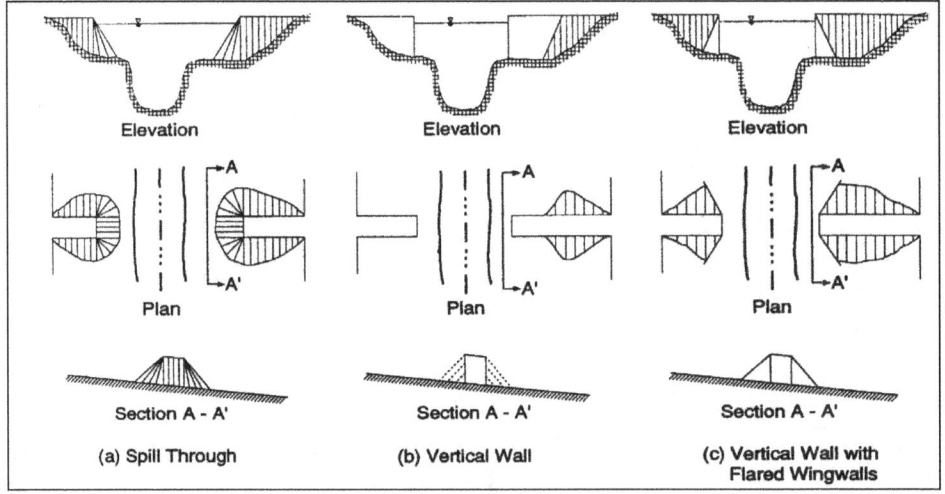

Figure 8.6. Abutment shape.

Table 8.1. Abutment Shape Coefficients.	
Description	K_1
Vertical-wall abutment	1.00
Vertical-wall abutment with wing walls	0.82
Spill-through abutment	0.55

8.6 ESTIMATING SCOUR AT ABUTMENTS

As a check on the potential depth of scour to aid in the design of the foundation and placement of rock riprap and/or guide banks, Froehlich's (TRB 1989) live-bed scour equation or the HIRE equation in HDS 6 (FHWA 2001a) can be used. A more recent alternative approach developed under NCHRP Project 24-20 is also presented for use in estimating scour at abutments (NCHRP 2010b).

8.6.1 Froehlich's Abutment Scour Equation

Froehlich (TRB 1989) analyzed 170 live-bed scour measurements in laboratory flumes by regression analysis to obtain the following equation:

$$\frac{y_s}{y_a} = 2.27 \, K_1 \, K_2 \left(\frac{L'}{y_a}\right)^{0.43} Fr^{0.61} + 1 \tag{8.1}$$

where:

- K_1 = Coefficient for abutment shape (Table 8.1)
- K_2 = Coefficient for angle of embankment to flow
- K_2 = $(\theta/90)^{0.13}$ (see Figure 8.5 for definition of θ)
  - $\theta<90°$ if embankment points downstream
  - $\theta>90°$ if embankment points upstream
- L' = Length of active flow obstructed by the embankment, ft (m)
- A_e = Flow area of the approach cross section obstructed by the embankment, ft^2 (m^2)
- Fr = Froude Number of approach flow upstream of the abutment = $V_e/(gy_a)^{1/2}$
- V_e = Q_e/A_e, ft/s (m/s)
- Q_e = Flow obstructed by the abutment and approach embankment, ft^3/s (m^3/s)
- y_a = Average depth of flow on the floodplain (A_e/L), ft (m)
- L = Length of embankment projected normal to the flow, ft (m)
- y_s = Scour depth, ft (m)

It should be noted that Equation 8.1 is not consistent with the fact that as L' tends to 0, y_s also tends to 0. The 1 was added to the equation so as to envelope 98 percent of the data. See Section 8.2.2 and Figure 8.4 for guidance on estimating L'.

8.6.2 HIRE Abutment Scour Equation

An equation based on field data of scour at the end of spurs in the Mississippi River (obtained by the USACE) can also be used for estimating abutment scour (FHWA 2001). This field situation closely resembles the laboratory experiments for abutment scour in that the discharge intercepted by the spurs was a function of the spur length. The modified equation, referred to herein as the HIRE equation, is applicable when the ratio of projected abutment length (L) to the flow depth (y_1) is greater than 25. This equation can be used to estimate scour depth (y_s) at an abutment where conditions are similar to the field conditions from which the equation was derived:

$$\frac{y_s}{y_1} = 4 \, Fr^{0.33} \frac{K_1}{0.55} K_2 \tag{8.2}$$

where:

- y_s = Scour depth, ft (m)
- y_1 = Depth of flow at the abutment on the overbank or in the main channel, ft (m)
- Fr = Froude Number based on the velocity and depth adjacent to and upstream of the abutment
- K_1 = Abutment shape coefficient (from Table 8.1)
- K_2 = Coefficient for skew angle of abutment to flow calculated as for Froehlich's equation (Section 8.7.1)

8.6.3 NCHRP 24-20 Abutment Scour Approach

NCHRP (2010b) developed abutment scour equations considering a range of abutment types, abutment locations, flow conditions, and sediment transport conditions. These equations use contraction scour as the starting calculation for abutment scour and apply a factor to account for large-scale turbulence that develops in the vicinity of the abutment. One important distinction regarding the contraction scour calculation is that the abutment creates a non-uniform flow distribution in the contracted section. The flow is more concentrated in the vicinity of the abutment and the contraction scour component is greater than for average conditions in the constricted opening. The three scour conditions illustrated in Figure 8.7 are (a) scour occurring when the abutment is in or close to the main channel, (b) scour occurring when the abutment is set back from the main channel, and (c) scour occurring when the embankment breaches and the abutment foundation acts as a pier. As illustrated in Figure 8.8, the NCHRP study also concluded that there is a limiting depth of abutment scour when the geotechnical stability of the embankment or channel bank is reached. The abutment scour computed from the NCHRP approach is total scour at the abutment; it is not added to contraction scour because it already includes contraction scour. The advantages of using the NCHRP abutment scour equations include (1) not using the effective embankment length, L', which is difficult to determine in many situations, (2) the equations are more physically representative of the abutment scour process, and (3) the equations predict total scour at the abutment rather than the abutment scour component that is then added to contraction scour. The scour equations for conditions (a) and (b) are:

$$y_{max} = \alpha_A y_c \text{ or } y_{max} = \alpha_B y_c \tag{8.3}$$

$$y_s = y_{max} - y_0 \tag{8.4}$$

where:

- y_{max} = Maximum flow depth resulting from abutment scour, ft (m)
- y_c = Flow depth including live-bed or clear-water contraction scour, ft (m)
- α_A = Amplification factor for live-bed conditions
- α_B = Amplification factor for clear-water conditions
- y_s = Abutment scour depth, ft (m)
- y_0 = Flow depth prior to scour, ft (m)

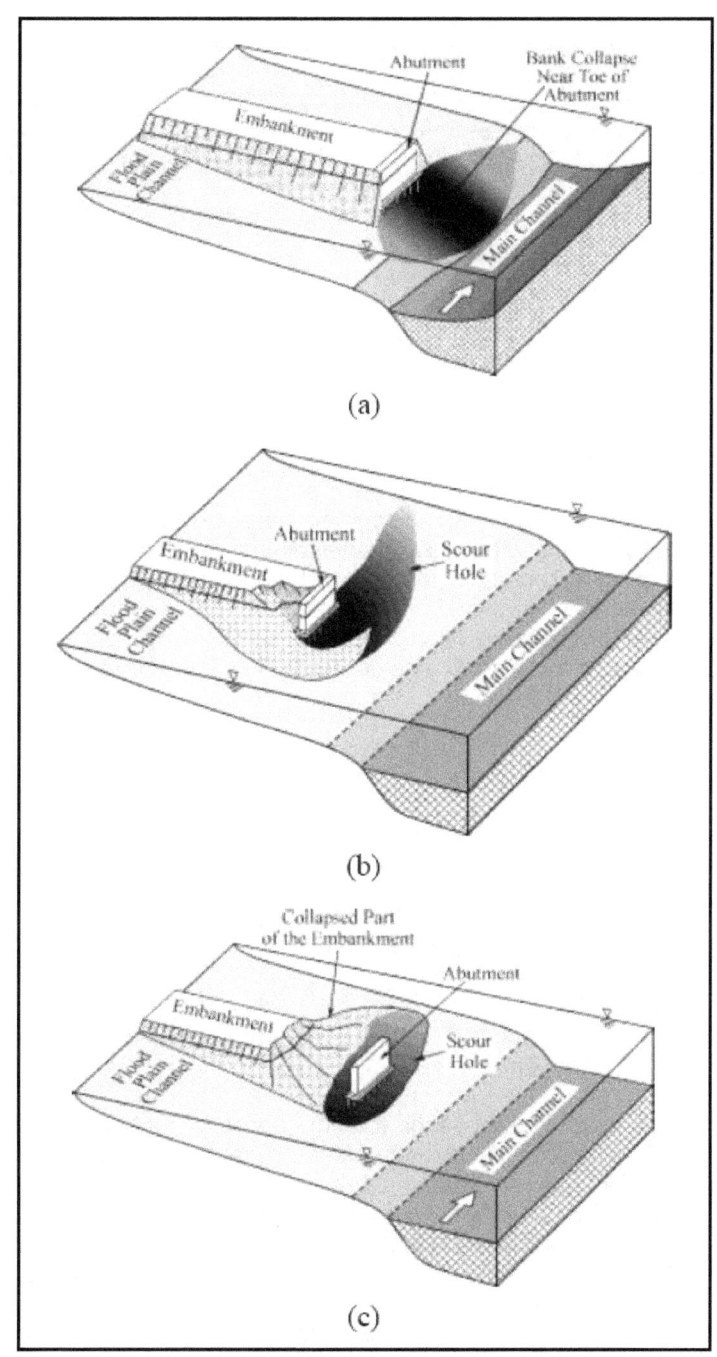

Figure 8.7. Abutment scour conditions (NCHRP 2010b).

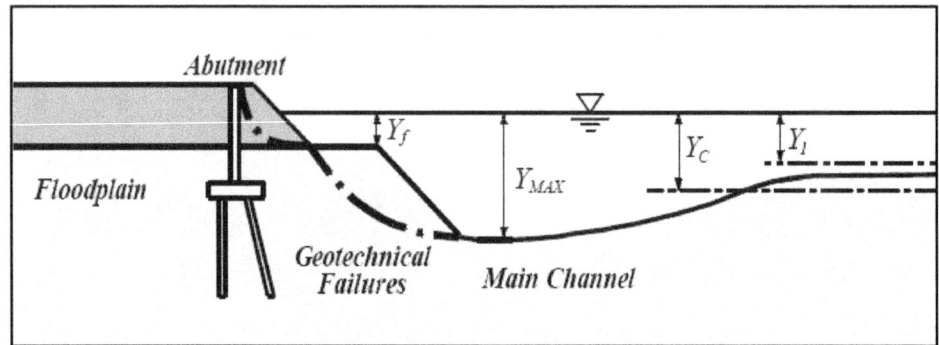

Figure 8.8. Conceptual geotechnical failures resulting from abutment scour (NCHRP 2010b).

Based on the NCHRP (2010b) study, if the projected length of the embankment, L, is 75 percent or greater than the width of the floodplain (B_f), scour condition (a) in Figure 8.7 occurs and the contraction scour calculation is performed using a live-bed scour calculation. The contraction scour equation is a simplified version of the live-bed contraction scour equation (see Chapter 6). The equation combines the discharge and width ratios due to the similarity of the exponents because other uncertainties are more significant. By combining the discharge and width, the live-bed contraction scour equation simplifies to the ratio of two unit discharges. Unit discharge (q) can be estimated either by discharge divided by width or by the product of velocity and depth. The contraction scour equation is:

$$y_c = y_1 \left(\frac{q_{2c}}{q_1} \right)^{6/7} \tag{8.5}$$

where:

y_c	=	Flow depth including live-bed contraction scour, ft (m)
y_1	=	Upstream flow depth, ft (m)
q_1	=	Upstream unit discharge, ft²/s (m²/s)
q_{2c}	=	Unit discharge in the constricted opening accounting for non-uniform flow distribution, ft²/s (m²/s)

The value of q_{2c} can be estimated as the total discharge in the bridge opening divided by the width of the bridge opening. The value of y_c is then used in Equation 8.3 to compute the total flow depth at the abutment. The value of α_A is selected from Figure 8.9 for spill through abutments and Figure 8.10 for wingwall abutments. The solid curves should be used for design. The dashed curves represent theoretical conditions that have yet to be proven experimentally. For low values of q_2/q_1, contraction scour is small, but the amplification factor is large because flow separation and turbulence dominate the abutment scour process. For large values of q_2/q_1, contraction scour dominates the abutment scour process and the amplification factor is small.

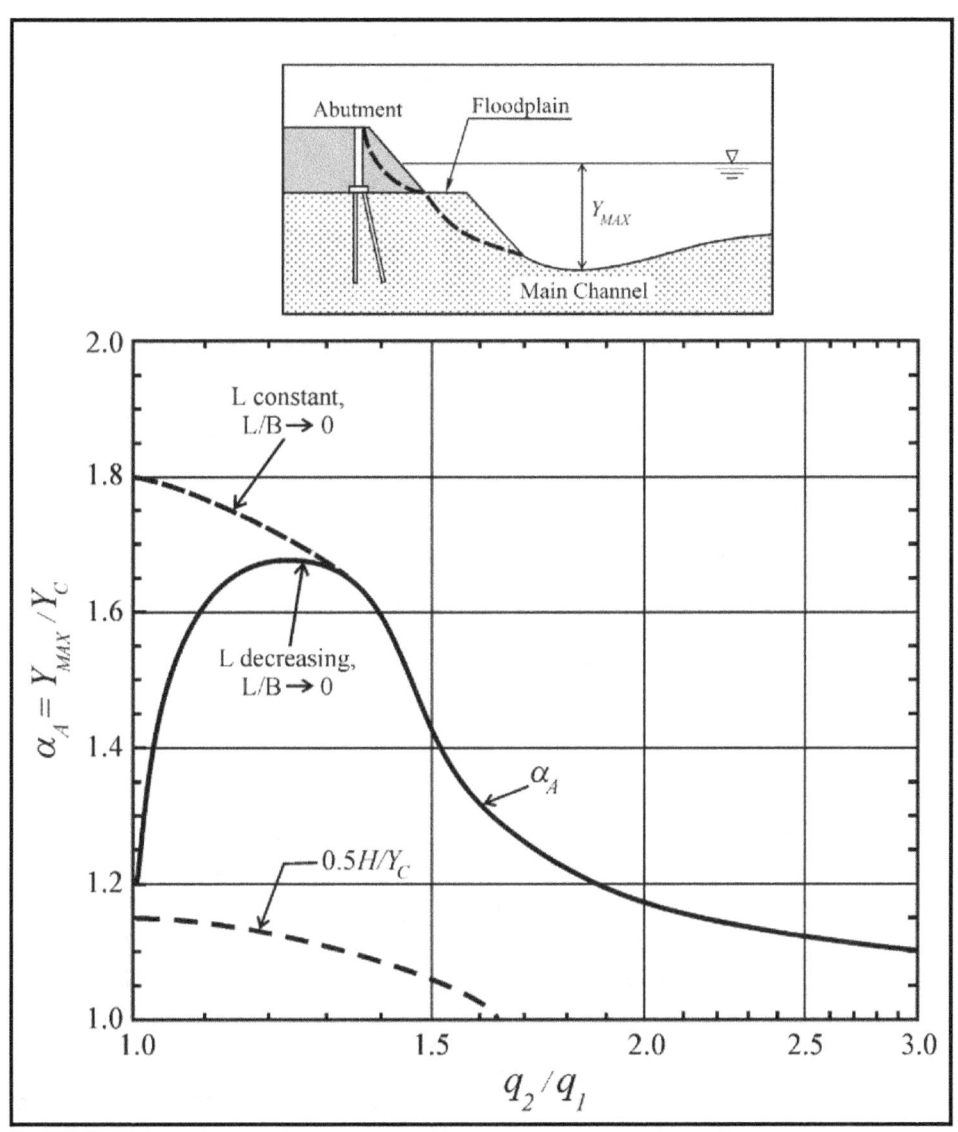

Figure 8.9. Scour amplification factor for spill-through abutments and live-bed conditions (NCHRP 2010b).

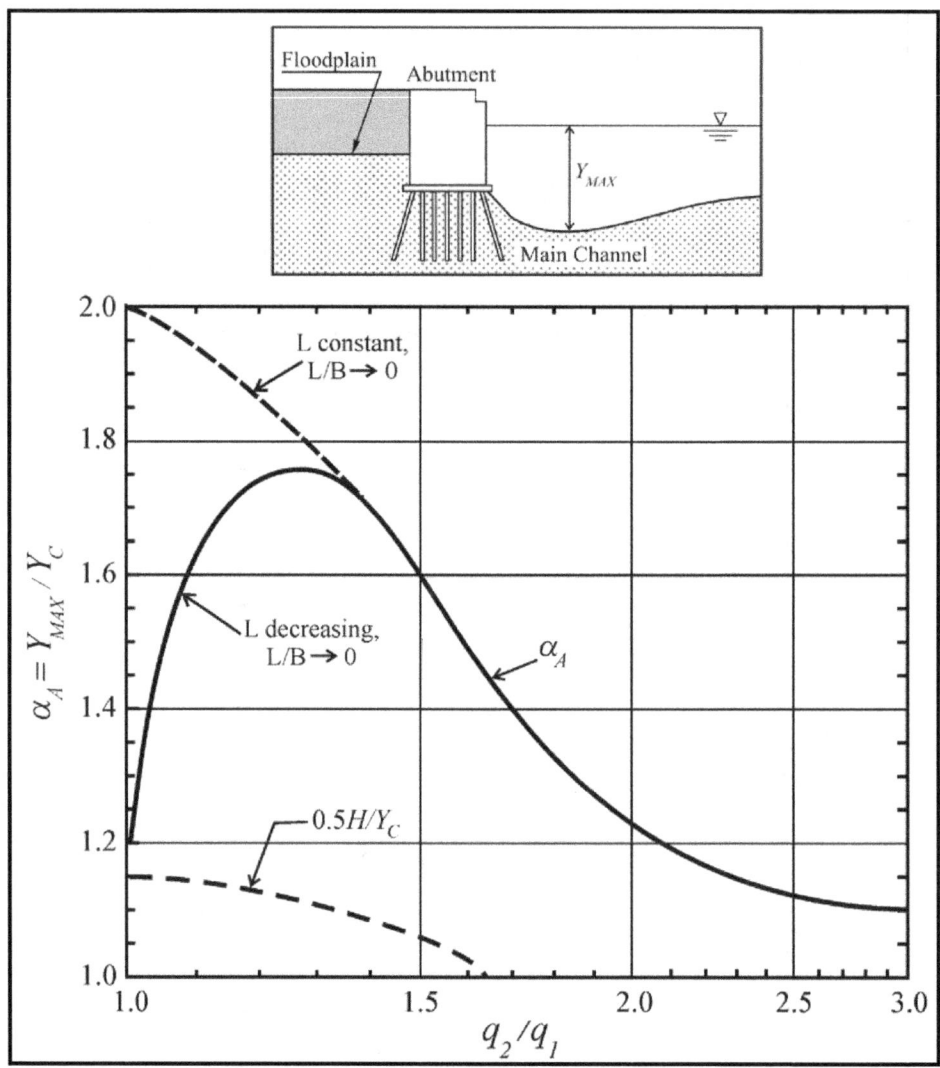

Figure 8.10. Scour amplification factor for wingwall abutments and live-bed conditions (NCHRP 2010b).

If the projected length of the embankment, L, is less than 75 percent of the width of the floodplain (B_f), scour condition (b) in Figure 8.7 occurs and the contraction scour calculation is performed using a clear-water scour calculation (see Chapter 6). The clear-water contraction scour equation also uses unit discharge (q), which can be estimated either by discharge divided by width or by the product of velocity and depth. Two clear-water contraction scour equations may be applied. The first equation is the standard equation based on grain size:

$$y_c = \left(\frac{q_{2f}}{K_u D_{50}^{1/3}} \right)^{6/7} \tag{8.6}$$

where:

y_c = Flow depth including clear-water contraction scour, ft (m)
q_{2f} = Unit discharge in the constricted opening accounting for non-uniform flow distribution, ft²/s (m²/s)
K_u = 11.17 English units
K_u = 6.19 SI
D_{50} = Particle size with 50 percent finer, ft (m)

Note that a lower limit of particle size of 0.2 mm is reasonable because cohesive properties limit the critical velocity and shear stress for cohesive soils. If the critical shear stress is known for a floodplain soil, then an alternative clear-water scour equation can be used:

$$y_c = \left(\frac{\gamma}{\tau_c}\right)^{3/7}\left(\frac{nq_{2f}}{K_u}\right)^{6/7} \tag{8.7}$$

where:

n = Manning n of the floodplain material under the bridge
τ_c = Critical shear stress for the floodplain material, lb/ft² (Pa)
γ = Unit weight of water, lb/ft³ (N/m³)
K_u = 1.486 English Units
K_u = 1.0 SI

The value of q_{2f} should be estimated including local concentration of flow at the bridge abutment. The value of q_f is the floodplain flow upstream of the bridge. The value of y_c is then used in Equation 8.3 to compute the total flow depth at the abutment. The value of α_B is selected from Figure 8.11 for spill through abutments and Figure 8.12 for wingwall abutments. The solid curves should be used for design. The dashed curves represent theoretical conditions that have yet to be proven experimentally. For low values of q_2/q_1, contraction scour is small, but the amplification factor is large because flow separation and turbulence dominate the abutment scour process. For large values of q_2/q_1, contraction scour dominates the abutment scour process and the amplification factor is small.

For scour estimates determined for either condition (a) or (b) the geotechnical stability of the channel bank or embankment should be considered. If the channel bank or embankment is likely to fail, then the limiting scour depth is the geotechnically stable depth and erosion will progress laterally. This may cause the embankment to breach and another scour estimate can be performed treating the abutment foundation as pier.

There are many uncertainties in determining the variables for these abutment scour equations. Determining the grain size or critical shear stress of the floodplain soils is one source of uncertainty. Determining the value of unit discharge near the abutment is another source of uncertainty. Two-dimensional models provide much better estimates of unit discharge throughout the bridge opening than one-dimensional models. This is illustrated in Figure 8.13. Unit discharge can be calculated at any point in the two-dimensional flow field by multiplying velocity and depth. Although two-dimensional modeling is strongly recommended for bridge hydraulic design, HEC-23 (FHWA, 2009) includes a method for estimating the velocity at an abutment. This method is used to size abutment riprap, but can also be used to determine the unit discharge at an abutment.

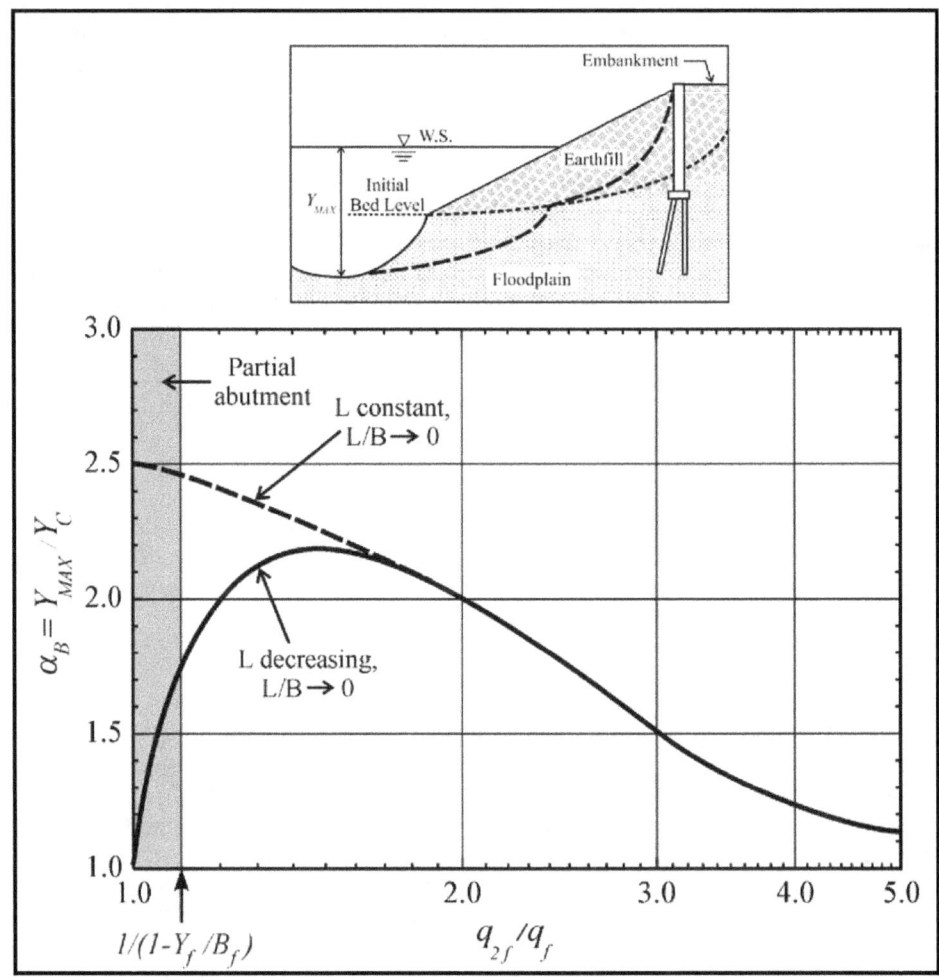

Figure 8.11. Scour amplification factor for spill-through abutments and clear-water conditions (NCHRP 2010b).

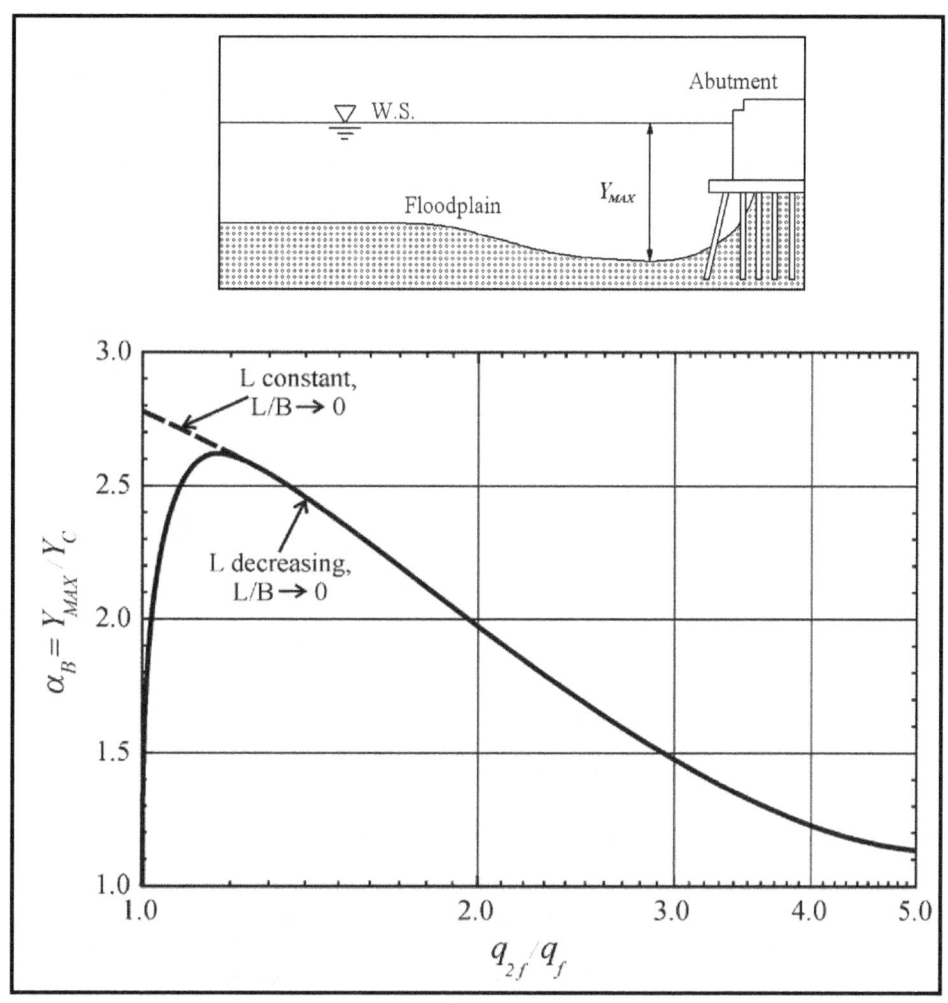

Figure 8.12. Scour amplification factor for wingwall abutments and clear-water conditions (NCHRP 2010b).

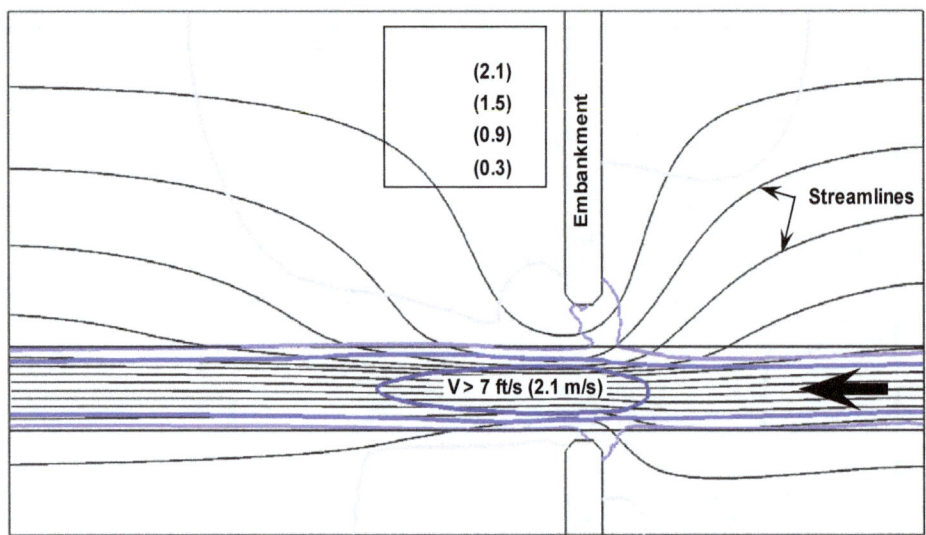

Figure 8.13. Velocity and streamlines at a bridge constriction (NCHRP 2010b).

The recommended procedure for selecting the velocity and unit discharge for abutment scour calculation is to use two-dimensional modeling. If one-dimensional modeling is used velocity and unit discharge are estimated as follows:

1. Determine the set-back ratio (SBR) of each abutment. SBR is the ratio of the set-back length to channel flow depth. The set-back length is the distance from the near edge of the main channel to the toe of abutment.

    SBR = Set-back length/average channel flow depth

    a. If SBR is less than 5 for both abutments (Figure 8.14), compute a velocity, Q/A, based on the entire contracted area through the bridge opening. This includes the total upstream flow, exclusive of that which overtops the roadway. Unit discharge in the channel is the computed velocity times channel flow depth. Unit discharge at the abutment is the computed velocity times the floodplain flow depth.

    b. If SBR is greater than 5 for an abutment (Figure 8.15), compute a velocity, Q/A, for the respective overbank flow only. Assume that the entire upstream overbank flow stays in the overbank section through the bridge opening. Unit discharge at the abutment is the computed velocity times the floodplain flow depth.

    c. If SBR for an abutment is less than 5 and SBR for the other abutment at the same site is more than 5 (Figure 8.16), a velocity determined from Step 1a for the abutment with SBR less than 5 may be unrealistically low. This would, of course, depend upon the opposite overbank discharge as well as how far the other abutment is set back. For this case, the velocity for the abutment with SBR less than 5 should be based on the flow area bounded by the abutment and the opposite channel bank. The appropriate discharge is the upstream channel flow and the upstream floodplain flow associated with that abutment. Unit discharge in the channel is the computed velocity times channel flow depth and unit discharge at the abutment is the computed velocity times the floodplain flow depth.

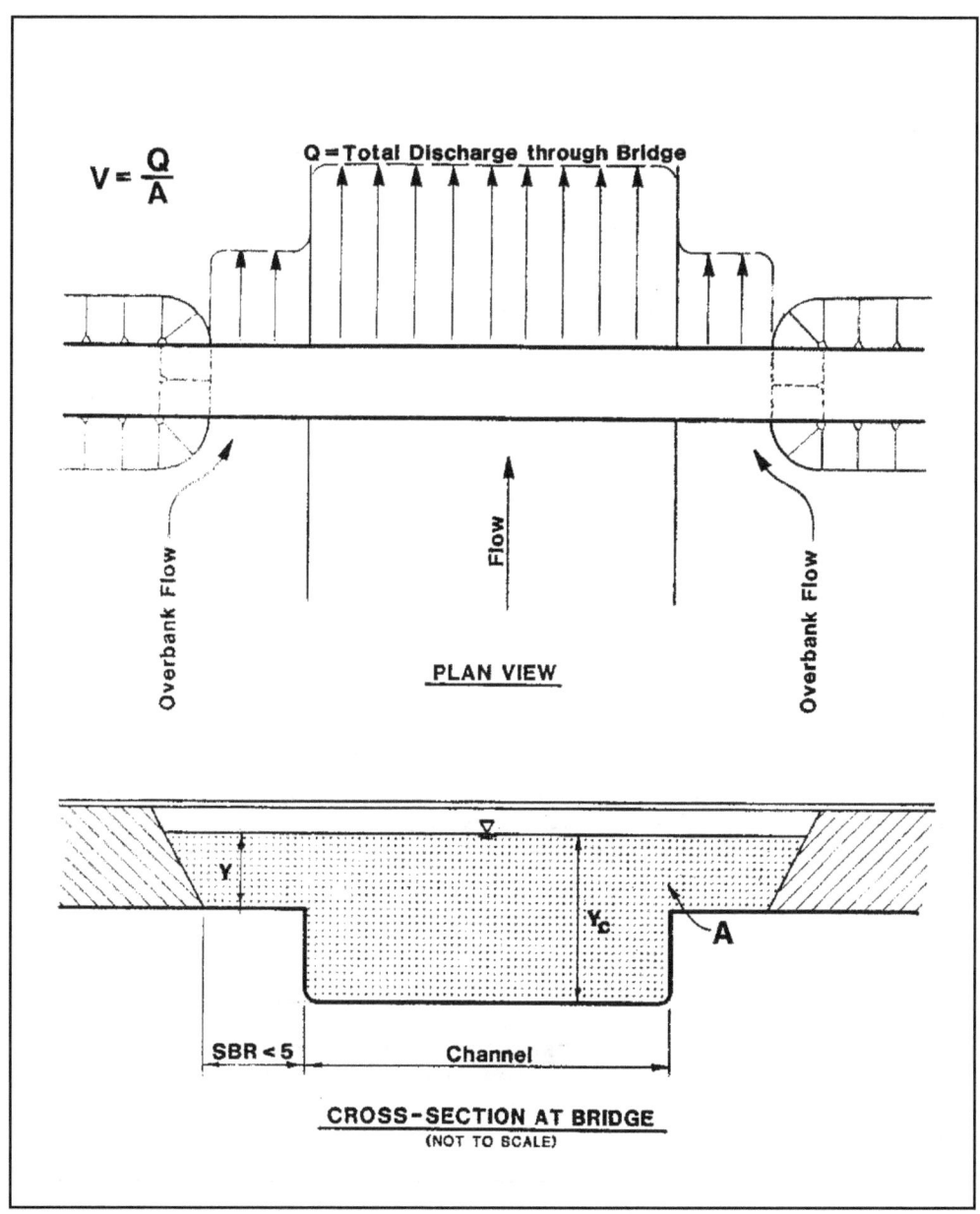

Figure 8.14. Velocity for SBR<5.

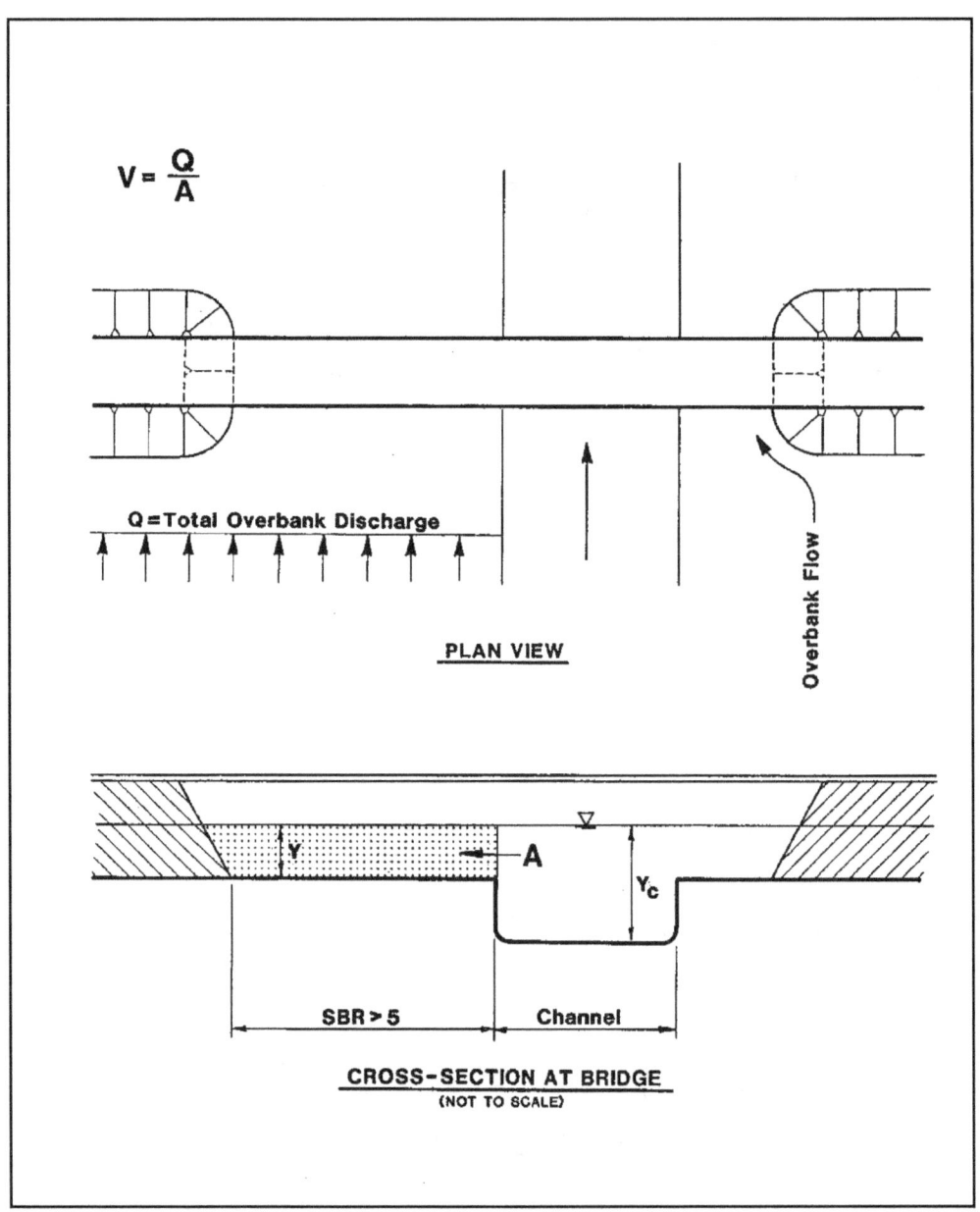

Figure 8.15. Velocity for SBR>5.

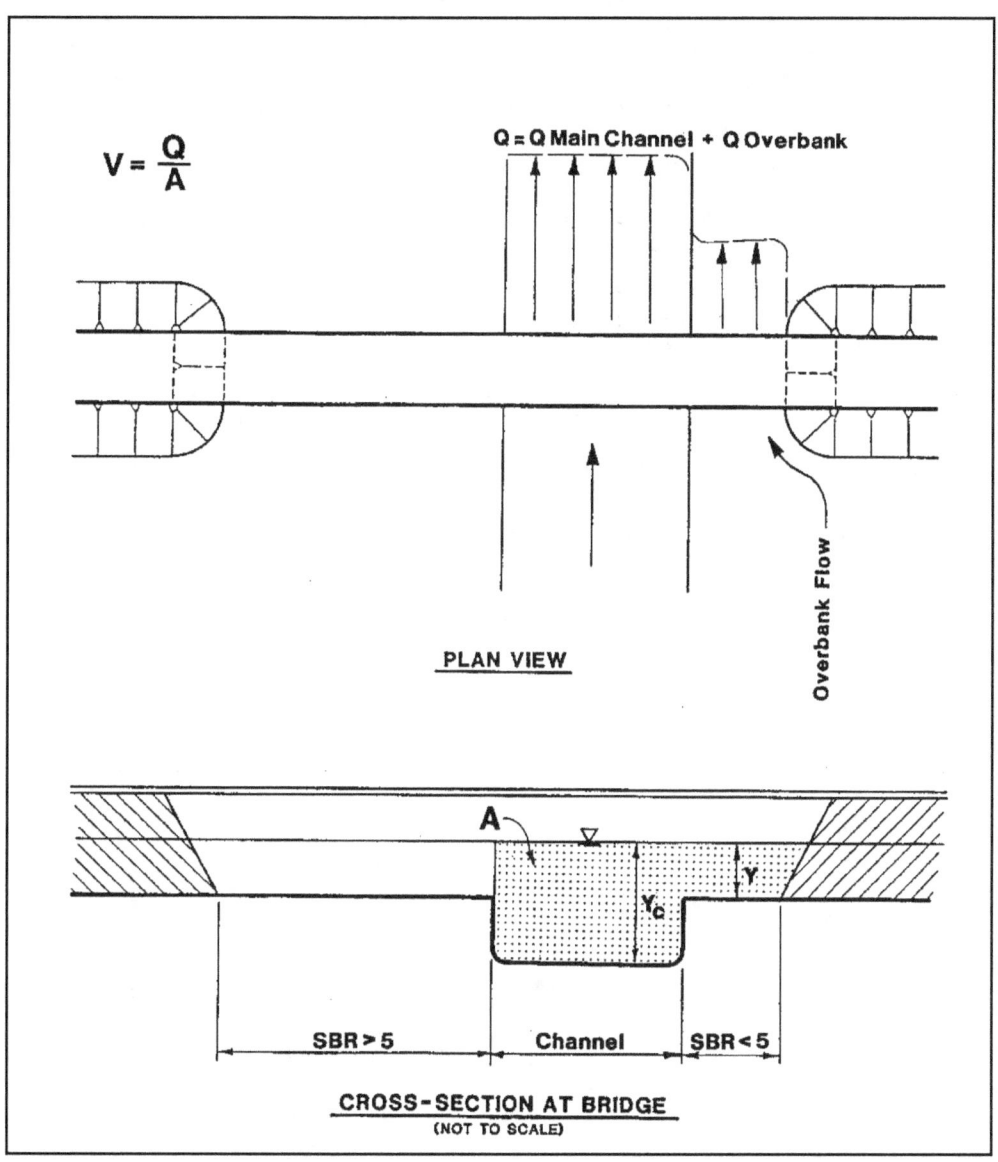

Figure 8.16. Velocity for SBR>5 and SBR<5.

2. Recent research results published by the Transportation Research Board as NCHRP Report 587, "Countermeasures to Protect Bridge Abutments from Scour," endorse the use of the SBR approach for sizing riprap at spill-through abutments (NCHRP 2007). NCHRP Report 568, "Riprap Design Criteria, Recommended Specifications, and Quality Control," recommends an additional criterion for selecting the velocity when applying the SBR method (NCHRP 2006). Based on the results of 2-dimensional computer modeling of a typical abutment configuration NCHRP Report 568 concludes:

   a. Whenever the SBR is less than 5, the average velocity in the bridge opening provides a good estimate for the velocity at the abutment.
   b. When the SBR is greater than 5, the recommended adjustment is to compare the velocity from the SBR method to the maximum velocity in the channel within the bridge opening and select the lower velocity.
   c. The SBR method is well suited for estimating velocity at an abutment if the estimated velocity does not exceed the maximum velocity in the channel.

8.7 ABUTMENT SCOUR EXAMPLE PROBLEMS

8.7.1 Example Problem 1 - Froehlich Equation

Determine abutment scour depth for the following conditions to aid in scour evaluation and design of countermeasures. The right approach embankment and abutment project 80 ft (24 m) onto the floodplain at an angle of 70° measured from the downstream floodplain edge. The left approach embankment and abutment project onto the floodplain 790 ft (240 m). The bridge abutment structures are vertical wall with wingwalls.

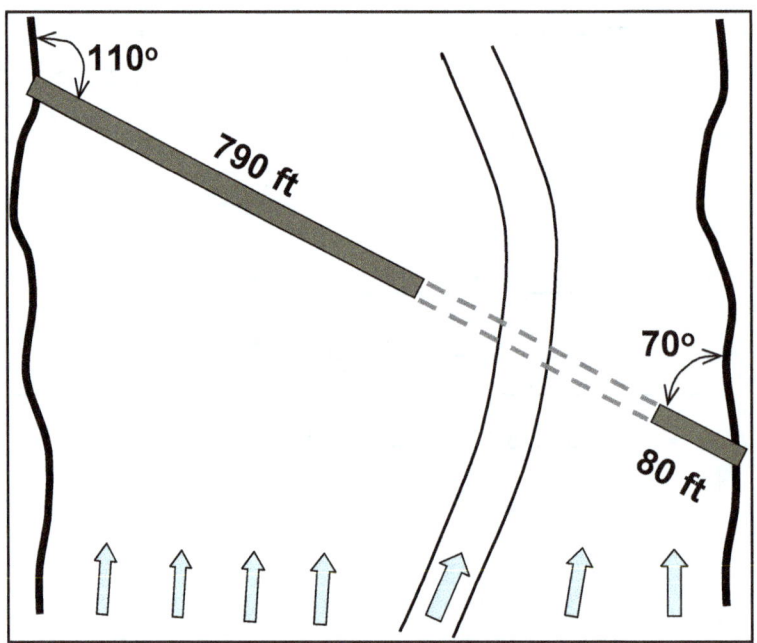

Plan view of abutment scour (Problems 1 and 2).

Given Data for Right Abutment:

Obstructed flow in right floodplain = 960 ft³/s (26.7 m³/s)
Average depth of floodplain flow upstream of embankment = 3.5 ft (1.6 m)

Determine:

The magnitude of abutment scour at the right abutment.

$L = 80 \cos(90° - 70°) = 75$ ft (22.8 m)

$y_a = 3.5$ ft

$\dfrac{L}{y_a} = \dfrac{75}{3.5} = 21.4 < 25$ therefore, use Froehlich Equation

$$\dfrac{y_s}{y_a} = 2.27 \, K_1 \, K_2 \left(\dfrac{L'}{y_a}\right)^{0.43} Fr^{0.61} + 1$$

$K_1 = 0.82$ for vertical wall with wingwalls

$$K_2 = \left(\dfrac{\theta}{90}\right)^{.13} = \left(\dfrac{70}{90}\right)^{0.13} = 0.97$$

$A_e = y_a \times L = 3.5 \times 75 = 262.5$ ft²

$V_e = \dfrac{Q_e}{A_e} = \dfrac{960}{262.5} = 3.65$ ft/s

$F_r = \dfrac{V_e}{\sqrt{gy_a}} = \dfrac{3.65}{\sqrt{(32.2)(3.5)}} = 0.34$

Calculate L', the length of active flow obstructed by the embankment:

From HEC-RAS stream tube upstream from abutment tip,

$V_{tube} = 4.6$ ft/s (1.4 m/s)

$y_{tube} = 5.0$ ft (1.5 m)

$q_{tube} = 4.6 \times 5.0 = 23.0$ ft³/s/ft (2.1 m³/s/m)

$L' = \dfrac{Q_e}{q_{tube}} = \dfrac{960}{23} = 42$ ft (12.7 m)

$$\frac{y_s}{y_a} = 2.27 \, K_1 \, K_2 \left(\frac{L'}{y_a}\right)^{0.43} Fr^{0.61} + 1$$

$$\frac{y_s}{3.5} = 2.27 \, (0.82)(0.97)\left(\frac{42}{3.5}\right)^{0.43} (0.34)^{0.61} + 1$$

$y_s = 3.5 \, (2.72 + 1) = 13.0$ ft (3.9 m)

8.7.2 Example Problem 2 - HIRE Equation

Given Data for Left Abutment:

See figure for Example Problems 1 and 2

Flow depth at abutment tip is 6.2 feet (1.9 m)
Velocity at tip of abutment $V_1 = 9.9$ ft/s (3.0 m/s) from HEC-RAS stream tube

Determine:

The magnitude of abutment scour at the left abutment.

$L = 790 \, Cos(110° - 90°) = 742$ ft

$y_1 = 6.2$ ft

$\dfrac{L}{y_1} = \dfrac{742}{6.2} = 120 > 25$ therefore, use HIRE equation

$$\frac{y_s}{y_1} = 4 F_r^{0.33} \frac{K_1}{0.55} K_2$$

$$F_r = \frac{V_1}{\sqrt{gy_1}} = \frac{9.9}{\sqrt{(32.2 \times 6.2)}} = 0.70$$

$K_1 = 0.82$

$$K_2 = \left(\frac{110}{90}\right)^{.13} = 1.03$$

$$\frac{y_s}{y_1} = 4(0.70)^{0.33} \frac{0.82}{0.55}(1.03) = 5.46$$

$y_s = 5.46 \times 6.2 = 33.9$ ft

8.7.3 Example Problem 3 - NCHRP Live-Bed Scour

Given:

The wingwall abutments are set near the channel such that $L/B_f = 0.85$.

Upstream channel unit discharge = 57.0 ft²/s (5.3 m²/s).
The bridge unit discharge = 78.6 ft²/s (7.3 m²/s).
Upstream flow depth (y_1) and bridge channel flow depth before scour (y_0) equal 10.0 ft (3 m)

Determine:

Compute abutment scour for NCHRP condition (a):

$q_{2c}/q_1 = 78.6/57.0 = 1.4$

$$y_c = y_1 \left(\frac{q_{2c}}{q_1}\right)^{6/7} = 10.0(1.4)^{6/7} = 13.3 \text{ ft } (4\text{m})$$

From Figure 8.10, the value of α_A is 1.7.

$y_{max} = \alpha_A y_c = 1.7 \times 13.3 = 22.6 \text{ ft} (6.9\text{m})$

$y_s = y_{max} - y_0 = 22.6 - 10.0 = 12.6 \text{ ft} (3.8\text{m})$

8.7.4 Example Problem 4 - NCHRP Clear-Water Scour (Particle Size)

Given:

The spill through abutment is set back from the channel such that $L/B_f = 0.6$.
Upstream floodplain unit discharge = 5.7 ft²/s (0.53 m²/s).
The abutment unit discharge = 10.1 ft²/s (0.94 m²/s).
Upstream floodplain flow depth (y_1) and abutment flow depth before scour (y_0) equal 3.5 ft (1.1 m) D_{50} = 0.3 mm (0.001 ft)

Determine:

Compute abutment scour for NCHRP condition (b):

$q_{2f}/q_f = 10.1/5.7 = 1.8$

$$y_c = \left(\frac{q_{2f}}{K_u D_{50}^{1/3}}\right)^{6/7} = \left(\frac{10.1}{11.17 \times .001^{1/3}}\right)^{6/7} = 6.6 \text{ ft } (2.0\text{m})$$

From Figure 8.11, the value of α_B is 2.1.

$y_{max} = \alpha_B y_c = 2.1 \times 6.6 = 13.9 \text{ ft} (4.2\text{m})$

$y_s = y_{max} - y_0 = 13.9 - 3.5 = 10.4 \text{ ft} (3.2\text{m})$

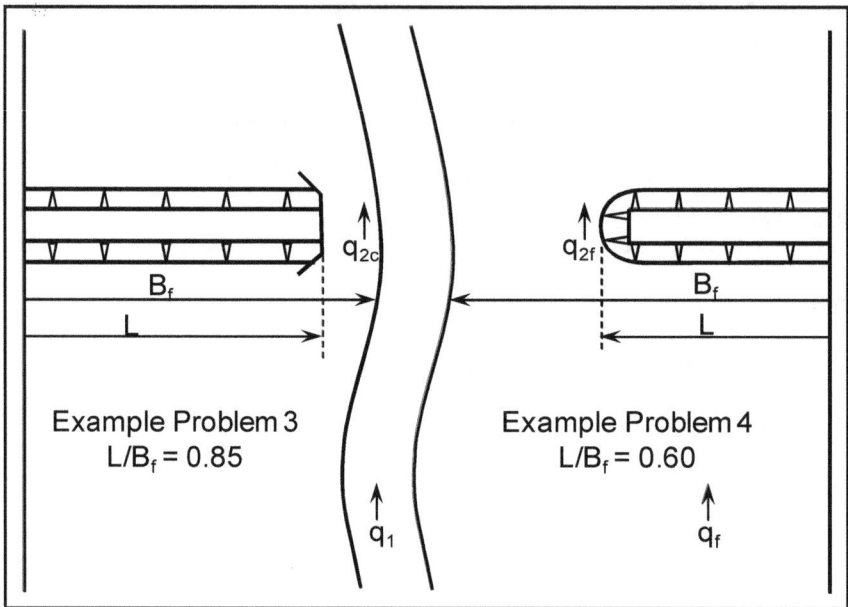

Plan view of abutment scour (Problems 3 and 4)

8.7.5 Example Problem 5 - NCHRP Clear-Water Scour (Shear Stress)

Given:

Same as Example Problem 4 except the critical shear stress is 0.04 lb/ft² (1.9 Pa)

Assume a Manning n roughness coefficient of 0.025.

Determine:

Compute abutment scour for NCHRP condition (b):

$q_{2f}/q_f = 10.1/5.7 = 1.8$

$$y_c = \left(\frac{\gamma}{\tau_c}\right)^{3/7} \left(\frac{nq_{2f}}{K_u}\right)^{6/7} = \left(\frac{62.4}{0.04}\right)^{3/7} \left(\frac{.025 \times 10.1}{1.486}\right)^{6/7} = 5.1 \text{ ft} (1.6 \text{ m})$$

From Figure 8.11, the value of α_B is 2.1.

$y_{max} = \alpha_B y_c = 2.1 \times 5.1 = 10.7$ ft (3.3 m)

$y_s = y_{max} - y_0 = 10.7 - 3.5 = 7.2$ ft (2.2 m)

8.7.6 Comprehensive Example

Additional abutment scour problems are included in the Comprehensive Example in Appendix D.

CHAPTER 9

SCOUR ANALYSIS FOR TIDAL WATERWAYS

9.1 INTRODUCTION

In the coastal region, scour at bridges over tidal waterways that are subjected to the effects of astronomical tides and storm surges is a combination of long-term degradation, contraction scour, local scour, and waterway instability. These are the same scour mechanisms that affect non-tidal (riverine) streams. Although many of the flow conditions are different in tidal waterways, the equations used to determine riverine scour are applicable if the hydraulic conditions (depth, discharge, velocity, etc.) are carefully evaluated.

This chapter presents an overview of methods for determining stream stability and scour at tidal inlets, tidal estuaries, bridge crossings to islands and streams affected by tides (tidal waterways). Other resources available to provide in-depth information, equations, and procedures to perform tidal scour analyses include the First and Second editions of HEC-25, "Tidal Hydrology, Hydraulics, and Scour at Bridges" (FHWA 2004), and "Highways in the Coastal Environment" (FHWA 2008) (see Sections 9.7 and 9.8).

Analysis of tidal waterways is very complex. The hydraulic analysis must consider the magnitude of the 100- and 500-year storm surge (storm tide - see Section 9.2 Glossary), the characteristics (geometry) of the tidal inlet, estuary, bay or tidal stream and the effect of any constriction of the flow due to the bridge. In addition, the analysis must consider the long-term effects of the normal tidal cycles on long-term aggradation or degradation, contraction scour, local scour, and stream instability. Coastal analyses require a synthesis of complex meteorological, bathymetric, geographical, statistical, and hydraulic disciplines and knowledge. The methods discussed in this chapter provide an overview of application of these elements in the context of tidal scour analyses.

A storm tide or storm surge in coastal waters results from astronomical tides, wind action, and rapid barometric pressure changes. In addition, the change in elevation resulting from the storm surge may be increased by resonance in harbors and inlets, whereby, the tidal range in an estuary, bay, or inlet is larger than on the adjacent coast.

The astronomical tidal cycle with reversal in flow direction can increase long-term degradation, contraction scour, and local scour. If sediment is being moved on the flood and ebb tide, there may be no net loss of sediment in a bridge reach because sediments are being moved back and forth. Consequently, no net long-term degradation may occur. However, local scour at piers and abutments can occur at both the inland and ocean side of the piers and abutments and will alternate with the reversal in flow direction. If, however, there is a loss of sediment in one or both flow directions, there will then be long-term degradation in addition to local scour. Also, the tidal cycles may increase bank erosion, migration of the channel, and thus, increase stream instability.

The complexity of the hydraulic analysis increases if the tidal inlet or the bridge constrict the flow and affect the amplitude of the storm surge (storm tide) in the bay or estuary so that there is a large change in elevation between the ocean and the estuary or bay. A constriction in the tidal inlet can increase the velocities in the constricted waterway opening, decrease interior wave heights and tidal range, and increase the phase difference (time lag) between exterior and interior water levels. Analysis of a constricted inlet or waterway may require the use of an orifice equation rather than tidal relationships (see FHWA 2004).

For the analysis of bridge crossings of tidal waterways, a three-level analysis approach similar to the approach outlined in HEC-20 is suggested (FHWA 2012b). Level 1 includes a qualitative evaluation of the stability of the inlet or estuary, estimating the magnitude of the tides, storm surges, and flow in the tidal waterway, and attempting to determine whether the hydraulic analysis depends on tidal or river conditions, or both. Level 2 represents the engineering analysis necessary to obtain the velocity, depths, and discharge for tidal waterways to be used in determining long-term aggradation, degradation, contraction scour, and local scour. The hydraulic variables obtained from the Level 2 analysis are used in the riverine equations presented in previous chapters to obtain total scour. Using these riverine scour equations, which are for steady-state equilibrium conditions for unsteady, dynamic tidal flow may result in estimating deeper scour depths than will actually occur, but this represents a conservative estimate for this level of analysis.

For complex tidal situations, Level 3 analysis using physical and 2-dimensional computer models may be required. This section will be limited to a discussion of Levels 1 and 2 analyses; however, HEC-25 First Edition (FHWA 2004) provides additional guidance on more advance tidal hydraulic modeling approaches. In Level 2 analyses, unsteady 1-dimensional or quasi 2-dimensional computer models may be used to obtain the hydraulic variables needed for the scour equations. The Level 1, 2, and 3 approaches are described in more detail in later sections.

9.2 OVERVIEW OF TIDAL PROCESS

9.2.1 Glossary

Bay. A body of water connected to the ocean with an inlet.

Celerity. See wave speed.

Diurnal tide. Tides with an approximate tidal period of 25 hours.

Ebb or ebb tide. Flow of water from the bay or estuary to the ocean.

Estuary. Tidal reach at the mouth of a river.

Fetch. The area over water where the wind is unobstructed with fairly uniform speed and direction.

Flood or flood tide. Flow of water from the ocean to the bay or estuary.

Hurricane. An intense type of tropical cyclone with well defined circulation and maximum sustained winds of 74 mph (120 kph) or higher.

Littoral transport or drift. Transport of beach material along a shoreline by wave action. Also, longshore sediment transport.

Littoral zone. The region that extends seaward from the coastline to just beyond the beginning of the breaking waves. Within this zone, waves and currents transport sediments. A current is generated by the incident waves within the littoral zone

Mean high water (MHW). The average of all high tides over a tidal epoch.

Passage. A tidal waterway between two islands or between the mainland and an island.

Run-up, wave. Height to which water rises above still-water elevation when waves meet a beach, wall, etc.

Semi-diurnal tide. Tides with an approximate tidal period of 12.5 hours.

Set-up, wave. Height to which water rises above still-water elevation as a result of storm wind effects.

Spring tide. Larger than normal tides that occur approximately twice per month at new and full moon when the sun and moon are aligned and the tidal forces are reinforced.

Still-water elevation. Flood height to which water rises as a result of barometric pressure changes occurring during a storm event.

Storm surge. Coastal flooding phenomenon resulting from wind and barometric changes. The storm surge is measured by subtracting the astronomical tide elevation from the total flood elevation (Hurricane surge).

Storm tide. Coastal flooding resulting from combination of storm surge and astronomical tide (often referred to as storm surge)

Tidal amplitude. Generally, half of tidal range.

Tidal cycle. One complete rise and fall of the tide.

Tidal day. Time of rotation of the earth with respect to the moon. Assumed to equal approximately 24.84 solar hours in length.

Tidal inlet. A channel connecting a bay or estuary to the ocean.

Tidal passage. A tidal channel connected with the ocean at both ends.

Tidal period. Duration of one complete tidal cycle. When the tidal period equals the tidal day (24.84 hours), the tide exhibits diurnal behavior. Should two complete tidal periods occur during the tidal day, the tide exhibits semi-diurnal behavior.

Tidal prism. Volume of water contained in a tidal bay, inlet or estuary between low and high tide levels.

Tidal range. Vertical distance between specified low and high tide levels.

Tidal waterways. A generic term which includes tidal inlets, estuaries, bridge crossings to islands or between islands, inlets to bays, crossings between bays, tidally affected streams, etc.

Tides, astronomical. Rhythmic diurnal or semi-diurnal variations in sea level that result from gravitational attraction of the moon and sun and other astronomical bodies acting on the rotating Earth.

Tsunami. Long-period ocean wave resulting from earthquake, other seismic disturbances or submarine land slides.

Waterway opening. Width or area of bridge opening at a specific elevation, measured normal to principal direction of flow.

Wave height. The vertical difference between successive wave crests and troughs.

Wave length. The horizontal difference between two successive wave crests or two successive wave troughs.

Wave period. Time interval between arrivals of successive wave crests at a point.

Wave speed or celerity. The travel speed of a wave equal to the wave length divided by the wave period.

Wave period. Time interval between arrivals of successive wave crests at a point.

9.2.2 Definition of Tidal and Coastal Processes

Typical bridge crossings of tidal waterways are sketched in Figure 9.1. From this figure, tidal flows can be defined as being between the ocean and a bay (or lagoon), from the ocean into an estuary, or through passages between islands.

Flow into (flood tide) and out of (ebb tide) a bay or estuary is driven by tides and by the discharge into the bay or estuary from upland areas. Assuming that the flow from upland areas is negligible, the ebb and flood in the bay or estuary will be driven solely by tidal fluctuations and storm surges. With no inflow of water from rivers and streams, the net flow of water into and out of the bay or estuary will be nearly zero. Increasing the discharge from rivers and streams will lead to a net outflow of water to the ocean.

Figure 9.2 illustrates the elevation and time variable nature of astronomical tides. For astronomical tides, maximum flood and ebb (or the time of maximum current and discharge) can be assumed to occur at the inflection point of (or halfway between) high tide and low tide, but actually can occur before or after the midtide level depending on the location. The addition of a storm surge to a high astronomical tide can lead to additional water surface elevations (High water, large tide plus storm surge in Figure 9.2), additional current, and associated flooding.

In the most conservative scenario, the greatest potential flood elevation would occur at the time where the high astronomical tide and maximum storm surge height coincide in time. In this circumstance, the maximum discharge would occur when the astronomical tidal period and the period associated with the storm surge event are the same value. The presence of any inland flood discharge would influence this discharge, particularly during the period when the flood levels recede (ebb).

Hydraulically, the above discussion presents two limiting cases for evaluation of the flow velocities in the bridge reach. With negligible flow from the upland areas, the flow through the bridge opening is based solely on the ebb and flood resulting from tidal fluctuations or storm surges. Alternatively, when the flow from the streams and rivers draining into the bay or estuary (inland flood) is large in relationship to the tidal flows (ebb and flood tide), the effects of tidal fluctuations are negligible. For this latter case, the evaluation of the hydraulic characteristics and scour can be accomplished using the methods described in previous chapters for inland rivers.

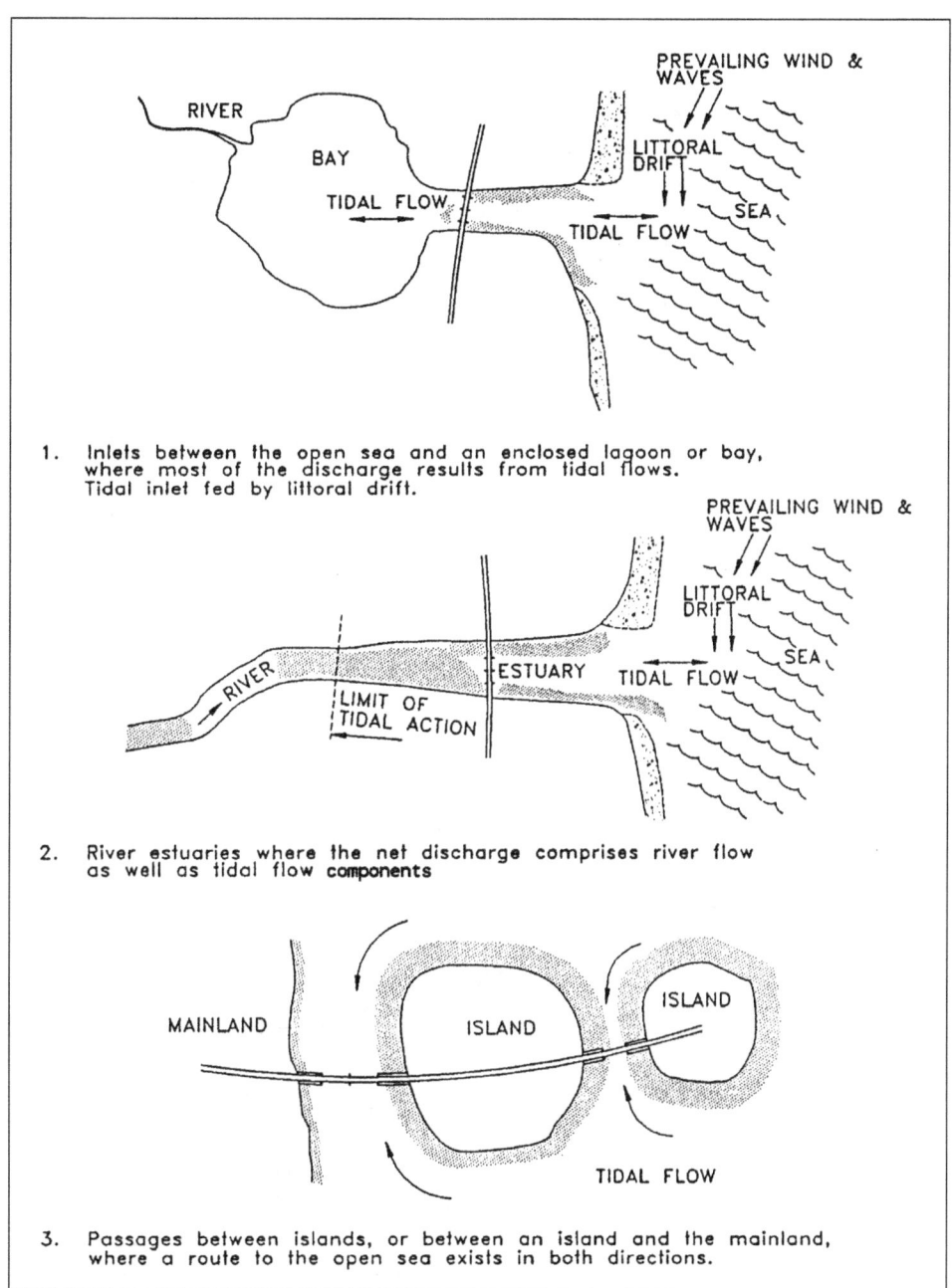

Figure 9.1. Types of tidal waterway crossings (after Neill 2004).

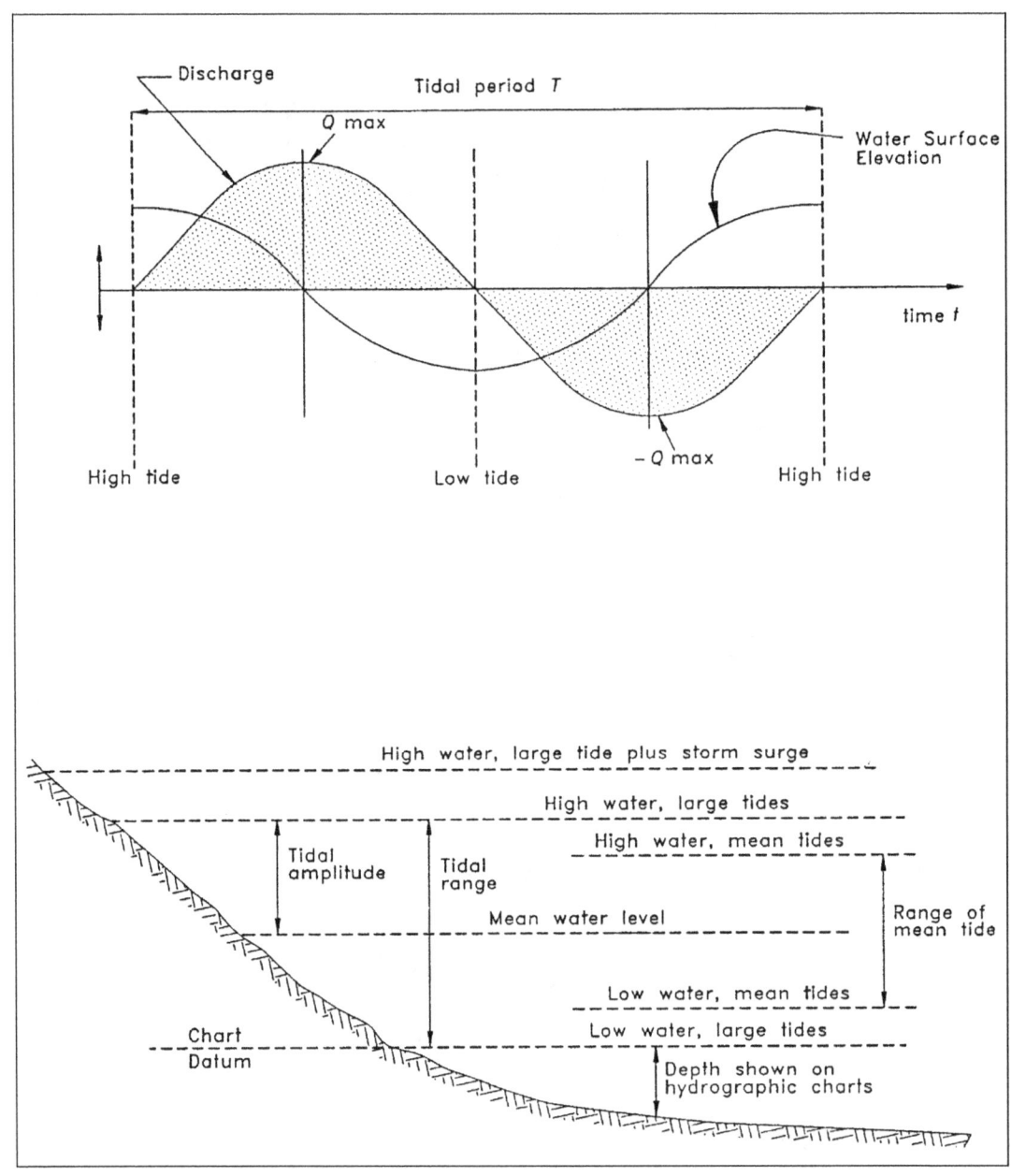

Figure 9.2. Principal tidal terms (after Neill 2004).

9.2.3 Aggradation, Degradation, and Scour in Tidal Waterways

Bridge scour in the coastal region results from the unsteady diurnal and semi-diurnal flows resulting from astronomical tides, large flows that can result from storm surges (hurricanes, nor'easters), and the combination of riverine and tidal flows. The forces which drive tidal fluctuations are, primarily, the result of the gravitational attraction of the sun and moon on the rotating earth (astronomical tides), wind and storm setup, and geologic disturbances (tsunamis). These different forces which drive tides produce varying tidal periods and amplitudes. In general semi-diurnal astronomical tides having tidal periods of approximately 12.5 hours occur in the lower latitudes while diurnal tides having tidal periods of approximately 25 hours occur in the higher latitudes. Typically, the storm surge period correlates with the associated storm type. Hurricane surges generally last from 12 to 15 hours. Nor'easters may produce a storm surge lasting several days. In general, storm surge periods may be assumed to be longer than astronomical tidal periods.

The continuous rise and fall of astronomical tides will usually influence long-term trends of aggradation or degradation, contraction and local scour. Worst-case hydraulic conditions for contraction and local scour are usually the result of infrequent tidal events such as storm surges and tsunamis. Storm surges and tsunamis are a single event phenomenon which, due to their magnitude, can present a significant threat to a bridge crossing in terms of scour. The hydraulic variables (discharge, velocity, and depths) and bridge scour in the coastal region can be determined with as much precision as riverine flows. These determinations are conservative and research is needed for both cases to improve scour determinations. Determining the magnitude of the combined flows can be accomplished by simply adding riverine flood flow to the maximum tidal flow, if the drainage basin is small, or routing the design riverine flows to the crossing and adding them to the storm surge flows.

The small size of the bed material (normally fine sand) as well as silts and clays with cohesion and littoral drift (transport of beach sand along the coast resulting from wave action) affect the magnitude of bridge scour. Mass density stratification of the water typically has a minor influence on bridge scour. Peak flows from storm surges may not have durations long enough to reach the ultimate scour depths determined from existing scour equations. Sediment transport equations can be used to compute the rate of contraction scour (see Section 9.5), but the time dependent characteristics of local scour require further research. Diurnal and semi-diurnal astronomical tides can cause long-term degradation if there is no source of sediment except at the crossing. At some locations, this has resulted in long-term degradation of 1.0 to 3.3 ft (0.3 to 1.0 m) per year with no indication of stopping (Butler and Lillycrop 1993, TRB 1993b). Existing scour equations can predict the magnitude of this scour, but not the time history (TRB 1995).

Mass density stratification (saltwater wedges), which can result when the denser more saline ocean water enters an estuary or tidal inlet with significant freshwater inflow, can result in larger velocities near the bottom than the average velocity in the vertical velocity profile. With careful evaluation, the correct velocity can be determined for use in the scour equations. With storm surges, mass density stratification will not normally occur. The density difference between salt and freshwater, except as it causes saltwater wedges, is not significant enough to affect scour equations. Density and viscosity differences between fresh and sediment-laden water can be much larger in riverine flows than the density and viscosity differences between salt and freshwater.

Salinity can affect the transport of silts and clays by causing them to flocculate and possibly deposit, which may affect stream stability and must be evaluated. Salinity may affect the erodibility of cohesive sediments, but this will only affect the rate of scour, not ultimate scour. Littoral drift is a source of sediment to a tidal waterway (TRB 1993a, Bruun 1966). An aggrading or stable waterway may exist if the supply of sediment to the bridge from littoral drift is large. This will have the effect of minimizing contraction scour, and possibly local scour. Conversely, long-term degradation, contraction scour and local scour can be exacerbated if the sediment from littoral drift is reduced or cut off. Evaluating the effect of littoral drift is a sediment transport problem involving historical information, future plans (dredging, jetties, etc.) for the waterway and/or the coast, sources of sediment, and other factors.

Evaluation of total scour at bridges crossing tidal waterways requires the assessment of long-term aggradation or degradation, local scour and contraction scour. Long-term aggradation or degradation estimates can be derived from a geomorphic evaluation coupled with computations of live-bed contraction scour if sediment transport is changed.

Although the hydraulics of flow for tidal waterways is complicated by the presence of two directional flow, the basic concept of sediment continuity is valid. Consequently, a clear understanding of the principle of sediment continuity is essential for evaluating scour at bridges spanning waterways influenced by tidal fluctuations. Technically, the sediment continuity concept states that the sediment inflow minus the sediment outflow equals the time rate of change of sediment volume in a given reach. More simply stated, during a given time period the amount of sediment coming into the reach minus the amount leaving the downstream end of the reach equals the change in the amount of sediment stored in that reach.

As with riverine scour, tidal scour can be characterized by either live-bed or clear-water conditions. In the case of live-bed conditions, sediment transported into the bridge reach will tend to reduce the magnitude of scour. Whereas, if no sediment is in transport to re-supply the bridge reach (clear-water), scour depths can be larger.

In addition to sediments being transported from inland areas, sediments are transported parallel to the coast by ocean currents and wave action. This littoral transport of sediment serves as a source of sediment supply to the inlet, bay or estuary, or tidal passage. During the flood tide, these sediments can be transported into the bay or estuary and deposited. During the ebb tide, these sediments can be re-mobilized and transported out of the inlet or estuary and either be deposited on shoals or moved further down the coast as littoral transport (Figure 9.3).

Sediment transported to the bay or estuary from the inland river system can also be deposited in the bay or estuary during the flood tide, and re-mobilized and transported through the inlet or estuary during the ebb tide. However, if the bay or estuary is large, sediments derived from the inland river system can deposit in the bay or estuary in areas where the velocities are low and may not contribute to the supply of sediment to the bridge crossing. The result is clear-water scour unless sediment transported on the flood tide (ocean shoals, littoral transport) is available on the ebb. Sediments transported from inland rivers into an estuary may be stored there on the flood and transported out during ebb tide. This would produce live-bed scour conditions unless the sediment source in the estuary was disrupted. Dredging, jetties or other coastal engineering activities can limit sediment supply to the reach and influence live-bed and clear-water conditions.

Application of sediment continuity involves understanding the hydraulics of flow and availability of sediment for transport. For example, a net loss of sediment in the inlet, bay or tidal estuary could be the result of cutting off littoral transport by means of a jetty projecting into the ocean (Figure 9.3). For this scenario, the flood tide would tend to erode sediment from the inlet and deposit sediment in the bay or estuary while the ensuing ebb tide would transport sediment out of the bay or estuary. Because the availability of sediment for transport into the bay is reduced, degradation of the inlet could result. As the cross sectional area of the inlet increases, the flow velocities during the flood tide increase, resulting in further degradation of the inlet. This can result in an unstable inlet which continues to enlarge as a result of sediment supply depletion.

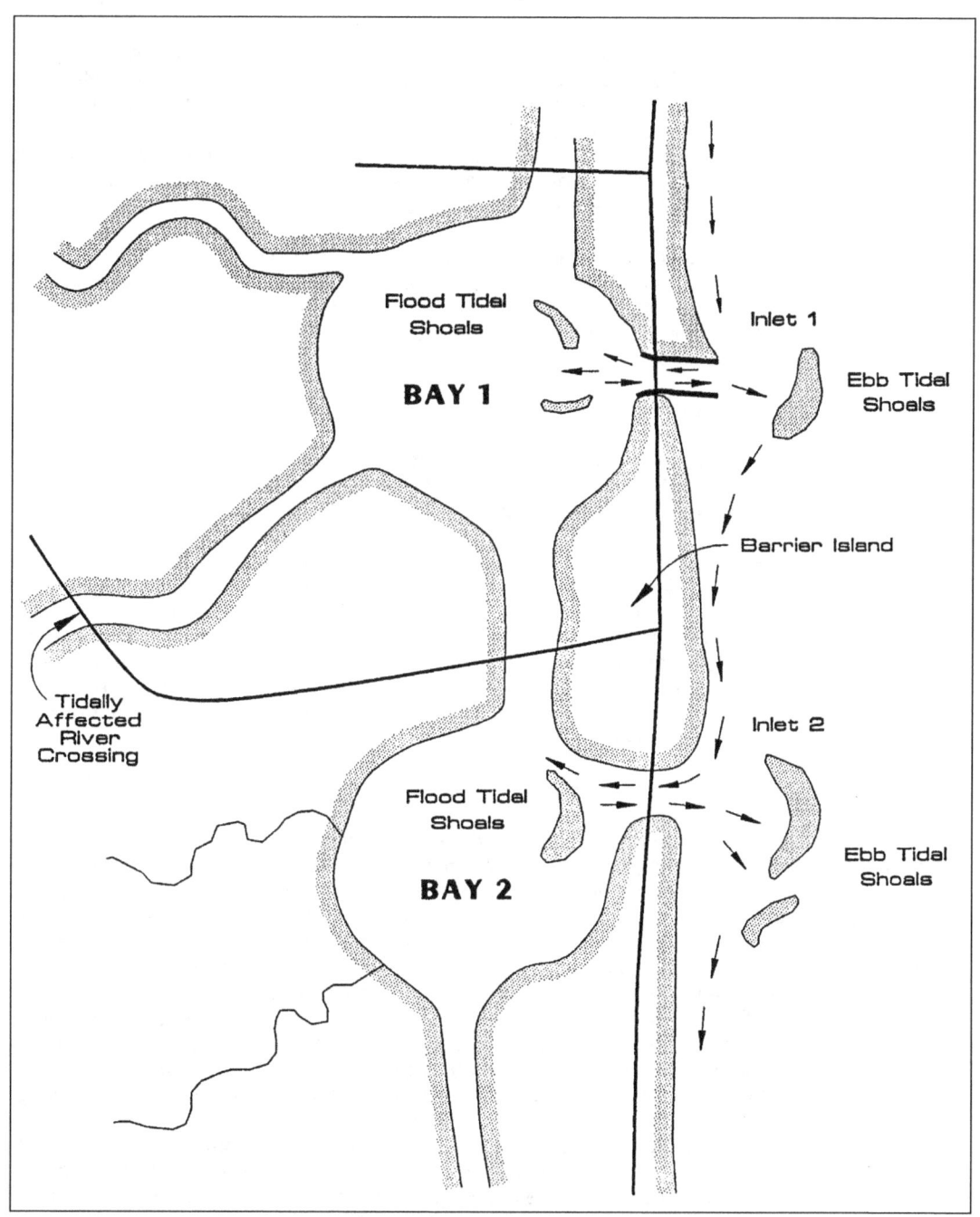

Figure 9.3. Sediment transport in tidal inlets (after TRB 1993a).

Although the above discussion would indicate that long-term degradation would continue indefinitely, this is not the case. As the scour depth increases there would be changes in the relationship between the incoming tide and the tide in the bay or estuary, and also between the tide in the bay and the ocean on the ebb tide. This could change the difference in elevation between the bay and ocean. At some level of degradation the incoming or outgoing tides could pick up sediment from either the bay or ocean which would then satisfy the transport capacity of the flow. Also, there could be other changes as scour progressed, such as accumulation of larger bed material on the surface (armor) or exposure of scour resistance rock which would decrease or stop the scour.

From the above discussion, it is clear that the concept of sediment continuity provides a valuable tool for evaluation of aggradation or degradation trends of a tidal waterway. Although this principle is not easy to quantify without direct measurement or hydraulic and sediment continuity modeling, the principle can be applied in a qualitative sense to assess long-term trends in aggradation or degradation.

9.3 LEVEL 1 ANALYSIS

The objectives of a Level 1 qualitative analysis are to determine the magnitude of the tidal effects on the crossing, the overall long-term stability of the crossing (vertical and lateral stability) and the potential for waterway response to change.

The first step in evaluation of highway crossings is to determine whether the bridge crosses a river which is influenced by tidal fluctuations (tidally affected river crossing) or whether the bridge crosses a tidal inlet, bay or estuary (tidally controlled). The flow in tidal inlets, bays and estuaries is predominantly driven by tidal fluctuations (with flow reversal), whereas, the flow in tidally affected river crossings is driven by a combination of river flow and tidal fluctuations. Therefore, tidally affected river crossings are not subject to flow reversal but the downstream tidal fluctuation acts as a cyclic downstream control. Tidally controlled river crossings will exhibit flow reversal.

9.3.1 Tidally Affected River Crossings

Tidally affected river crossings are characterized by both river flow and tidal fluctuations. From a hydraulic standpoint, the flow in the river is influenced by tidal fluctuations which result in a cyclic variation in the downstream control of the tail water in the river estuary. The degree to which tidal fluctuations influence the discharge at the river crossing depends on such factors as the relative distance from the ocean to the crossing, riverbed slope, cross-sectional area, storage volume, and hydraulic resistance. Although other factors are involved, relative distance of the river crossing from the ocean can be used as a qualitative indicator of tidal influence (see Figure 9.1(2)). At one extreme, where the crossing is located far upstream, the flow in the river may only be affected to a minor degree by changes in tailwater control due to tidal fluctuations. As such, the tidal fluctuation downstream will result in only minor fluctuations in the depth, velocity, and discharge through the bridge crossing.

As the distance from the crossing to the ocean is reduced, again assuming all other factors are equal, the influence of the tidal fluctuations increases. Consequently, the degree of tail water influence on flow hydraulics at the crossing increases. A limiting case occurs when the magnitude of the tidal fluctuations is large enough to reduce the discharge through the bridge crossing to zero at high tide. River crossings located closer to the ocean than this limiting case have two directional flows at the bridge crossing, and because of the storage of the river flow at high tide, the ebb tide will have a larger discharge and velocities than the flood tide.

For the Level 1 analysis, it is important to evaluate whether the tidal fluctuations will significantly affect the hydraulics at the bridge crossing. If the influence of tidal fluctuations is considered to be negligible, then the bridge crossing can be evaluated based on the procedures outlined for inland river crossings presented previously in this document. If not, then the hydraulic flow variables must be determined using dynamic tidal flow relationships. This evaluation should include extreme events such as the influence of storm surges and inland floods.

From historical records of the stream at the highway crossing, determine whether the worst-case conditions of discharge, depths and velocity at the bridge are created by tides and storm surge, or by inland floods or a combination of the two. Historical records could consist of tidal and stream flow data from Federal Emergency Management Agency (FEMA), National Oceanic and Atmospheric Administration (NOAA), USACE, and USGS records; aerial photographs of the area; maintenance records for the bridge or bridges in the area; newspaper accounts of previous high tides and/or flood flows; and interviews in the local area.

If the primary hazard to the bridge crossing is from inland flood events, then scour can be evaluated using the methods given previously in this circular and in HEC-20 (FHWA 2012b). If the primary hazard to the bridge is from tide and storm surge or tide, storm surge and inland flood runoff, then use the analyses discussed in the following sections on tidal waterways and presented in detail in HEC-25 First and Second Editions (FHWA 2004, 2008). If it is unclear whether the worst hazard to the bridge will result from a storm surge, maximum tide, or from an inland flood, it may be necessary to evaluate scour considering each of these scenarios and compare the results.

9.3.2 Tidal Inlets, Bays, and Estuaries

For tidal inlets, bays and estuaries, the goal of the Level 1 analysis is to determine the stability of the inlet and identify and evaluate long-term trends at the location of the highway crossing. This can be accomplished by careful evaluation of present and historical conditions of the tidal waterway and anticipating future conditions or trends.

Existing cross-sectional and sounding data can be used to evaluate the stability of the tidal waterway at the highway crossing and to determine whether the inlet, bay or estuary is increasing or decreasing in size, or is relatively stable. For this analysis it is important to evaluate these data based on past and current trends. The data for this analysis could consist of aerial photographs, cross section soundings, location of bars and shoals on both the ocean and bay sides of an inlet, magnitude and direction of littoral drift, and longitudinal elevations through the waterway. It is also important to consider the possible impacts (either past or future) of the construction of jetties, breakwaters, or dredging of navigation channels.

Sources of data would be USACE, FEMA, USGS, U.S. Coast Guard (USCG), NOAA, local Universities, oceanographic institutions and publications in local libraries. For example, publications by Bruun contain information on many tidal inlets on the east coast of the United States (Bruun 1966, 1990).

A site visit is recommended to gather such data as the conditions of the beaches (ocean and bay side); location and size of any shoals or bars; seasonal direction of ocean waves; magnitude of the currents in the bridge reach at mean water level (midway between high and low tides); and size of the sediments. Sounding the channel both longitudinally and in cross section using a conventional "fish finder" sonic fathometer is usually sufficiently accurate for this purpose.

Observation of the tidal inlet to identify whether the inlet restricts the flow of either the incoming or outgoing tide is also recommended. If the inlet or bridge restricts the flow, there will be a noticeable drop in head (change in water surface elevation) in the channel during either the ebb or flood tide. If the tidal inlet or bridge restricts the flow, an orifice equation may need to be used to determine the maximum discharge, velocities and depths (see FHWA 2004).

Velocity measurements in the tidal inlet channel along several cross sections, several positions in the cross section and several locations in the vertical can also provide useful information for verifying computed velocities. Velocity measurements should be made at maximum discharge (Q_{max}). Maximum discharge usually occurs around the midpoint in the tidal cycle between high and low tide (Figure 9.2), although constricted inlets usually cause peak discharge to occur closer to high and low tides.

The velocity measurements can be made from a boat or from a bridge located near the site of a new or replacement bridge. If a bridge exists over the channel, a recording velocity meter could be installed to obtain measurements over several tidal cycles. Currently, there are instruments available that make velocity data collection easier. For example, broad-band acoustic Doppler current profiles and other emerging technologies will greatly improve the ability to obtain and use velocity data.

In order to develop adequate hydraulic data for the evaluation of scour, it is recommended that recording water level gages located at the inlet, at the proposed bridge site and in the bay or estuary upstream of the bridge be installed to record tide elevations at 15-minute intervals for several full tidal cycles. This measurement should be conducted during one of the spring tides where the amplitude of the tidal cycle will be largest. The gages should be referenced to the same datum and synchronized. The data from these recording gages are necessary for calibration of unsteady models such as UNET (HEC-RAS), FST2DH, and RMA-2V (USACE 1985, 1993b, 1997; FHWA 1994, 2003b). These data are also useful for calibration of HEC River Analysis System (RAS) when the bridge crosses tidally affected channels (USACE 2010a).

The data and evaluations suggested above can be used to estimate whether present conditions are likely to continue into the foreseeable future and as a basis for evaluating the hydraulics and total scour for the Level 2 analysis. A stable inlet could change to one which is degrading if the channel is dredged or jetties are constructed on the ocean side to improve the entrance, since dredging or jetties could modify the supply of sediment to the inlet. In addition, plans or projects which might interrupt existing conditions of littoral drift should be evaluated.

It should be noted that in contrast to an inland river crossing, the discharge at a tidal inlet is not fixed. In inland rivers, the design discharge is fixed by the runoff and is virtually unaffected by the waterway opening. In contrast, the discharge at a tidal inlet can increase as the area of the tidal inlet increases, thus increasing long-term aggradation or degradation and local scour. Also, as Neill points out, constriction of the natural waterway opening may modify the tidal regime and associated tidal discharge (Neill 2004).

9.4 LEVEL 2 ANALYSIS

9.4.1 Introduction

Level 2 analysis involves the basic engineering assessment of scour problems at highway crossings. Scour equations developed for inland rivers are recommended for use in estimating and evaluating scour for tidal flows. However, in contrast to the evaluation of scour at inland river crossings, the evaluation of the hydraulic conditions at the bridge crossing using a 1-dimensional steady flow model is only suitable for tidally affected crossings where tidal fluctuations result in a variable tailwater control without flow reversal. Other methods are recommended for tidally affected and tidally controlled crossings where the tidal fluctuation has a significant influence on the tidal hydraulics. Several methods to obtain hydraulic characteristics of tidal flows at the bridge crossing are available. These range from simple procedures such as a tidal prism or orifice equation approach to more complex 2-dimensional and quasi 2-dimensional unsteady flow models. An overview of the unsteady flow models which are suitable for modeling tidal hydraulics at bridge crossings is presented in the First Edition of HEC-25 (FHWA 2004).

9.4.2 Evaluation of Hydraulic Characteristics

The velocity, depth and discharge at the bridge waterway are the most significant variables for evaluating bridge scour in tidal waterways. Direct measurements of the value of these variables for the design storm are seldom available. Therefore, it is usually necessary to develop the hydraulic and hydrographic characteristics of the tidal waterway, estuary or bay, and calculate the discharge, velocities, and depths in the crossing using unsteady flow modeling and coastal engineering methods. These values can then be used in the scour equations given in previous sections to calculate long-term aggradation or degradation, contraction scour, and local scour.

Although unsteady flow models are suitable for determining the hydraulic conditions, their use requires careful application and calibration. The effort required to utilize these models may be more than is warranted for many tidal situations. As such, the use of these models may be more applicable under a Level 3 analysis. However, these models could be used in the context of a Level 2 analysis, if deemed necessary, to better define the hydraulic conditions at the bridge crossing.

9.4.3 Design Storm and Storm Tide

Normally, long-term aggradation or degradation at a tidal inlet or estuary are influenced primarily by the periodic tidal fluctuations associated with astronomical tides. Therefore, flow hydraulics at the bridge should be determined considering the tidal range as depicted in Figure 9.2 for evaluation of long-term aggradation or degradation.

Extreme events associated with inland floods and storm tides should be used to determine the hydraulics at the bridge to evaluate local and contraction scour (see Table 2.1). Difficulty arises in determining whether the storm tide, inland flood or the combination of storm tide and inland flood should be considered controlling. The effect of the inland flood discharges (if any), would be most significant during the period when storm tide floodwaters recede (ebb), as those discharges would likely add to, and increase the storm tide associated discharges.

When inland flood discharges are small in relationship to the magnitude of the storm tide and are the result of the same storm event, then the flood discharge can be added to the discharge associated with the design tidal flow, or the volume of the runoff hydrograph can be added to the volume of the tidal prism. If the inland flood and the storm tide may result from different storm events, then, a joint probability approach may be warranted to determine the magnitude of the extreme events.

In some cases there may be a time lag between the storm tide discharge and the stream flow discharge at the bridge crossing. For this case, stream flow-routing methods such as the USACE HEC-HMS model can be used to estimate the timing of the flood hydrograph derived from runoff of the watersheds draining into the bay or estuary (USACE 2010b).

For cases where the magnitude of the inland flood is much larger than the magnitude of the storm tide, evaluation of the hydraulics reduces to using the equations and procedures recommended for inland rivers. The selection of the method to use to combine inland flood and storm tide flows is a matter of judgment and must consider the characteristics of the site and the storm events.

9.4.4 Scour Evaluation Concepts

The total scour at a bridge crossing can be evaluated using the scour equations recommended for inland rivers and the hydraulic characteristics determined using the procedures outlined in the previous sections. However, it should be emphasized that the scour equations and subsequent results need to be carefully evaluated considering other (Level 1) information from the existing site, other bridge crossings, or comparable tidal waterways or tidally affected streams in the area.

Evaluation of long-term aggradation or degradation at tidal highway crossings, as with inland river crossings, relies on a careful evaluation of the past, existing and possible future condition of the site. This evaluation is outlined under Level 1 and should consider the principles of sediment continuity. A longitudinal sonic sounder survey of a tide inlet is useful to determine if bed material sediments can be supplied to the tidal waterway from the bay, estuary or ocean. When available, historical sounding data should also be used in this evaluation. Factors which could limit the availability of sediment should also be considered.

Over the long-term in a stable tidal waterway, the quantity of sediment being supplied to the waterway by ocean currents, littoral transport and inland flows and being transported out of the tidal waterway are nearly the same. If the supply of sediment is reduced either from the ocean or from the bay or estuary, a stable waterway can be transformed into a degrading waterway. In some cases, the rate of long-term degradation has been observed to be large and deep. An estimate of the maximum depth that this long-term degradation can achieve can be made by employing the clear-water contraction scour equation to the inlet (see Chapter 6). For this computation the flow hydraulics should be developed based on the range of mean tide as described in Figure 9.2. **It should be noted that the use of this equation would provide an estimate of the worst case long-term degradation which could be expected assuming no sediments were available to be transported to the tidal waterway from the ocean or inland bay or estuary**. As the waterway degrades, the flow conditions and storage of sediments in shoals will change, ultimately developing a new equilibrium. The presence of scour resistant rock would also limit the maximum long-term degradation.

Potential contraction scour for tidal waterways also needs to be carefully evaluated using hydraulic characteristics associated with the storm surge or inland flood as described in the previous section. For highway crossings of estuaries or inlets to bays, where either the channel narrows naturally or where the channel is narrowed by the encroachment of the highway embankments, the live-bed or clear water contraction scour equations can be utilized to estimate contraction scour.

Soil boring or sediment data are needed in the waterway upstream, downstream, and at the bridge crossing in order to determine if the scour is clear-water or live-bed and to support scour calculations if clear-water contraction scour equations are used. Equation 6.1 and the ratio of V_*/ω can be used to assess whether scour would be clear-water or live-bed.

A mitigating factor which could limit contraction scour concerns sediment delivery to the inlet or estuary from the ocean due to a storm surge or from an inland flood. A surge may transport large quantities of sediment into the inlet or estuary during the flood tide. Likewise, inland floods can also transport sediment to an estuary during extreme floods. Thus, contraction scour during extreme events may be classified as live-bed because of the sediment being delivered to the inlet or estuary from the combined effects of the storm surge and inland flood. **The magnitude of contraction scour must be carefully evaluated using engineering judgment which considers the geometry of the crossing, estuary or bay, the magnitude and duration of the discharge associated with the storm surge or inland flood, the basic assumptions for which the contraction scour equations were developed, and mitigating factors which would tend to limit contraction scour.**

Evaluation of local scour at piers can be made by using the equations in Chapter 7 as recommended for inland river crossings. These equations can be applied to piers in tidal flows in the same manner as given for inland bridge crossings. However, the flow velocity and depth will need to be determined considering the design flow event and hydraulic characteristics for tidal flows.

9.5 TIME DEPENDENT CHARACTERISTICS OF TIDAL SCOUR

In tidal areas, hurricane storm surges often produce extreme hydraulic conditions. Computing ultimate contraction scour amounts for these conditions may not be reasonable based on the short duration (often less than 3 hours) of the flow produced by the surge. Based on equations in the Dutch Scour Manual (Hoffmans and Verheij 1997), (see also Transportation Research Board Research Results Digest (TRB 1999)), the time development of scour holes can be estimated. To provide confirmation of these results, the Yang (1996) sediment transport equation was used to compute contraction scour hole development based on the erosion of the scour hole equal to the transport capacity in the contracted bridge opening. The scour rates for this situation are shown on Figures 9.4 and 9.5. Figure 9.4 shows the complete development of scour with time plotted on a logarithmic axis and Figure 9.5 shows the first 100 hours of development with time plotted on an arithmetic axis. The scour rates predicted by the two methods are extremely similar and indicate that the scour that could be generated in a few hours during a storm surge is significantly less than the ultimate contraction scour condition.

Also shown in Figures 9.4 and 9.5 is the development of a pier scour hole for the same hydraulic conditions. The pier scour hole reaches 90 percent of ultimate scour in the first 20 hours while the clear-water contraction scour reaches only about 30 percent of ultimate scour.

The Dutch equations are based on clear-water scour and the conditions used to test the Yang equation were close to clear-water. The Dutch Scour Manual (Hoffmans and Verheij 1997) indicates that under live-bed conditions scour reaches ultimate conditions more rapidly and that the ultimate scour is less than the equivalent clear-water case which is consistent with current U.S. guidance. Figure 9.6 shows the development of contraction scour (using the Yang equation) under varying amounts of upstream sediment supply relative to the transport capacity in the bridge opening. This approach involves a basic sediment continuity analysis as outlined in HEC-20 (FHWA 2012b). For the case shown, if the upstream channel is supplying 50 percent of the contracted section transport capacity, the scour hole reaches the ultimate depth in approximately one hour. Based on this review, it appears that under storm surge conditions contraction scour should be analyzed on a case-by-case basis to assess the level of contraction scour that could occur over a short time. It also suggests that local scour occurs more rapidly and time dependence is a less significant factor.

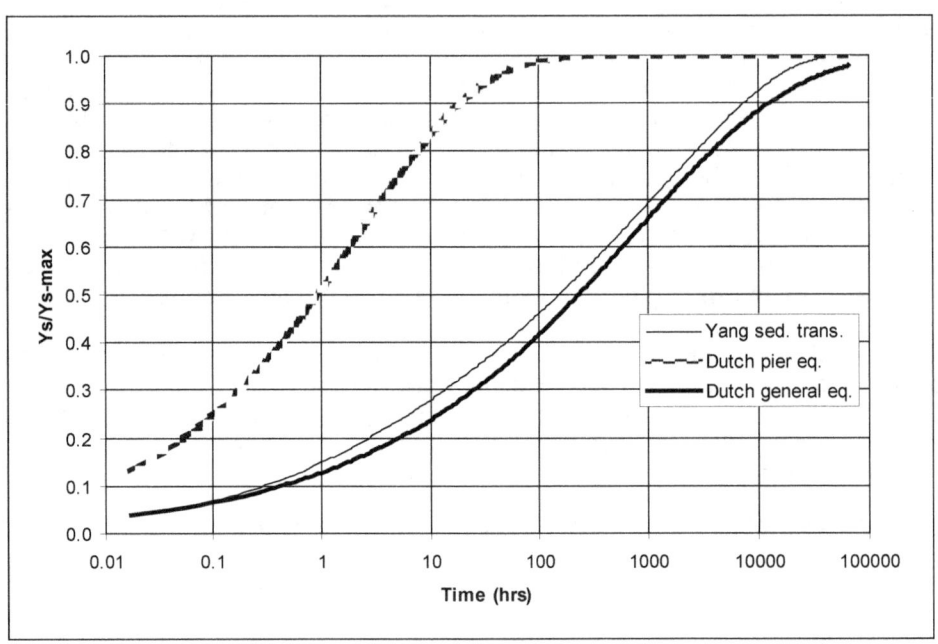

Figure 9.4. Time development of clear-water scour.

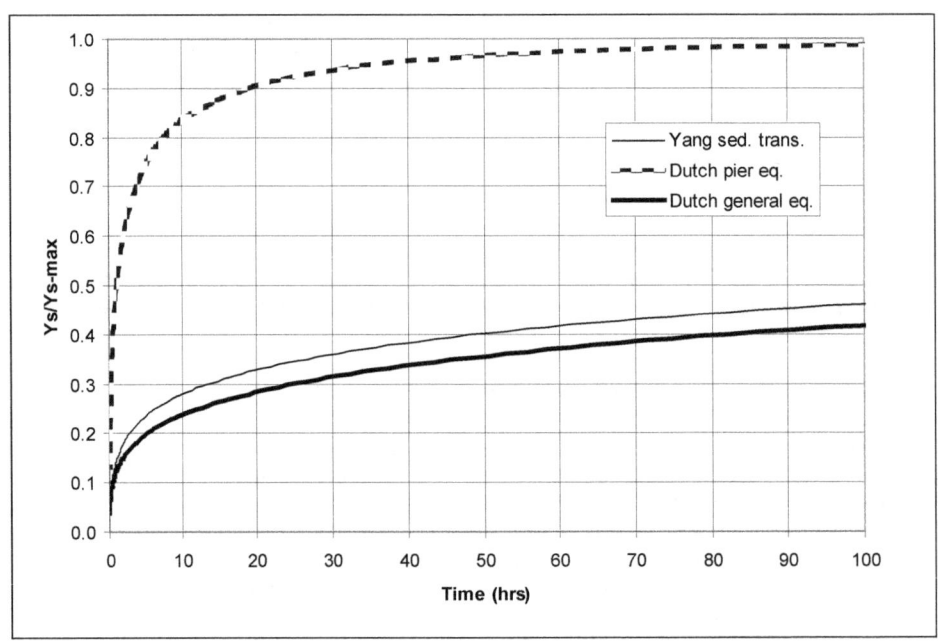

Figure 9.5. Initial clear-water scour development.

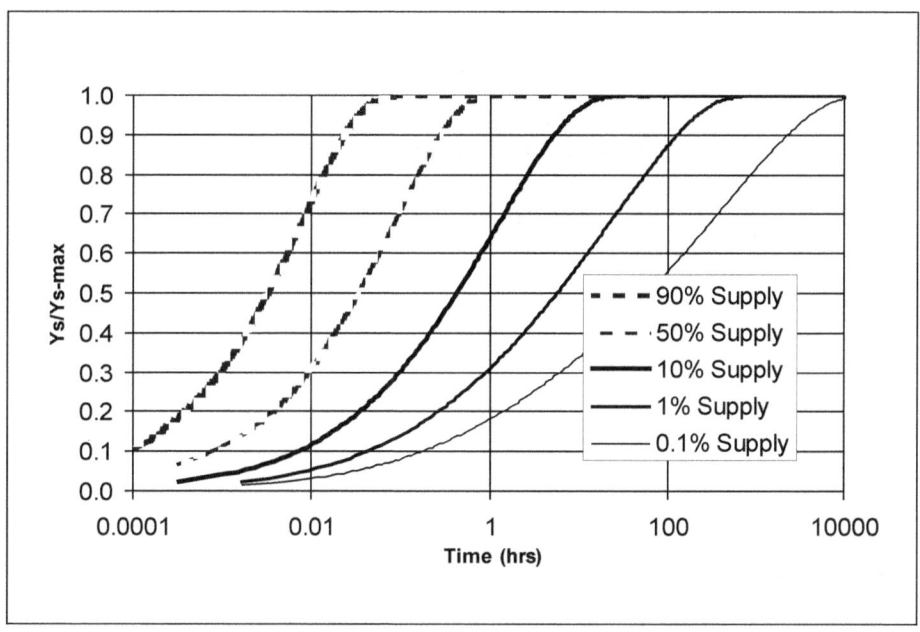

Figure 9.6. Contraction scour development with sediment supply.

9.6 LEVEL 3 ANALYSIS

As discussed in HEC-20, Level 3 analysis involves the use of physical models or more sophisticated computer models for complex situations where Level 2 analysis techniques have proven inadequate (FHWA 2012b). In general, crossings that require Level 3 analysis will also require the use of qualified hydraulic/coastal engineers. Level 3 analysis by its very nature is specialized and beyond the scope of this manual.

9.7 TIDAL HYDROLOGY, HYDRAULICS, AND SCOUR AT BRIDGES

Additional information to support the analysis of scour for bridges crossing tidal waterways can be found in HEC-25 First Edition (FHWA 2004).

The purpose of the First Edition is to provide guidance on hydraulic modeling for bridges over tidal waterways. This manual includes descriptions of: (1) common physical features that affect transportation projects in coastal areas, (2) tide causing astronomical and hydrologic processes, (3) approaches for determining hydraulic conditions for bridges in tidal waterways, (4) applying the hydraulic analysis results to provide scour estimates. It is expected that using the methods in the First Edition, better predictions of bridge hydraulics and scour in tidal waterways will result. In many cases, simplified tidal hydraulic methods will provide adequate results. However, when the simplified methods yield overly conservative results, use of the recommended modeling approaches will provide more realistic predictions and hydraulic variables and scour.

The First Edition of HEC-25 provides guidance on the following topics:

- Tidal Hydrology and Boundary Conditions

    - Astronomical Tides and Tidal Currents
    - Storms and Other Climatologic Conditions
    - Storm Surges
    - Predicting Storm Surge Hydrology
    - Storm Tides
    - Wind Considerations
    - Upland Runoff

- Basic Tidal Hydraulic Methods

    - Introduction
    - Tidal Prism
    - Orifice Approach
    - Routing Approach

- Tidal Hydraulic Modeling

    - Model Extent
    - One-Dimensional Modeling
    - Two-Dimensional Modeling
    - Model Selection
    - Model Calibration and Troubleshooting
    - Physical Modeling in Coastal Engineering

- Tidal Scour

    - Bridge Scour Analysis for Tidal Waterways
    - Time Dependent Contraction Scour
    - Time Dependent Local Scour

- Data Requirement and Sources

    - Tide Gages
    - Tidal Benchmarks and Vertical Datums
    - Hurricane Surge Data
    - Wind Data
    - Bathymetric and Topographic Data
    - Aerial Photography and Mapping

- Other Considerations

    - Coastal Zones and Beach Dynamics
    - Wave Analysis
    - Shore Protection Countermeasures

9.8 HIGHWAYS IN THE COASTAL ENVIRONMENT

The Second Edition of HEC-25 (FHWA 2008) provides guidance for the analysis, planning, design and operation of highways in the coastal environment. The focus is on roads near the coast that are always, or occasionally during storms, influenced by coastal tides and waves. It is estimated that there are over 60,000 road miles in the United States that can be called "coastal highways." A primary goal of this manual is the integration of coastal engineering principles and practices in the planning and design of coastal highways. Some of the physical coastal science concepts and modeling tools that have been developed by the coastal engineering community, and are applicable to highways, are briefly summarized. This includes engineering tools for waves, water levels, and sand movement. Applications to several of the highway and bridge planning and design issues that are unique to the coastal environment are also summarized. This includes coastal revetment design, planning and alternatives for highways that are threatened by coastal erosion, roads that overwash in storms, and coastal bridge issues including wave loads on bridge decks.

The Second Edition of HEC-25 provides guidance on the following topics:

- Coastal Highways

    - What are Coastal Highways?
    - Estimating the Extent of Coastal Roads and Bridges
    - Societal Demand for Coastal Highways
    - Natural Coastal Processes Impacting Highways
    - Coastal Highway Planning and Design
    - Coastal Engineering as a Specialty Area
    - Coastal Engineering in the Highway Community

- Tides, Storm Surge and Water Levels

    - Astronomical Tides
    - Storm Surge
    - Sea Level Rise
    - Lake Water Level Fluctuations

- Waves

    - Definitions, Theories, and Properties of Waves
    - Wave Transformation and Breaking
    - Irregular Waves
    - Wave Generation
    - Tsunamis
    - Ship Wakes

- Coastal Sediment Processes

    - Overview of Coastal Geomorphology
    - Beach Terminology
    - Coastal Sediment Characteristics
    - Cross-Shore Sand Transport and Dune Erosion Modeling
    - Longshore Sand Transport and Shoreline Change Modeling

- Tidal Inlets
- Physical Models in Coastal Engineering

- Coastal Revetments for Wave Attack

  - Types of Revetments and Seawalls
  - Hudson's Equation for Armor Stone Size
  - Design Wave Heights for Revetment Design
  - Practical Issues for Coastal Revetment Design

- Roads in Areas of Receding Shorelines

  - Quantifying Shoreline Change Rates
  - Estimating Future Shoreline Positions
  - Vulnerability Studies for Coastal Roads and Bridges
  - Relocation Considerations
  - Coastal Structures

- Highway Overwashing

  - Description of Issue
  - The Coastal Weir-Flow-Damage Mechanism
  - Strategies for Roads that Overwash

- Coastal Bridges

  - Locations of Coastal Bridges
  - Coastal Bridge Scour
  - Coastal Bridge Wave Forces
  - Selection of Design Storm Surge and Design Wave Heights

CHAPTER 10

SCOUR EVALUATION, INSPECTION, AND PLAN OF ACTION

10.1 INTRODUCTION

Approximately 500,000 bridges in the National Bridge Inventory are built over waterways. Statistically, we can expect hundreds of these bridges to experience floods on the order of a 100-year flood or greater each year. Because it is not economically feasible to construct all bridges to resist all conceivable floods, or to install scour countermeasures at all existing bridges to ensure absolute invulnerability from scour damage, some risks of failure from future floods may have to be accepted. **However, every bridge over water, whether existing or under design, should be assessed for its vulnerability to floods in order to determine the prudent measures to be taken.**

State departments of transportation (DOTs) have been conducting scour evaluations of their bridges over water in accordance with the 1991 FHWA Technical Advisory T 5140.23 (see FHWA 1988a and USDOT 1988). The evaluation is to be conducted by an interdisciplinary team of hydraulic, geotechnical and structural engineers who can make the necessary engineering judgments to determine the vulnerability of a bridge to scour. In general, the program consisted of screening all bridges over water to determine their scour vulnerability, and setting priorities for their evaluation. Each DOT structured its own evaluation program using guidelines furnished by FHWA. The screening and evaluation has helped bridge owners in rating each bridge in the National Bridge Inventory (NBI) using rating factors for Item 113, Scour Critical Bridges. A description of Item 113 rating factors is given in FHWA (1995) along with the other codes for rating bridge foundations, i.e., Item 60 - Substructure, Item 61 - Channel and Channel Protection, Item 71 - Waterway Adequacy, Item 92 - Critical Feature Inspection, Item 93 - Critical Feature Inspection Date.

At the present time, virtually all existing bridges in the United States have received an initial screening and more than 90 percent of all bridges had been evaluated for scour. More than half of the DOTs have reported a 90 percent or better completion percentage for the evaluation of all their bridges over waterways.

There are two main objectives to be accomplished in inspecting bridges for scour:

1. Accurately record the present condition of the bridge and the stream, and

2. Identify conditions that are indicative of potential problems with scour and/or stream instability for further review and evaluation by others.

In order to accomplish these objectives, the inspector needs to recognize and understand the interrelationship between the bridge, the stream, and the floodplain. Typically, a bridge spans the main channel of a stream and perhaps a portion of the floodplain. The road approaches to the bridge are typically on embankments which obstruct flow on the floodplain. This overbank or floodplain flow must, therefore, return to the stream at the bridge, flow through relief structures (culverts or relief bridges) and/or overtop one or both approach roadways.

Where overbank flow is forced to return to the main channel at the bridge, zones of turbulence are established and scour is likely to occur at the bridge abutments. Further, piers and abutments may present obstacles to flood flows in the main channel, creating conditions for local scour because of the turbulence around the bridge foundations. After flowing through the bridge, the flood water will expand back to the floodplain, creating additional zones of turbulence and scour.

The following sections present guidance for bridge inspectors in developing an understanding of the overall flood flow patterns at each bridge inspected. Guidance on the use of this information for rating the present condition of the bridge and evaluating the potential for damage from scour is also presented. When an actual or potential scour problem is identified by a bridge inspector, the bridge should be further evaluated by an Interdisciplinary Team as described in Technical Advisory T 5140.23 (FHWA 1988a and USDOT 1988). The results of this evaluation should be recorded under Item 113 of the "Recording and Coding Guide" (FHWA 1995 and USDOT 2001).

If the bridge is determined to be scour critical, a Plan of Action (Section 10.6) must be developed for establishing bridge-specific inspection type and frequency, installing scour countermeasures, and providing other critical guidance such as identifying flood conditions that will trigger closing of the bridge to reduce the risk to the traveling public. Also, the rating factor for the bridge substructure (Item 60 of the Recording and Coding Guide) should be consistent with the rating factor of Item 113 when scour around or underneath a substructure element has been observed.

10.2 OFFICE REVIEW

It is desirable to conduct an office review of bridge plans and previous inspection reports prior to performing the bridge inspection. Information obtained from the office review provides a better basis for inspecting the bridge and the stream. Items for consideration in the office review include:

1. Has an engineering scour evaluation study been made? If so, is the bridge scour-critical?
2. If the bridge is scour-critical, has a Plan of Action been developed?
3. What do comparisons of streambed cross sections taken during successive inspections reveal about the streambed? Is it stable? Degrading? Aggrading? Moving laterally? Are there scour holes around piers and abutments?
4. What equipment is needed (rods, poles, sounding lines, sonar, etc.) to measure streambed elevations so that a cross section diagram can be prepared?
5. Are there sketches and/or aerial photographs to indicate the planform location of the stream and determine whether the main channel is migrating or the flow direction is changing at the bridge?
6. What type of bridge foundation was constructed? (Spread footings, piles, drilled shafts, etc.) Are footing and pile tip elevations known? Do the foundations appear to be vulnerable to scour? What are the sub-surface soil conditions? (sand, gravel, silt, clay, rock?)
7. Do special conditions exist that require particular methods and equipment (divers, boats, electronic gear for measuring stream bottom, etc.) for underwater inspections?
8. Are there special items that should be looked at? (Examples might include damaged riprap, stream channel presenting a skewed angle of flow as it approaches the substructure elements, past problems with debris, etc.)

10.3 BRIDGE INSPECTION

10.3.1 Safety Considerations

The bridge inspection team should understand and practice prudent safety precautions during the conduct of the bridge inspection. Warning signs should be set up at the approaches to the bridge to alert motorists of the activity on the bridge. This is particularly important if streambed measurements are to be taken from the bridge, since most bridges have minimal clearances between the parapet and the edge of the travel lane. Inspectors should wear brightly colored vests so that they are conspicuous to motorists.

When measurements are made in the stream, the inspector should be secured by a safety line whenever there is deep or fast flowing water and a boat should be available in case of emergency. If waders become overtopped, they will fill and may drag the inspector downstream and under water in a matter of a few seconds.

The inspection team should leave word with their office regarding their schedule of work for the day. The team should also carry a cell phone with them so that they can get immediate help in the event of an emergency.

10.3.2 FHWA Recording and Coding Guide

During the bridge inspection, the condition of the bridge waterway opening, substructure, channel protection, and scour countermeasures should be evaluated, along with the condition of the stream.

FHWA Report No. FHWA-PD-96-001, "Recording and Coding Guide for the Structure Inventory and Appraisal of the Nation's Bridges," (FHWA 1995). The Recording and Coding Guide for bridges contains 116 items that are specific to each bridge. Only about half of the items are "condition ratings" that might change from one inspection to the next; six of those are related to hydraulic and scour/stream instability issues:

1. Item 60:   Substructure
2. Item 61:   Channel and Channel Protection
3. Item 71:   Waterway Adequacy
4. Item 92:   Critical Feature Inspection
5. Item 93:   Critical Feature Inspection Date
6. Item 113: Scour Critical Bridges

The guidance in the Recording and Coding Guide for rating the present condition of Items 61, 71, and 113 is set forth in detail. Items 92 and 93 are used to alert inspectors to specific elements that require special attention, for example countermeasures or monitoring devices that have been installed to address scour problems previously encountered at a particular bridge.

Guidance for rating the present condition of Item 60, Substructure, is general and does not include specific details for scour; however, the rating given to Item 60 should be consistent with the one given for Item 113 whenever a rating of 2 or below is determined for Item 113. Current policy and technical guidance for Items 60 and 113 are provided in the FHWA memorandum dated April 27, 2001 (USDOT 2001).

The following sections present approaches to evaluating the present condition of the bridge foundation for scour and the overall scour potential at the bridge.

10.3.3 General Site Considerations

In order to appreciate the relationship between the bridge and the river it is crossing, observation should be made of the conditions of the river up- and downstream of the bridge:

- Is there evidence of general degradation or aggradation of the river channel resulting in unstable bed and banks?
- Is there evidence of on-going development in the watershed and particularly in the adjacent floodplain that could be contributing to channel instability?
- Are there active gravel or sand mining operations in the channel near the bridge?
- Are there confluences with other streams? How will the confluence affect flood flow and sediment transport conditions?
- Is there evidence at the bridge or in the up- and downstream reaches that the stream carries large amounts of debris? Are the bridge superstructure and substructure elements streamlined to pass debris, or is it likely that debris will hang up on the bridge and create adverse flow patterns with resulting scour?
- The best way of evaluating flow conditions through the bridge is to look at and photograph the bridge from the up- and downstream channel. Is there a significant angle of attack of the flow on a pier or abutment?

10.3.4 Assessing the Substructure Condition

Item 60, Substructure, is the key item for rating the bridge foundations for vulnerability to scour damage. When a bridge inspector finds that a scour problem has already occurred, it should be considered in the rating of Item 60. Both existing and potential problems with scour should be reported so that a scour evaluation can be made by an interdisciplinary team. The scour evaluation is reported on Item 113 in the Recording and Coding Guide (FHWA 1995). If the bridge is determined to be scour critical and has an Item 113 rating factor of 2 or less, the rating factor for Item 60 should be consistent with that of Item 113 to ensure that existing scour problems have been considered and reported. The following items are recommended for consideration in inspecting the present condition of bridge substructure elements for scour-related problems:

1. Evidence of movement of piers and abutments;

    - Rotational movement (check with plumb line)
    - Settlement (check lines of substructure and superstructure, bridge rail, etc., for discontinuities; check for structural cracking or spalling)
    - Check bridge seats for excessive movement

2. Damage to scour countermeasures protecting the foundations (riprap, guide banks, sheet piling, sills, etc.). Examples of damage could include riprap placed around piers and/or abutments that has been displaced or replaced with river bed material. A common cause of damage to abutment riprap protection is runoff from the ends of the bridge deck which flows down to the riprap and undermines it. This condition can be corrected by installing bridge-end drains.

3. Changes in streambed elevation at foundations (undermining of footings, exposure of piles), and

4. Changes in streambed cross section at the bridge, including location and depth of scour holes.

   - Note and measure any depressions around piers and abutments
   - Note the approach flow conditions. Is there an angle of attack of flood flow on piers or abutments?

In order to evaluate the conditions of the foundations, the inspector should measure the elevation of the streambed to a common bench mark at the bridge cross section during each inspection. These cross-section elevations should be plotted to a common datum and successive cross sections compared. Careful observations and measurements should be made of scour holes at piers and abutments, including probing soft material in scour holes to determine the location of a firm bottom where possible. If equipment or conditions do not permit measurement of the stream bottom, this condition should be noted for further action.

10.3.5 Assessing the Condition of Countermeasures

Countermeasures associated with scour and stream instability problems can be classified according to three types (FHWA 2009):

1) River training countermeasures (typically used to mitigate stream instability problems);
2) Armoring countermeasures (used to protect bed and banks in the immediate vicinity of piers and abutments)
3) Monitoring devices (used to detect scour during flood events to provide early warning and trigger bridge closure; these devices do not actually correct the scour problem).

FHWA Publication No. FHWA-NHI-08-106, "Stream Stability, Bridge Scour, and Countermeasures: A Field Guide for Bridge Inspectors" provides detailed guidance regarding the inspection of these countermeasure types. Over the last several decades, a wide variety of countermeasure structures, armoring materials, and monitoring devices have been used at existing bridges to mitigate scour and stream stability problems. While most bridge inspectors are familiar with standard countermeasures such as riprap, it is unlikely that they are knowledgeable of the full spectrum of countermeasures currently available and in use.

The Field Guide was developed as a pocket-sized document that inspectors can easily take to the bridge to help identify other countermeasure types and aid in the assessment of their condition. Inspection tips are provided for many different types of countermeasures, including:

1) River training countermeasures

   - Spurs (both permeable and impermeable)
   - Bendway weirs
   - Guide banks
   - Drop structures and check dams

2) Armoring countermeasures

   - Rock riprap
   - Grouted riprap
   - Concrete slope paving
   - Articulating concrete blocks
   - Gabion mattresses
   - Grout-filled mats

3) Monitoring devices
- Fixed sonar devices
- Sounding rods
- Buried/driven rods (e.g., sliding magnetic collar devices)
- Other buried devices (e.g., float-out transmitters)

As an example of the type of guidance that inspectors can find in the Field Guide, the inspection tips for standard riprap are shown in Table 10.1. Inspectors are encouraged to consult the Field Guide where inspection tips for other countermeasure types can be found.

Table 10.1. Tips for Inspecting Riprap.

1. Riprap should be **angular and interlocking** (old bowling balls would not make good riprap). Flat sections of broken concrete paving do not make good riprap.
2. Riprap should have a **granular or geotextile filter** between the rock and the subgrade to prevent loss of the finer subgrade material, whether on the bed or the bank.
3. Riprap should be **well graded** (a wide range of rock sizes). The maximum rock size should be no greater than about twice the median (d_{50}) size.
4. When inspecting riprap, the following would be strong indicators of problems:
    - Have riprap stones been **displaced** downstream?
    - Has the riprap blanket **slumped** down the slope?
    - Has angular riprap material been **replaced** over time by smoother river run material?
    - Has the riprap material physically **deteriorated**, **disintegrated**, or been **abraded** over time?
    - Are there **holes** in the riprap blanket where the filter has been exposed or breached?
5. **Riprap revetment** must have an adequate burial depth at the toe (toe down) to prevent it from being undermined. Toe down should be deeper than the expected long-term degradation and contraction scour.
6. For **piers and abutments**, riprap should generally extend up to the bed elevation so that the top of the riprap is visible to the inspector during and after floods.

10.3.6 Assessing Scour Potential at Bridges

The items listed in Table 10.2 are provided for bridge inspectors' consideration in assessing the adequacy of the bridge to resist scour. In making this assessment, inspectors need to understand and recognize the interrelationships between Item 60 (Substructure), Item 61 (Channel and Channel Protection), Item 71 (Waterway Adequacy), and 113 (Scour-Critical Bridges). Items requiring additional attention at a particular bridge are listed in Item 92 (Critical Feature Inspection) and 93 (Critical Feature Inspection Date). As noted earlier, additional follow-up by an interdisciplinary team should be made utilizing Item 113 - Scour Critical Bridges when the bridge inspection reveals a potential problem with scour (USDOT 2001).

Table 10.2. Assessing the Scour Potential at Bridges.

1.   UPSTREAM CONDITIONS

a.   <u>Banks</u>

    <u>STABLE</u>:   Natural vegetation, trees; Bank stabilization measures such as riprap, paving, gabions; Channel stabilization measures such as dikes and jetties.

    <u>UNSTABLE</u>:   Bank sloughing, undermining, evidence of lateral movement, damage to stream stabilization measures, etc.

b.   <u>Main Channel</u>

- Clear and open with good approach flow conditions, or meandering or braided with main channel at an angle to the orientation of the bridge.
- Existence of islands, bars, debris, cattle guards, fences that may affect flow.
- Aggrading or degrading streambed.
- Evidence of movement of channel with respect to bridge (make sketches, take pictures).
- Evidence of ponding of flow.

<u>Floodplain</u>

- Evidence of significant flow on floodplain.
- Floodplain flow patterns - does flow overtop road and/or return to main channel?
- Existence and hydraulic adequacy of relief bridges (if relief bridges are obstructed, they will affect flow patterns at the main channel bridge).
- Extent of floodplain development and any obstruction to flows approaching the bridge and its approaches.
- Evidence of overtopping approach roads (debris, erosion of embankment slopes, damage to riprap or pavement, etc.).
- Evidence of ponding of flow.

<u>Debris</u>

- Extent of debris in upstream channel.

<u>Other Features</u>

- Existence of upstream tributaries, bridges, dams, or other features that may affect flow conditions at bridges.

Table continues

Table 10.2. Assessing the Scour Potential at Bridges (continued).

2. CONDITIONS AT BRIDGE

a. Substructure

- Is there evidence of scour at piers?
- Is there evidence of scour at abutments (upstream or downstream sections)?
- Is there evidence of scour at the approach roadway (upstream or downstream)?
- Are piles, pile caps or footings exposed?
- Is there debris on the piers or abutments?
- If riprap has been placed around piers or abutments, is it still in place and in good condition?

Superstructure

- Evidence of overtopping by flood water (Is superstructure tied down to substructure to prevent displacement during floods?)
- Obstruction to flood flows (Does superstructure collect debris or present a large area that would obstruct the flow?)
- Design (Is superstructure vulnerable to collapse in the event of foundation movement, e.g., simply-supported spans and nonredundant design for load transfer?)

Channel Protection and Scour Countermeasures

- Riprap (Is riprap adequately toed into the streambed or is it being undermined and washed away? Is riprap pier protection intact, or has riprap been removed and replaced by bed-load material? Can displaced riprap be seen in streambed beneath or downstream of the bridge?)
- Guide banks (Are guide banks in place? Have they been damaged by scour and erosion?)
- Stream and streambed (Is main current impinging upon piers and abutments at an angle? Is there evidence of scour and erosion of streambed and banks, especially adjacent to piers and abutments? Has stream cross section changed since last measurement? In what way?)

d. Waterway Area Does waterway area at the bridge appear small in relation to the stream and floodplain? Is there evidence of scour across a large portion of the streambed at the bridge? Do bars, islands, vegetation, and debris constrict the flow and concentrate it in one section of the bridge or cause it to attack piers and abutments? Do the superstructure, piers, abutments, and fences, etc., collect debris and constrict flow? Are approach roads regularly overtopped? If waterway opening is inadequate, does this increase the scour potential at bridge foundations?

Table continues

Table 10.2 Assessing the Scour Potential at Bridges (continued).

3. DOWNSTREAM CONDITIONS

a. Banks

   STABLE: Natural vegetation, trees; Bank stabilization measures such as riprap, paving, gabions; Channel stabilization measures such as dikes and jetties.

   UNSTABLE: Bank sloughing, undermining, evidence of lateral movement, damage to stream stabilization measures, etc.

b. Main Channel

   - Clear and open with smooth exit conditions, vs. meandering or braided with bends, islands, bars, cattle guards, debris, and fences that retard and obstruct flow.
   - Aggrading or degrading streambed.
   - Evidence of movement of channel with respect to the bridge (make sketches and take pictures).
   - Evidence of extensive bed erosion.

c. Floodplain

   - Clear and open so that contracted flow at bridge will expand and return smoothly to the floodplain downstream of the bridge, vs. restricted and blocked by dikes, development encroachment, trees, debris, or other obstructions.
   - Evidence of scour and erosion due to downstream turbulence.

d. Other Features

   - Downstream dams or confluences with larger streams which may cause variable tailwater depths. (This may create conditions for high velocity flow through bridge.)

10.3.7 Underwater Inspections

Perhaps the single most important aspect of inspecting the bridge for actual or potential damage from scour is measuring and plotting of stream bottom elevations in relation to the bridge foundations. Where conditions are such that the stream bottom cannot be accurately measured by rods, poles, weighted sounding lines or other means, other arrangements, such as underwater inspections, need to be made to determine the stream bottom elevation around the foundations and to determine the condition of the foundations. Other approaches to determining the cross section of the streambed at the bridge include:

- Use of divers
- Use of scour monitoring devices (see Section 10.4)

For the purpose of evaluating resistance to scour of the substructure under Item 60 of the Recording and Coding Guide, the questions remain essentially the same for foundations in deep water as for foundations in shallow water:

- Is the foundation footing, pile cap, or the piling exposed to or undetermined by the stream flow, and if so, what is the extent and probably consequences of this condition?
- Has riprap around piers or abutments been damaged, displaced, or removed?

Technical Advisory T5140.21 contains additional guidance for underwater inspections by divers.

10.3.8 Notification Procedures

A positive means of promptly communicating inspection findings to the appropriate agency personnel must be established. **Any condition that a bridge inspector considers to be of an emergency or potentially hazardous nature should be reported immediately.** That information as well as other conditions which do not pose an immediate hazard, but still warrant further action, should be conveyed to the interdisciplinary team for review.

A report form is therefore needed to communicate pertinent problem information to the hydraulic, structural, and geotechnical engineers. An existing report form may currently be used by bridge inspectors within a DOT to advise maintenance personnel of specific needs. Regardless of whether an existing report is used or a new one is developed, a bridge inspector should be provided the means of advising the interdisciplinary team of problems in a timely manner.

10.4 MONITORING BRIDGES FOR SCOUR

10.4.1 General

Periodic inspections of all bridges serve as the foundation for a bridge owner's management plan to ensure public safety. This includes underwater inspection of foundations located in deep water. A river and its floodplain are constantly changing, whereas the bridge and its foundation are fixed. A measuring system may be necessary to track the lateral and vertical movement of the channel bed over time. The measurements help to determine whether changes are random and within acceptable tolerances, or whether definite trends are occurring which may threaten the stability of the bridge.

Scour critical bridges are typically inspected more frequently than the routine biennial interval, and often include inspection during and immediately after flood events. Special attention should be given to the condition of foundation elements during these inspections. In many cases, special scour monitoring efforts should be put into effect at scour critical bridges as necessary to ensure that these bridges remain stable; if scour begins to approach an unstable condition during a flood event, monitoring devices can provide early warning so that the bridge can be closed until the danger has passed.

A wide range of monitoring procedures and devices can be used, depending on the nature and circumstances of the scour criticality condition at any particular bridge. The Plan of Action prepared for each scour critical bridge serves as the basis for (1) selecting the appropriate monitoring procedures and (2) providing special instructions to the bridge inspector regarding the procedures, as discussed in detail in Section 10.6. Monitoring may include:

- Stationing inspectors at the bridge during and immediately after flood events and providing them with portable equipment to measure scour depths
- Installing permanent scour monitoring devices at bridge piers and/or abutments
- Closing the bridge to traffic when conditions become unsafe

It should be noted that a monitoring program involves more than just instrumentation. It must describe specific actions to be taken immediately when an unsafe scour condition has been detected. In some cases, a properly designed scour monitoring program can be an acceptable countermeasure by itself. **As noted, however, monitoring does not fix the scour problem, and therefore, does not allow changing the Item 113 coding on a scour-critical bridge.** In other cases, a monitoring program allows time to implement hydraulic or structural countermeasures.

During a flood, scour is generally not visible and during the falling stage of a flood, scour holes typically fill in. Visual monitoring during a flood and inspection after a flood cannot fully determine that a bridge is safe. Instruments to measure or monitor scour can resolve this uncertainty. Using monitoring devices as a countermeasure for a scour critical bridge involves two basic categories of instruments: portable instruments and fixed instruments.

The selection of fixed or portable instruments in a scour monitoring program depends on many different factors. Each instrument has advantages and limitations that influence when and where they should be used. The idea of a toolbox, with various instruments that can be used under specific conditions, best illustrates the strategy to use when trying to select instrumentation for a scour monitoring program. Specific factors to consider include the frequency of data collection desired, the physical conditions at the bridge and stream channel, and traffic safety issues. Hydraulic Engineering Circular No. 23 (FHWA 2009) provides detailed information on advantages and limitations of various monitoring technologies in relation to an overall bridge management program.

10.4.2 Portable Monitoring Devices

Portable monitoring devices are usually preferred when only occasional measurements are required, such as during and after a flood, or where many different bridges must be monitored on a relatively infrequent basis. By definition, portable instruments require an inspector to operate the device and to take it from point to point along a bridge, or from one bridge to another. Portable devices provide readings at key locations at a specific point in time, and therefore are not used to document the time history of scour at a particular location.

Portable devices provide flexibility and the capability to respond quickly to flood conditions; however, collecting data may become very labor intensive and costly to implement where a large number of bridges are involved. The physical conditions at a bridge, such as height off the water and type of superstructure, can influence the decision to use fixed vs. portable equipment. For example, bridges that are very high off the water, or that have large deck overhang or projecting geometries, would complicate using portable measurements from the bridge deck. Traffic safety issues include the need for traffic control or lane closures when inspectors need to make portable measurements from the bridge deck.

The two main categories of portable monitoring devices are: 1) physical probes (sounding weights, extensible rods or poles), and 2) sonar instruments (e.g., fathometers or "fish finders"). Typically, portable devices are deployed from the bridge deck either manually or with a truck equipped with a boom or small crane. Both physical probes and sonar devices can be deployed from the water surface using an inspection boats, but hazardous conditions during flood flows usually prevent this type of use during flood watch activities. Sonars have been used successfully with tethered kneeboards or remote-controlled boats to avoid this safety concern.

10.4.3 Fixed Monitoring Devices

Fixed instrumentation is most often used when frequent measurements or regular, ongoing monitoring (e.g., weekly, daily, or continuous) are required. As the name implies, fixed devices are installed at a bridge at selected locations that are of primary concern from a scour perspective (typically at an abutment exposed to attack during floods, or in front of one or more piers). The devices can be read onsite by an inspector, and are often equipped with a data logger that can be downloaded to a laptop computer at the time of the inspection. Within the last decade, a popular choice for data retrieval involves a telemetered signal that can be observed in real time using land line, cell phone, satellite phone, or radio frequency.

During flood watch activities, an obvious advantage of fixed instruments is their ability to measure and report conditions during a flood event without the need for an inspector to be physically present. Fixed instruments typically record the maximum depth of scour that occurred during the flood, even if the scour hole refills with sediment as floodwaters recede; some fixed instruments, such as sonar devices, can track the refill process as well. Limitations include higher initial cost (including installation at the bridge), and measurement limited to one location, and potential damage from ice and/or debris.

There are several main categories of fixed scour monitoring instruments, including (NCHRP 1997):

- Sounding rods (typically fixed to bridge piers or abutments)
- Sonars
- Buried or driven rods (e.g., sliding magnetic collar)
- Other buried devices (e.g., float-out sensors)
- Tilt sensors
- Time domain reflectometers

In many cases, monitoring devices are deployed not for measuring the depth of a scour hole, but to detect the movement of countermeasures during a flood. If a device detects movement of riprap at a pier (for example), inspectors can be mobilized to the bridge with portable instruments and, if necessary, the bridge can be closed until the danger has passed.

Both fixed and portable devices are discussed in detail in NCHRP Synthesis 20-05, Topic 36-02, "Monitoring Scour Critical Bridges (NCHRP 2009) and are summarized in HEC-23 (FHWA 2009).

10.4.4 Selection and Maintenance of Monitoring Devices

Selecting the appropriate monitoring devices is specific to each individual bridge, and should consider the nature and location of the scour problem(s), accessibility issues created by the bridge superstructure and substructure elements, desired monitoring frequency, and cost over the remaining life of the bridge.

Monitoring devices are considered a type of scour countermeasure, and even though they do not correct the scour problem, they reduce the risk to the traveling public by providing early warning and allow for timely bridge closure when dangerous conditions develop. Like any other type of countermeasure, monitoring devices must be inspected and maintained to ensure that they function properly when they are needed.

10.5 CASE HISTORIES OF BRIDGE INSPECTION PROBLEMS

10.5.1 Introduction

Since 1987 there have been three bridge failures with loss of life that illustrate the importance of bridge inspections. In two of the failures, inspectors failed to observe changed conditions that if corrected may have saved the bridge. In one case, the inspectors documented the changes, but there was no follow-up action to evaluate the changes and to protect the bridge. In the following sections, the inspection problems associated with these bridge failures are described and issues related to inspection are highlighted.

10.5.2 Schoharie Creek Bridge Failure

On April 5, 1987 the New York State Thruway Authority Bridge (I-90) over Schoharie Creek collapsed killing 10 persons. The National Transportation Safety Board (NTSB 1988) investigated the collapse and gave as the probable cause as:

> "............the failure of the New York State Thruway Authority to maintain adequate rip rap around the bridge piers, which led to severe erosion in the soil beneath the spread footings. Contributing to the accident were ambivalent plans and specifications used for construction of the bridge, an inadequate NYSTA bridge inspection program, and inadequate oversight by the New York State Department of Transportation and the Federal Highway Administration. Contributing to the severity of the accident was the lack of structural redundancy in the bridge."

The bridge was built in 1953 on piers with spread footings and no piles. The footings were 5 ft (1.5 m) deep, 18 ft (5.5 m) wide and 82 ft (25 m) long. The tops of the footings were at the streambed and embedded into a substrate consisting of ice contact stratified draft (glacial till). The footings were protected by riprap. In 1955 the bridge survived a larger flood (2084 73,600 cfs (2,084 m^3/s)) than the 1987 flood (62,100 cfs (1,760 m^3/s)). However, from 1953 to 1987 the bridge was subjected to many floods which progressively removed riprap from the piers, enabling the spread footings to be undermined during the April 1987 flood (Figures 10.1 and 10.2).

The NYSTA inspected the bridge annually or biennially with the last inspection on April 1, 1986. A 1979 inspection by a consultant hired by NYSDOT indicated that most of the riprap around the piers was missing (Figures 10.1 and 10.2); however, the 1986 inspection failed to detect any problems with the condition of the riprap at the piers (Richardson et al. 1987). Based on the Safety Board findings, the conclusions from this failure are that inspectors and their supervisors must recognize that riprap does not necessarily make a bridge safe from scour, and inspectors must be trained to recognize when riprap is missing and the significance of this condition.

Figure 10.1. Photograph of riprap at pier 2, October 1956.

Figure 10.2. Photograph of riprap at pier 2, August 1977 (flow is from right to left).

10.5.3 Hatchie River Bridge Failure

On April 1, 1989 the northbound U.S. Route 51 bridge over the Hatchie River in Tennessee collapsed killing eight persons. The National Transportation Safety Board (NTSB 1990) investigated the collapse and gave as the probable cause:

> ".........the northward migration of the main river channel which the Tennessee Department of Transportation failed to evaluate and correct. Contributing to the severity of the accident was the lack of redundancy in the design of the bridge spans."

A 2-lane bridge on Route 51 was opened to traffic in 1936. It was 4,000 ft (1,219 m) long and spanned the main channel (approximately 300 ft (91 m)) and the majority of the floodplain. In 1974 a second 2-lane (southbound) bridge was added. Its length was 1,000 ft 305 m) and centered approximately on the main channel downstream from the northbound bridge. The earthfill approaches to the new southbound bridge blocked the floodplain flow that had formerly moved through the open bents of the 1936 (northbound) bridge. This concentrated the flow in both bridges and caused the main channel to move northward and into the floodplain bents of the northbound bridge.

Each of the floodplain bents of the 1936 (northbound) bridge was on a pile cap (bottom elevation 237.9 ft) supported by five untreated wooden piles 20 ft (6 m) long. The main channel bridge was on piers with a pile cap (bottom elevation 223.67 ft) supported on 20 ft (6 m) long precast concrete piles. The northward movement of the channel exposed the piles of the bent next to the channel to local pier scour and it collapsed, dropping three spans. The channel migration was documented by Tennessee DOT and U.S. Army Corps of Engineers (USACE). At the time of the collapse the flow was not large (8,620 cfs (244 m^3/s) but the flow was overbank and of long duration. By comparison, the maximum flood peak for the 1989 flood season was 28,700 cfs (813 m^3/s) with a 3-year recurrence interval (USGS 1989).

Since 1975, the bridge had been inspected at intervals of 24 to 26 months and the last inspection was in September 1987. The NTSB report stated "the 1979, 1985, and 1987 inspection reports accurately identified the channel migration around column bent 70," (the floodplain bent that failed). The report further stated "....on-site inspections of the northbound U.S. 51 Bridge adequately identified the exposure of the column bent footings and piles due to the northward migration of the Hatchie River channel." The report also noted that the inspectors did not have design or as-built plans with them during the inspection. Because of this, the inspectors were mistaken in the thickness of the pile cap and calculated that 1 ft (0.3 m) of the piles was exposed. Whereas, the piles were actually exposed 3 ft (0.9 m) in 1987. The Safety Board noted other (unrelated) bridge collapses where inspectors did not have design or as-built plans, and as a result, deficiencies were overlooked that contributed to bridge failures. Therefore, the Safety Board believes that "it is essential for inspectors to have available bridge design or as-built plans during the on-site bridge inspection."

The NTSB noted that although TDOT inspectors measured the streambed depth at each substructure element and the USACE maintained historical channel profile (cross section) data at the bridge "a channel profile of the river was not being maintained by TDOT." As a result the TDOT evaluator of the inspection report used only the 1985 and 1987 measurements and was not able to determine the extent of channel migration. In other words, if the profiles had been plotted, the evaluator should have easily detected the lateral migration.

The Safety Board also noted that an underwater inspection did not occur in 1987 because the bridge foundation was submerged less than 10 ft (3 m), TDOT criteria at that time. In 1990, TDOT changed the criteria to 3.5 ft (1 m). The Safety Board stated "a diver inspection of the bridge should have been conducted following the 1987 inspection because of the exposure of the untreated timber piles noted in the inspection report."

In conclusion, inspectors should have design or as-built plans on site during an inspection and should measure and plot a profile of the river cross section at the bridge. Submerged bridge elements that can not be examined visually or by feel should have an underwater inspection. Good communication must be established between inspectors, evaluators, and decision makers. Changes in the river need to be evaluated through comparisons of successive channel cross sections to determine whether the changes are (1) random and insignificant or (2) represent a significant pattern of change to the channel which may endanger the stability of the bridge.

10.5.4 Arroyo Pasajero Bridge Failure

On March 10, 1995 the two I-5 bridges over Los Gatos Creek (Arroyo Pasajero) in the California Central Valley near Coalinga collapsed killing seven persons and injuring one. CALTRANS retained a team of engineers from FHWA, USGS, and private consultants to investigate the accident. No report was prepared by CALTRANS but three of the investigators, in the interest of bridge engineering, prepared a paper which was published by ASCE (Richardson et al. 1997). The probable cause of the failure was:

> "The minimum scour depth from long-term degradation 10 ft (3 m) from inspection records, contraction scour 8.5 ft (2.6 m) calculated using Laursen's live bed equation, and local pier scour 6.7 ft (2 m) determined from a model study, exposed 8.9 ft (2.7 m) of the cast in place columns below the point where there was steel reinforcement. The force of the flood waters (at an angle of attack of 15 to 26 degrees) on the unreinforced columns, with their area increase by a web wall and debris, caused the bridge to fail."

The bridges, built in 1967, were 122 ft (37 m) long, with vertical wall abutments (with wing walls) and three piers. Each pier consisted of six 16 inch (406 mm) cast in place concrete columns. The columns were spaced 7.5 ft (2.3 m) on centers. They were embedded 41 ft (12.5 m) below original ground surface but only had steel reinforcing for 17 ft (5.2 m) below the original ground surface. The abutments were on pile-supported footings and the piles were 36.7 ft (11.3 m) long. A flood in 1969 lowered the bed 6 ft (1.83 m) and damaged one column. In repairing the damage CALTRANS maintenance constructed a web wall 8 or 12 ft (2.4 or 3.6 m) high, 38 ft (11.6 m) long and 2 ft (0.6 m) wide around the columns to reinforce them. The elevation of the bottom of the web wall was unknown.

Los Gatos Creek is an ephemeral stream (dry most of the time) which drains from the eastern side of the coastal range onto an alluvial fan whose head is approximately 2 mi (3.2 km) upstream of the two bridges. About 1,800 ft (548 m) upstream of the bridges Chino Creek (also ephemeral) joins Los Gatos Creek. At the time of construction Chino Creek spread over and infiltrated into its alluvial fan. Some time after construction a channel was constructed connecting the two streams and increasing the drainage area of Los Gatos Creek by about 33 percent.

The Los Gatos Creek channel upstream of the bridge is from 300 to 400 ft (91 to 122 m) wide, but only 150 to 250 ft (46 to 76 m) wide downstream. The 122 ft (37 m) wide bridge severely constricts the channel and the March 10, 1995 flood ponded upstream of the bridge. From 1955 to 1995, differential land subsidence between bench marks approximately 1.5 mi (2.4 km) upstream and 5.3 mi (8.5 km) downstream was measured as 11.5 ft (3.5 m). The bed of the stream is sand and the bedform is plane bed. Discharges are hard to quantify for this stream. For the 1995 flood, the USGS using slope-area methods determined that the discharge ranged from 16,300 to 40,300 cfs (462 to 1,141 m^3/s) and the most probable discharge was 27,300 cfs (773 m^3/s) with a recurrence interval of 75 years based on historical data.

The factors involved in the I-5 bridge failure were:

- Increase in channel slope caused by regional subsidence
- Change in the original design by maintenance forces adding a web wall between columns to repair damage from an earlier flood. With an angle of attack from 15 to 26 degrees this action potentially increased local pier scour depth by a factor of 3.6 to 4.4
- Increase in drainage area of 33 percent above the bridge by land use change and the construction of a channel to link two streams (Chino Creek to Los Gatos Creek)
- Long-term degradation of 10 ft (3 m) since the bridge was built
- Significant contraction of the flow, i.e., channel width of 300 to 400 ft (91 to 122 m) wide to a bridge width of 122 ft (37 m)

In conclusion, the various factors that contributed to this failure illustrate the complexities of inspection and the need for all elements of a DOT (inspection, maintenance, design and management) to be involved in the process. Inspectors must continually observe the conditions at the bridge, and the stream channel above and below the bridge, and communicate actions, conditions, and changes in the bridge and stream to the different sections of the organization.

10.5.5 Conclusions

These three cases illustrate the difficulty and necessity for inspection of bridges. They also illustrate the need for good communication between DOT inspection, maintenance, design and management. Inspectors must have design or as-built plans on site; must measure, plot, and compare cross sections of the channel at the bridge, and they must observe and carefully document the conditions of the bridge and the channel upstream and downstream. Maintenance personnel must inform inspectors, designers and others when they make changes to a bridge or channel. Communication is very important. Designers need to inform inspection and maintenance personnel of design assumptions and what to look for. Inspectors and maintenance personnel, because they are the "eyes" of the DOT team, must look for changes and inform others.

10.6 PLAN OF ACTION

10.6.1 Background

Scour related deficiencies are the leading cause of serious bridge failures and closures. A national scour evaluation program as an integral part of the National Bridge Inspection Standards was established in 1988 by Technical Advisory T 5140.20, published following the April 1987 collapse of New York's Schoharie Bridge due to scour. T 5140.20 was superseded in 1991 by T 5140.23, "Evaluating Scour at Bridges." This Technical Advisory provides more guidance on the development and implementation of procedures for evaluating bridge scour to meet the requirements of 23 CFR 650, Subpart C. T 5140.23 states that a Plan of Action (POA) should be developed for each existing bridge found to be scour critical. Subsequent to that technical advisory, 23 CFR 650.313(e)(3), enacted January 13, 2005, makes a POA mandatory for scour critical bridges.

Simply put, the goal of a POA is to provide guidance for inspectors and engineers that can be implemented for scour critical bridges before, during, and after flood events to protect the traveling public. The two primary components of a Plan of Action are:

1. Instructions regarding type and frequency of inspections to be made at the bridge
2. A schedule for timely design and construction of scour countermeasures

The Technical Advisory further recommends appropriate training and instruction for bridge inspectors in scour issues. These include issues such as collection and comparison of cross section data, identification of conditions indicative of potential scour problems, and effective notification procedures when an actual or potential problem is identified at or in the vicinity of the bridge.

Developing a POA for a scour critical bridge involves much more than simply establishing a schedule for inspection and installation of countermeasures. Bridge management and inspection strategies need to be considered. Countermeasure alternatives need to be evaluated. A monitoring program must be developed that addresses issues related to the type and frequency of inspection, possible use of instrumentation, and flood monitoring. Bridge closure instructions and planning for detour routes are also necessary. Finally, consideration must be given to implementation and maintenance of the POA, including points of contact, responsibilities, communications, coordination with other agencies, and notification of the public if closure or detour actions are necessary (USDOT 2003).

10.6.2 Developing a Plan of Action

Many state DOTs have developed a standardized template for POAs that is appropriate for state-specific hydrologic, hydraulic, structural and geotechnical conditions and that recognizes the relationships between the DOT, other bridge owners, and cooperating state and local government agencies.

In addition, FHWA has developed an on-line training seminar titled "Plan of Action (POA) for Scour Critical Bridges." This POA training seminar is intended for individuals at the Federal, State and local level who are involved in the planning, development, implementation and maintenance of POAs for bridges determined to be scour critical. This seminar was developed as a collaborative effort between the FHWA's National Hydraulics Team (NHT) and the National Highway Institute (NHI) and is designated as Course Number FHWA-NHI 135085.

The POA training seminar features three lessons:

- Guidance and Regulations - Provides among other details, an expanded guidance on management and inspection strategies, and components of a POA Standard Template developed by the NHT.
- Riverine Case Study - Presents a riverine case study overview of a scour critical bridge.
- Details of POA Standard Template and Riverine Case Study POA - Presents detailed guidance for completing the POA Standard Template with the information provided in the riverine case study.

Other features of the training seminar include expanded guidance for each section of the POA Standard Template, a monitoring case study, resources (technical references), and a glossary. The training seminar is available on line at no cost on the FHWA website.

The FHWA's POA Standard Template consists of 10 sections, summarized as follows:

Section	Title	Description
1	General Information	This section provides basic information about the bridge (structure number, route carried, name of water way crossed, year built, ADT etc.).
2	Responsibility for POA	POA author; responsible agency and personnel contact information.
3	Scour Vulnerability	Current rating factor for Item 113, description of scour evaluation, and summary of scour history.
4	Recommended Action(s)	Status of recommended actions (any/all): Increased inspection frequency, scour monitoring devices, flood monitoring program, countermeasures.
5	NBIS Coding Information	Current and previous rating factors for Items 60, 61, 71 and 113.
6	Monitoring Program	Details and personnel contact information for scour monitoring procedures, including scour alert and/or scour critical trigger conditions and actions required if those conditions occur.
7	Countermeasure Recommendations	Prioritized list of countermeasure alternatives, cost estimates, anticipated schedule for installation, and personnel contact information.
8	Bridge Closure Plan	Conditions requiring bridge closure, personnel contact information, and conditions under which the bridge may be re-opened to the public.
9	Detour Route	Description of detour route, identification of other bridges on detour route, traffic control equipment and location (signs, barriers), and instructions for notifying other agencies and public media.
10	Attachments	Supporting data and information including maps, photos, sketches, as-built drawings, scour calculations, preliminary countermeasure designs, etc.

10.6.3 Maintaining a Plan of Action

Once the initial POA for a bridge has been developed, it must be updated both on a periodic basis and as warranted by changes to the bridge or waterway (e.g., channel changes or bridge scour due to a flood event, installation of countermeasures, etc.) Section 2 of the POA template identifies the agency contact person responsible for the POA, as well as the planned frequency of periodic updates.

In general, periodic updates will typically be administrative in nature, for example, revising the schedule for countermeasure implementation, or changing the contact information for key personnel. Because a POA involves multi-agency coordination and communication, periodic updates are essential if a flood event were to trigger incident management and response.

In contrast, significant changes to the bridge or waterway may result in a recoding of NBIS Item 113, which requires updating the POA immediately. Inspections conducted during or after a flood event would typically be the activity that reveals the need for an immediate POA update based on an observed condition.

Other elements of periodic POA maintenance include:

- Preparedness: Preparedness involves an integrated combination of planning, training, exercises, personnel qualification and certification standards, equipment acquisition and upgrades, and publications. The updated POA must include information to reflect any changes in resource availability, communications equipment and protocols, management structure, etc. This not only applies to changes within the DOT itself, but changes that have occurred within other agencies involved in coordinated incident management activities, for example:
    - State patrol and local law enforcement agencies
    - Public works and utilities departments
    - Federal, State, local and tribal public safety organizations
    - Nongovernmental organizations and contractors

- Supporting technologies: Advances in technology may be phased into POAs, for example, in bridge monitoring programs, or communications and data-sharing systems. To the extent that DOTs and other agencies associated with a POA incorporate these advances, the POA must be updated to reflect the new technology. The management of communications and information using GIS, video conferencing, and Internet-based systems for emergency response are examples of emerging technologies applicable to POA development.

CHAPTER 11

LITERATURE CITED

Ahmad, M., 1953, "Experiments on Design and Behavior of Spur Dikes," Proceedings of the International Association of Hydraulic Research, American Society of Civil Engineers Joint Meeting, University of Minnesota, August.

American Association of State Highway and Transportation Officials, 1992a, "Standard Specifications for Highway Bridges," Fifteenth Edition, Washington, D.C.

American Association of State Highway and Transportation Officials, 1992b, "Highway Drainage Guidelines, Vol. VII, Hydraulic Analyses for the Location and Design of Bridges," Washington, D.C.

American Association of State Highway and Transportation Officials, 2010, "AASHTO LRFD Bridge Design Specifications," Customary U.S. Units, 5th Edition, with 2010 Interim Revisions," American Association of State Highway and Transportation Officials, Washington, D.C.

Annandale, G.W., 1995, "Erodibility," Journal of Hydraulic Research, Vol. 33, p. 471-494.

Annandale, G.W., 2006, "Scour Technology," McGraw-Hill, New York.

Arneson, L.A., 1998, "Vertical Contraction Scour at Bridges With Water Flowing Under Pressure Conditions," Compendium on Stream Stability and Scour Papers from ASCE Water Resources Conferences (1991 to 1998), Richardson and Lagasse (eds.), ASCE, Reston, VA.

Bieniawski, Z.T., 1989, "Engineering Rock Mass Classifications," New York: Wiley.

Briaud, J.L., F.C.K. Ting, H.C. Chen, R. Gudavaiii, S. Perugu, G. Wei, 1999a, "SRICOS: Prediction of Scour Rate in Cohesive Soils at Bridge Piers," ASCE Journal of Geotechnical and Geoenvironmental Engineering, Vol. 125, No. 4, pp 237-246, Reston, VA.

Briaud, J.L., F.C.K. Ting, H.C. Chen, R. Gudavaiii, K. Kwak, B. Philogene, S.W. Han., S. Perugu, G. Wei, P. Nurtjahyo, Y. Cao, Y. Li, 1999b, "SRICOS Prediction of Scour Rate at Bridge Piers," Report 2937-F, Texas Depart. of Transportation, Texas A&M University, Civil Engineering, College Station, TX 77843-3136.

Briaud, J.L., H.C. Chen, K.A. Chang, S.J. Oh, S. Chen, J. Wang, Y. Li, K. Kwak, P. Nartjaho, R. Gudaralli, W. Wei, S. Pergu, Y.W. Cao, and F. Ting, 2011, "The Sricos – EFA Method" Summary Report, Texas A&M University.

Bruun, P., 1966, "Tidal Inlets and Littoral Drift," Vol. 2, Washington, D.C.

Bruun, P., 1990, "Tidal Inlets on Alluvial Shores," Chapter 9, Vol. 2, Port Engineering, 4th edition, Gulf Publishing, Houston, TX.

Butler, H.L. and J. Lillycrop, 1993, "Indian River Inlet: Is there a Solution?" Hydraulic Engineering, Proc. of the 1993 National Conference, ASCE, Vol. 2, pp. 1218-1224.

Chang, F. and S.R. Davis, 1999a, "Maryland SHA Procedure for Estimating Scour at Bridge Abutments, Part I - Live Bed Scour," ASCE Compendium, Stream Stability and Scour at Highway Bridges, Richardson and Lagasse (eds.), Reston, VA.

Chang, F. and S.R. Davis, 1999b, "The Maryland State Highway Administration ABSCOUR Program," Maryland SHA.

Costa, J.E. and J.E. O'Connor, 1995, "Geomorphically Effective Floods, in Natural and Anthropogenic Influences in Fluvial Geomorphology - Wolman Volume," Costa, J.E., Miller, A.J., Potter, K.W., and Wilcock, P.R. eds., Washington D.C., American Geophysical Union Geophysical Monograph 89, p. 45-56.

Dickenson, S.E. and M.W. Baillie, 1999, "Predicting Scour in Weak Rock of the Oregon Coast Range," unpublished research report, Department of Civil, Construction, and Environmental Engineering, Oregon State University, Corvallis, OR, Final Report SPR 382, Oregon Department of Transportation and Report No. FHWA-OR-RD-00-04.

Ettema, R., 1980, "Scour at Bridge Piers," Report 215, Dept. of Civil Engineering, University of Auckland, Auckland, New Zealand.

Federal Highway Administration, 1973, "A Statistical Summary of the Cause and Cost of Bridge Failures," Federal Highway Administration, U.S. Department of Transportation, Washington, D.C. (Chang, F.F.M.).

Federal Highway Administration, 1978, "Countermeasures for Hydraulic Problems at Bridges," Vol. 1 and 2, FHWA/RD-78-162&163, Federal Highway Administration, U.S. Department of Transportation, Washington, D.C. (Brice, J.C. and J.C. Blodgett).

Federal Highway Administration, 1979, "Scour Around Bridge Piers at High Froude Numbers," Federal Highway Administration, Report No. FHWA-RD-79-104, U.S. Department of Transportation, Washington, D.C., April (Jain, S.C. and R.E. Fischer).

Federal Highway Administration, 1988a, "Scour at Bridges," Technical Advisory T5140.20, updated by Technical Advisory T5140.23, October 28, 1991, "Evaluating Scour at Bridges," U.S. Department of Transportation, Washington, D.C.

Federal Highway Administration 1988b, "Interim Procedures for Evaluating Scour at Bridges," U.S. Department of Transportation, Washington, D.C.

Federal Highway Administration, 1988c, "Revisions to the National Bridge Inspection Standards (NBIS), T5140.21, U.S. Department of Transportation, Washington, D.C.

Federal Highway Administration, 1991, "Scourability of Rock Formations," U.S. Department of Transportation Memorandum, HNG-31, Washington, D.C.

Federal Highway Administration, 1994, "Development of Hydraulic Computer Models to Analyze Tidal and Coastal Hydraulic Conditions at Highway Structures, Phase I Report," FHWA-SC-94-4, Federal Highway Administration, Washington, D.C. (Richardson, E.V., B.L. Edge, L.W. Zevenbergen, J.R. Richardson, P.F. Lagasse, J.S. Fisher, and R. Greneir).

Federal Highway Administration, 1995, "Recording and Coding Guide for the Structure Inventory and Appraisal of the Nation's Bridges, Report No. FHWA-PD-96-001, U.S. Department of Transportation, Washington, D.C.

Federal Highway Administration, 1999a, "Abutment Scour Studies for Compound Channels," U.S. Department of Transportation, Federal Highway Administration, September (Sturm, T.W.).

Federal Highway Administration, 1999b, "Predicting Scour in Weak Rock of the Oregon Coast Range," Report No. FHWA-OR-RD-00-04, Washington, D.C., October (Dickenson, S.E. and M.W. Baillie).

Federal Highway Administration, 2001, "River Engineering for Highway Encroachments - Highways in the River Environment," FHWA NHI 01-004, Federal Highway Administration, Hydraulic Design Series No. 6, Washington, D.C. (Richardson, E.V., D.B. Simons, and P.F. Lagasse).

Federal Highway Administration, 2002, "Evaluation of Soil and Rock Properties," Geotechnical Engineering Circular No. 5, FHWA Publication No. FHWA-IF-0-034, Washington, D.C. (Sabatini, P.J., R.C. Bachus, P.W. Mayne, J.A. Schneider, T.E. Zettler).

Federal Highway Administration, 2003a, "Bottomless Culvert Scour Study: Phase I Laboratory Report," Federal Highway Administration, Report No. FHWA-RD-02-078 (Kerenyi, K., J.S. Jones, and S. Stein).

Federal Highway Administration, 2003b, "Finite Element Surface-Water Modeling System: Two-Dimensional Flow in a Horizontal Plane," FESWMS-2DH, Version 2, User's Manual, U.S. Department of Transportation, Research, Development, and Technology, Turner-Fairbank Highway Research Center, McLean, VA (Froehlich, D.C.).

Federal Highway Administration, 2004, "Tidal Hydrology, Hydraulics and Scour at Bridges," Hydraulic Engineering Circular No. 25, First Edition, available as a web document, FHWA, Washington, D.C. (Zevenbergen, L.W., P.F. Lagasse, and B.L. Edge).

Federal Highway Administration, 2005, "Design and Construction of Driven Pile Foundations," Volumes I and II, FHWA-NHI-05-042 and 043, Washington, D.C.

Federal Highway Administration, 2006, "Soils and Foundations Workshop - Reference Manual Volume 1," FHWA-06-088, Federal Highway Administration, Washington, D.C., December.

Federal Highway Administration, 2007, "Bottomless Culvert Scour Study: Phase II Laboratory Report," Federal Highway Administration, Report No. FHWA-HRT-07-026 (Kerenyi, K., J.S., Jones, and S. Stein).

Federal Highway Administration, 2008, "Highways in the Coastal Environment," Hydraulic Engineering Circular No. 25, Second Edition, Publication No. FHWA-NHI-07-096, Washington, D.C. (Douglass, S.L. and J. Krolak).

Federal Highway Administration, 2009, "Bridge Scour and Stream Instability Countermeasures - Experience, Selection, and Design Guidelines, <u>Hydraulic Engineering Circular No. 23</u>, Third Edition, FHWA-NHI 09-111 (Vol. 1), FHWA-NHI-09-112 (Vol. 2), Federal Highway Administration, Washington, D.C. (Lagasse, P.F., P.E. Clopper, J.E. Pagán-Ortiz, L.W. Zevenbergen, L.A. Arneson, J.D. Schall, and L.G. Girard).

Federal Highway Administration, 2010, "Drilled Shafts: Construction Procedures and LRFD Design Methods," Geotechnical Engineering Circular No. 10, FHWA Publication No. FHWA-NHI-10-0016, Washington, D.C. (Brown, D.A., J.P. Turner, and R.J. Castelli).

Federal Highway Administration, 2012a, "Hydraulic Design of Safe Bridges," Report FHWA-HIF-12-018, <u>Hydraulic Design Series No. 7</u>, Washington, D.C. (Zevenbergen, L.W., L.A. Arneson, J.H. Hunt, and A.C. Miller).

Federal Highway Administration, 2012b, "Stream Stability at Highway Structures," <u>Hydraulic Engineering Circular No. 20</u>, Fourth Edition, HIF-FHWA-12-004, Federal Highway Administration, Washington, D.C. (Lagasse, P.F., L.W. Zevenbergen, W.J. Spitz, and L.A. Arneson).

Federal Highway Administration, 2012c, "Submerged-Flow Bridge Scour under Clear-Water Condition," Federal Highway Administration, Report No. FHWA-HRT-12-034 (Suaznabar, O., H. Shan, Z. Xie, J. Shen, and K. Kerenyi).

Federal Highway Administration, 2012d, "Pier Scour in Clear-Water Conditions with Nonuniform Bed Materials," Federal Highway Administration, Report No. FHWA-HRT-12-022 (Guo, J., O. Suaznabar, H. Shan, Z. Xie, J. Shen, and K. Kerenyi).

Florida Department of Transportation, 2011, "Bridge Scour Manual," Tallahassee, FL.

Hoffmans, G.J.C.M. and H.J. Verheij, 1997, "Scour Manual," A.A. Balkema: Rotterdam, Brookfield.

International Society for Rock Mechanics, 1981, "ISRM Report on Teaching of Rock Mechanics," International Society for Rock Mechanics, ISRM Secretariat, Lisbon, Portugal.

Jones, J.S. and D.M. Sheppard, 2000, "Local Scour at Complex Pier Geometries," Proceedings of the ASCE 2000 Joint Conference on Water Resources Engineering and Water Resources Planning and Management, July 30 - August 2, Minneapolis, MN.

Kerenyi, K. and J. Pagán-Ortiz, 2007, "Testing Bottomless Culverts," Public Roads, Vol. 70, No. 6, May/June 2007.

Kirsten, H.A.D., 1982, "A Classification System for Excavation in Natural Materials," The Civil Engineer in South Africa, pp. 292-308, July.

Landers, M.N., D.S. Mueller, and E.V. Richardson, 1999, "U.S. Geological Survey Field Measurements of Pier Scour," ASCE Compendium, Stream Stability and Scour at Bridges, Richardson and Lagasse (eds.), Reston, VA.

Laursen, E.M., 1960, "Scour at Bridge Crossings," Journal Hydraulic Division, American Society of Civil Engineers, Vol. 86, No. HY 2.

Laursen, E.M., 1963, "An Analysis of Relief Bridge Scour," Journal Hydraulic Division, American Society of Civil Engineers, Vol. 89, No. HY3.

Laursen, E.M., 1980, "Predicting Scour at Bridge Piers and Abutments," General Report No. 3, Arizona Department of Transportation, Phoenix, AZ.

Liu, H.K., Chang, F.M., and M.M. Skinner, 1961, "Effect of Bridge Constriction on Scour and Backwater," Department of Civil Engineering, Colorado State University, Fort Collins, CO.

McWhorter, D.B. and D.K. Sunada, 1977, "Ground-Water Hydrology and Hydraulics," Water Resources Publications, Fort Collins, CO.

Melville, B.W., 1992, "Local Scour at Bridge Abutments," Journal of Hydraulic Engineering, American Society of Civil Engineers, Hydraulic Division, Vol. 118, No. 4.

Melville, B.W. and A.J. Sutherland, 1988, "Design Method for Local Scour at Bridge Piers," American Society of Civil Engineers, Journal Hydraulic Division, Vol. 114, No. 10, October.

Melville, B.W. and S.E. Coleman, 2000, "Bridge Scour," Water Resources Publications, LLC, Highlands Ranch, CO.

Mueller, D.S., 1996, "Local Scour at Bridge Piers in Nonuniform Sediment Under Dynamic Conditions," Dissertation in partial fulfillment of the requirements for the Degree of Doctor of Philosophy, Colorado State University, Fort Collins, CO.

National Cooperative Highway Research Program, 1997, "Instrumentation for Measuring Scour at Bridge Piers and Abutments," NCHRP Report 396, Transportation Research Board, National Academy of Science, Washington, D.C. (Lagasse, P.F., E.V. Richardson, J.D. Schall, and G.R. Price).

National Cooperative Highway Research Program, 2004, "Pier and Contraction Scour in Cohesive Soils," NCHRP Report 516, Transportation Research Board, National Academy of Science, Washington, D.C. (Briaud, J.L., H.C. Chen, Y. Li, P. Nurtjahyo, and J. Wang).

National Cooperative Highway Research Program, 2006, "Riprap Design Criteria, Recommended Specifications and Quality Control, NCHRP Report 568, Transportation Research Board, National Academy of Science, Washington, D.C. (Lagasse, P.R., P.E. Clopper, L.W. Zevenbergen, and J.F. Ruff).

National Cooperative Highway Research Program, 2007, "Countermeasures to Protect Bridge Abutments from Scour," NCHRP Report No. 587, Transportation Research Board, National Academy of Science, Washington, D.C. (Barkdoll, B.D., R. Ettema, and B.W. Melville).

National Cooperative Highway Research Program, 2009, "Monitoring Scour Critical Bridges: A Synthesis of Highway Practice," NCHRP Synthesis Report 396, Transportation Research Board, National Academy of Science, Washington, D.C. (Hunt, B.E.).

National Cooperative Highway Research Program, 2010a, "Effects of Debris on Bridge Pier Scour," NCHRP Report 653, Transportation Research Board, National Academy of Science, Washington, D.C. (Lagasse, P.F., P.E. Clopper, L.W. Zevenbergen, W.J. Spitz, and L.G. Girard).

National Cooperative Highway Research Program, 2010b, "Estimation of Scour Depth at Bridge Abutments," NCHRP Project 24-20, Draft Final Report, Transportation Research Board, National Academy of Science, Washington, D.C. (Ettema, R., T. Nakato, and M. Muste).

National Cooperative Highway Research Program, 2011a, "Evaluation of Bridge Pier Scour Research: Scour Processes and Prediction," NCHRP Project 24-27(01), Transportation Research Board, National Academy of Science, Washington, D.C. (Ettema, R., Constantinescu, G., and B.W. Melville).

National Cooperative Highway Research Program, 2011b, "Evaluation of Bridge-Scour Research: Abutment and Contraction Scour Processes and Prediction," NCHRP Project 24-27(02), Transportation Research Board, National Academy of Science, Washington, D.C., (Sturm, T., Melville, B.W., and R. Ettema).

National Cooperative Highway Research Program, 2011c, "Scour at Wide Piers and Long Skewed Piers," NCHRP Report 682, Transportation Research Board, National Academy of Science, Washington, D.C., (Sheppard, D.M., Melville, B.W., and H. Deamir).

National Cooperative Highway Research Program, 2011d, "Estimation of Scour Depth at Bridge Abutments," NCHRP Project 24-20, Transportation Research Board, National Academy of Science, Washington, D.C., (Ettema, R., Nagato, T., and M. Muste).

National Cooperative Highway Research Program, 2011e, "Scour at Bridge Foundations on Rock," Final Report, NCHRP Project 24-29, Transportation Research Board, National Academy of Science, Washington, D.C. (Keaton, J.R., S.K. Mishra, and P.E. Clopper).

National Transportation Safety Board, 1988, "Collapse of the New York Thruway (I-90) Bridge over the Schoharie Creek, Near Amsterdam, New York, April 5, 1987," NTSB/HAR-88/02, NTSB, Washington, D.C.

National Transportation Safety Board, 1990, "Collapse of the Northbound U.S. Route 51 Bridge Spans over the Hatchie River near Covington, Tennessee," April 1, 1989, NTSB/HAR-90/01, National Transportation Safety Board, Washington, D.C.

Neill, C.R. (Editor), 2004, "Guide to Bridge Hydraulics," Second Edition prepared by Project Committee on Bridge Hydraulics Roads and Transportation Association of Canada (TAC), University of Toronto Press, Toronto, Canada.

Raudkivi, A.J., 1986, "Functional Trends of Scour at Bridge Piers," American Society of Civil Engineers, Journal of the Hydraulics Division, Vol. 112, No. 1.

Richardson, E.V. and J.R. Richardson, 1992, Discussion of Melville, B.W., 1992, "Local Scour at Bridge Abutments," American Society of Civil Engineers, Journal of Hydraulics Division, September.

Richardson, E.V. and L. Abed, 1993, "Topwidth of Pier Scour Holes in Free and Pressure Flow," ASCE Hydraulic Engineering, Proc. 1993 National Conference, San Francisco, CA, August.

Richardson, E.V. and P.F. Lagasse (eds.), 1999, "Stream Stability and Scour at Highway Bridge - Compendium of Papers, ASCE Water Resources Engineering Conferences 1991-1998," American Society of Civil Engineers, Reston, VA.

Richardson, E.V., J.S. Jones, and J.C. Blodgett, 1997, "Findings of the I-5 Bridge Failure," ASCE Hydraulic Engineering Proceedings of Theme A, 27th IAHR Congress, San Francisco, CA.

Richardson, E.V., P.F. Lagasse, J.D. Schall, J.F. Ruff, T.E. Brisbane, and D.M. Frick, 1987, "Hydraulic, Erosion and Channel Stability Analysis of the Schoharie Creek Bridge Failure, New York," Resource Consultants, Inc. and Colorado State University, Fort Collins, CO.

Salim, M. and J.S. Jones, 1995, "Effects of Exposed Pile Foundations on Local Pier Scour," Proceedings ASCE Water Resources Engineering Conference, San Antonio, TX.

Salim, M. and J.S. Jones, 1996, "Scour Around Exposed Pile Foundations," Proceedings ASCE North American and Water and Environment Congress, '96, Anaheim, CA (also issued as FHWA Memo).

Salim, M. and J.S. Jones, 1999, Scour Around Exposed Pile Foundations," ASCE Compendium, Stream Stability and Scour at Highway Bridges, Richardson and Lagasse (eds.), Reston, VA.

Sheppard, D.M., 2001, "A Methodology for Predicting Local Scour Depths Near Bridge Piers with Complex Geometries," unpublished design procedure, University of Florida, Gainesville, FL.

Sheppard, D.M. and W. Miller, 2006, "Live-bed Local Pier Scour Experiments," Journal of Hydraulic Engineering - ASCE, 132(7), 635-642.

Smith, W.L., 1999, "Local Structure-Induced Sediment Scour at Pile Groups," M.S. Thesis, University of Florida, Gainesville, FL.

Sowers, G.B. and G.F. Sowers, 1970, "Introductory Soil Mechanics and Foundations," Macmillian Publishing Col, NY.

Sturm, T.W., 1999, "Abutment Scour in Compound Channels," ASCE Compendium, Stream Stability and Scour at Highway Bridges, Richardson and Lagasse (eds.), Reston, VA.

Sturm, T.W. and A. Chrisochoides, 1998, "Abutment Scour in Compound Channel for Variable Setbacks," Water Resources Engineering, Proc. of the International Water Resources Engineering Conference, ASCE, Memphis, 1, 174-179.

Transportation Research Board, 1983, "Comparison of Prediction Equations for Bridge Pier and Abutment Scour," Transportation Research Record 950, Second Bridge Engineering Conference, Vol. 2, Transportation Research Board, Washington, D.C. (Jones, J.S.).

Transportation Research Board, 1984, "Case Histories of Scour Problems at Bridges," Transportation Research Record 950, Second Bridge Engineering Conference, Vol. 2, Transportation Research Board, Washington, D.C. (Davis, S.R.).

Transportation Research Board, 1989, "Abutment Scour Prediction," Presentation, Transportation Research Board, Washington, D.C. (Froehlich, D.C.).

Transportation Research Board, 1993a, "Bridge Scour in Tidal Waters," Transportation Research Board, Washington, D.C. (Sheppard, D.M.).

Transportation Research Board, 1993b, "Tidal Inlet Bridge Scour Assessment Model," Transportation Research Record 1420, TRB, National Research Council, Washington, D.C., pp. 7-13 (Vincent, M.S., M.A. Ross, and B.E. Ross).

Transportation Research Board, 1994, "Scour Around Wide Piers in Shallow Water," Transportation Research Board Record 1471, Transportation Research Board, Washington, D.C. (Johnson, P.A. and E.F. Torrico).

Transportation Research Board, 1995, "Bridge Scour in the Coastal Region," Proc. Fourth International Bridge Conference, Transportation Research Board, Washington, D.C. (Richardson, J.R., E.V. Richardson, and B.L. Edge).

Transportation Research Board, 1998a, "Highway Infrastructure Damage Caused by the 1993 Upper Mississippi River Basin Flooding," Transportation Research Board Report 417, Washington, D.C. (Parola, A.C., D.J. Hagerty, and S. Kamojjala).

Transportation Research Board, 1998b, "Vertical contraction scour at bridges with water flowing under pressure conditions." Transportation Research Record. 1647, 10-17 (Arneson, L.A. and S.R. Abt).

Transportation Research Board, 1999, "1998 Scanning Review of European Practice for Bridge Scour and Stream Instability Countermeasures," National Cooperative Highway Research Program, Research Results Digest, Number 241, Washington, D.C.

Transportation Research Board, 2009, "Joint Workshop on Abutment Scour: Present Knowledge and Future Needs," Research Results Digest 334, National Cooperative Research Program, TRB, Washington, D.C.

Umbrell, E.R., G.K. Young, S.M. Stein, and J.S. Jones, 1998, "Clear-water contraction scour under bridges in pressure flow," J. Hydraul. Engrg. 124(2), 236-240.

U.S. Army Corps of Engineers, 1981, "Final Report to Congress, The Streambank Erosion Control Evaluation and Demonstration Act of 1974," Washington, D.C.

U.S. Army Corps of Engineers, 1983, "Streambank Protection Guidelines," U.S. Army Engineers Waterways Experiment Station, Vicksburg, MS (Keown, M.P.).

U.S. Army Corps of Engineers, 1985, "Users Manual for the Generalized Computer Program System: Open Channel Flow and Sedimentation, TABS-2," U.S. Army Engineers Waterways Experiment Station, Vicksburg, MS, 671 pp. (Thomas, W.A. and W.H. McAnally).

U.S. Army Corps of Engineers, 1993a, EM1110-2-1601, ELT1110-2-120, Vicksburg, MS.

U.S. Army Corps of Engineers, 1993b, "UNET - One Dimensional Unsteady Flow Through a Full Network of Open Channels," Report CPD-66, U.S. Army Corps of Engineers, Hydrologic Engineering Center, Davis, CA (Barkau, R.L.).

U.S. Army Corps of Engineers, 1997, "Users Guide to RMA2 WES Version 4.3," U.S. Army Corps of Engineers - Waterways Experiment Station, Barbara Donnell, ed., Vicksburg, MS.

U.S. Army Corps of Engineers, 2010a, "River Analysis System," HEC-RAS, User's Manual Version 4.1, Hydrologic Engineering Center, Davis, CA.

U.S. Army Corps of Engineers, 2010b, "Hydrologic Modeling System HEC-HMS," User's Manual Version 3.5, Hydrologic Engineering Center, Davis, CA.

U.S. Department of Agriculture, 1993, "Soil Survey Manual," Soil Conservation Service, USDA Handbook 18, Washington, D.C., October.

U.S. Department of Transportation, Federal Highway Administration, 1988, "Scour at Bridges," Technical Advisory T5140.20, updated by Technical Advisory T5140.23, October 28, 1991, "Evaluating Scour at Bridges," U.S. Department of Transportation, Washington, D.C.

U.S. Department of Transportation, 1989, "Laboratory Studies of the Effects of Footings and Pile Groups on Bridge Pier Scour," U.S. Interagency Sedimentation Committee Bridge Scour Symposium, Washington, D.C. (Jones, J.S.).

U.S. Department of Transportation, Federal Highway Administration, 2001, "Revision of Coding Guide, Item 113 - Scour Critical Bridges," Memorandum, HIBT-30, April 27, Washington, D.C.

U.S. Department of Transportation, Federal Highway Administration, 2003, "Compliance with National Bridge Inspection Standards; Plan of Action for Scour Critical Bridges," Memorandum, HIBT-20, July 24, Washington, D.C.

U.S. Department of Transportation, Federal Highway Administration, 2004, "National Bridge Inspection Standards," Federal Register, Volume 69, No. 239, 23CFR Part 650, FHWA Docket No. FHWA-2001-8954, Final Rule, December 14, 2004, effective January 13, 2005, Washington, D.C.

U.S. Geological Survey, 1989, "Channel Evolution of the Hatchie River near the U.S. Highway 51 Crossing in Lauderdale and Tipton Counties, West Tennessee," USGS Open-File Report 89-598, Nashville, TN (Bryan, B.S.).

U.S. Geological Survey, 2011, "Investigation of Pier Scour in Coarse-Bed Streams in Montana, 2001 through 2007," In cooperation with the Montana Department of Transportation, Scientific Investigations Report 2011-5107 (Holnbeck, S.R.).

Yang, C.T., 1996, "Sediment Transport: Theory and Practice," B.J. Clark and J.M. Morris (eds.), McGraw-Hill Companies, Inc.

(page intentionally left blank)

APPENDIX A

METRIC SYSTEM, CONVERSION FACTORS, AND WATER PROPERTIES

(page intentionally left blank)

APPENDIX A

Metric System, Conversion Factors, and Water Properties

The following information is summarized from the Federal Highway Administration, National Highway Institute (NHI) Course No. 12301, "Metric (SI) Training for Highway Agencies." For additional information, refer to the Participant Notebook for NHI Course No. 12301.

In SI there are seven base units, many derived units and two supplemental units (Table A.1). Base units uniquely describe a property requiring measurement. One of the most common units in civil engineering is length, with a base unit of meters in SI. Decimal multiples of meter include the kilometer (1000m), the centimeter (1m/100) and the millimeter (1 m/1000). The second base unit relevant to highway applications is the kilogram, a measure of mass which is the inertial of an object. There is a subtle difference between mass and weight. In SI, mass is a base unit, while weight is a derived quantity related to mass and the acceleration of gravity, sometimes referred to as the force of gravity. In SI the unit of mass is the kilogram and the unit of weight/force is the newton. Table A.2 illustrates the relationship of mass and weight. The unit of time is the same in SI as in the English system (seconds). The measurement of temperature is Centigrade. The following equation converts Fahrenheit temperatures to Centigrade, $°C = 5/9 (°F - 32)$.

Derived units are formed by combining base units to express other characteristics. Common derived units in highway drainage engineering include area, volume, velocity, and density. Some derived units have special names (Table A.3).

Table A.4 provides useful conversion factors from English to SI units. The symbols used in this table for metric units, including the use of upper and lower case (e.g., kilometer is "km" and a newton is "N") are the standards that should be followed. Table A.5 provides the standard SI prefixes and their definitions.

Table A.6 provides physical properties of water at atmospheric pressure in SI system of units. Table A.7 gives the sediment grade scale and Table A.8 gives some common equivalent hydraulic units.

Table A.1. Overview of SI Units.

	Base Units	Units	Symbol
Base units	length	meter	m
	mass	kilogram	kg
	time	second	s
	temperature*	kelvin	K
	electrical current	ampere	A
	luminous intensity	candela	cd
	amount of material	mole	mol
Supplementary units	angles in the plane	radian	rad
	solid angles	steradian	sr

*Use degrees Celsius (°C), which has a more common usage than kelvin.

Table A.2. Relationship of Mass and Weight.

System	Mass	Weight or Force of Gravity	Force
English	slug, pound-mass	pound, pound-force	pound, pound-force
metric	kilogram	newton	newton

Table A.3. Derived Units With Special Names.			
Quantity	Name	Symbol	Expression
Frequency	hertz	Hz	s^{-1}
Force	newton	N	kg • m/s^2
Pressure, stress	pascal	Pa	N/m^2
Energy, work, quantity of heat	joule	J	N • m
Power, radiant flux	watt	W	J/s
Electric charge, quantity	coulomb	C	A • s
Electric potential	volt	V	W/A
Capacitance	farad	F	C/V
Electric resistance	ohm	Ω	V/A
Electric conductance	siemens	S	A/V
Magnetic flux	weber	Wb	V • s
Magnetic flux density	tesla	T	Wb/m^2
Inductance	henry	H	Wb/A
Luminous flux	lumen	lm	cd • sr
Illuminance	lux	lx	lm/m^2

Table A.4. Useful Conversion Factors.			
Quantity	From English Units	To Metric Units	Multiply by *
Length	mile	km	1.609
	yard	m	0.9144
	foot	m	0.3048
	inch	mm	25.40
Area	square mile	km^2	2.590
	acre	m^2	4047
	acre	hectare	0.4047
	square yard	m^2	0.8361
	square foot	m^2	0.09290
	square inch	mm^2	645.2
Volume	acre foot	m^3	1233
	cubic yard	m^3	0.7646
	cubic foot	m^3	0.02832
	cubic foot	L (1000 cm^3)	28.32
	100 board feet	m^3	0.2360
	gallon	L (1000 cm^3)	3.785
	cubic inch	cm^3	16.39
Mass	lb	kg	0.4536
	kip (1000 lb)	metric ton (1000 kg)	0.4536
Mass/unit length	plf	kg/m	1.488
Mass/unit area	psf	kg/m^2	4.882
Mass density	pcf	kg/m^3	16.02
Force	lb	N	4.448
	kip	kN	4.448
Force/unit length	plf	N/m	14.59
	klf	kN/m	14.59
Pressure, stress, modulus of elasticity	psf	Pa	47.88
	ksf	kPa	47.88
	psi	kPa	6.895
	ksi	MPa	6.895
Bending moment, torque	ft-lb	N · m	1.356
	ft-kip	kN · m	1.356
Moment of mass	lb · ft	m	0.1383
Moment of inertia	lb · ft^2	kg · m^2	0.04214
Second moment of area	in^4	mm^4	416200
Section modulus	in^3	mm^3	16390
Power	ton (refrig)	kW	3.517
	Btu/s	kW	1.054
	hp (electric)	W	745.7
	Btu/h	W	0.2931
*4 significant figures; underline denotes exact conversion			

Table A.4. Useful Conversion Factors (continued).			
Quantity	From English Units	To Metric Units	Multiply by *
Volume rate of flow	ft^3/s	m^3/s	0.02832
	cfm	m^3/s	0.0004719
	cfm	L/s	0.4719
	mgd	m^3/s	0.0438
Velocity, speed	ft/s	m/s	<u>0.3048</u>
Acceleration	f/s^2	m/s^2	<u>0.3048</u>
Momentum	lb · ft/sec	kg · m/s	0.1383
Angular momentum	lb · ft^2/s	kg · m^2/s	0.04214
Plane angle	degree	rad	0.01745
	degree	mrad	17.45

*4 significant figures; underline denotes exact conversion

Table A.5. Prefixes.					
Submultiple Name	Submultiple Factor	Submultiple Symbol	Multiple Name	Multiple Factor	Multiple Symbol
deci	10^{-1}	d	deka	10^1	da
centi	10^{-2}	c	hecto	10^2	h
milli	10^{-3}	m	kilo	10^3	k
micro	10^{-6}	µ	mega	10^6	M
nano	10^{-9}	n	giga	10^9	G
pica	10^{-12}	p	tera	10^{12}	T
femto	10^{-15}	f	peta	10^{15}	P
atto	10^{-18}	a	exa	10^{18}	E
zepto	10^{-21}	z	zetta	10^{21}	Z
yocto	10^{-24}	y	yotto	10^{24}	Y

Table A.6. Physical Properties of Water at Atmospheric Pressure in SI Units.

Temp-erature	Temp-erature	Density	Specific weight	Dynamic Viscosity	Kinematic Viscosity	Vapor Pressure	Surface Tension[1]	Bulk Modulus
Centigrade	Fahrenheit	kg/m^3	N/m^3	N.s/m^2	m^2/s	N/m^2 abs.	N/m	GN/m^2
0	32	1,000	9,810	1.79 x 10^{-3}	1.79 x 10^{-6}	611	0.0756	1.99
5	41	1,000	9,810	1.51 x 10^{-3}	1.51 x 10^{-6}	872	0.0749	2.05
10	50	1,000	9,810	1.31 x 10^{-3}	1.31 x 10^{-6}	1,230	0.0742	2.11
15	59	999	9,800	1.14 x 10^{-3}	1.14 x 10^{-6}	1,700	0.0735	2.16
20	68	996	9,790	1.00 x 10^{-3}	1.00 x 10^{-6}	2,340	0.0728	2.20
25	77	997	9,781	8.91 x 10^{-4}	8.94 x 10^{-7}	3,170	0.0720	2.23
30	86	996	9,771	7.97 x 10^{-4}	8.00 x 10^{-7}	4,250	0.0712	2.25
35	95	994	9,751	7.20 x 10^{-4}	7.24 x 10^{-7}	5,630	0.0704	2.27
40	104	992	9,732	8.53 x 10^{-4}	6.58 x 10^{-7}	7,380	0.0696	2.28
50	122	988	9,693	5.47 x 10^{-4}	5.53 x 10^{-7}	12,300	0.0679	
60	140	983	9,843	4.68 x 10^{-4}	4.74 x 10^{-7}	20,000	0.0662	
70	158	978	9,694	4.04 x 10^{-4}	4.13 x 10^{-7}	31,200	0.0644	
80	176	972	9,535	3.54 x 10^{-4}	3.64 x 10^{-7}	47,400	0.0626	
90	194	965	9,467	3.15 x 10^{-4}	3.26 x 10^{-7}	70,100	0.0607	
100	212	958	9,398	2.82 x 10^{-4}	2.94 x 10^{-7}	101,300	0.0589	

[1]Surface tension of water in contact with air

Table A.7. Physical Properties of Water at Atmospheric Pressure in English Units.

Temperature	Temperature	Density	Specific Weight	Dynamic Viscosity	Kinematic Viscosity	Vapor Pressure	Surface Tension[1]	Bulk Modulus
Fahrenheit	Centigrade	Slugs/ft^3	Weight lb/ft^3	lb-sec/ft^2 x 10^{-4}	ft^2/sec x 10^{-5}	lb/in^2	lb/ft	lb/in^2
32	0	1.940	62.416	0.374	1.93	0.09	0.00518	287,000
39.2	4.0	1.940	62.424					
40	4.4	1.940	62.423	0.323	1.67	0.12	0.00514	296,000
50	10.0	1.940	62.408	0.273	1.41	0.18	0.00508	305,000
60	15.6	1.939	62.366	0.235	1.21	0.26	0.00504	313,000
70	21.1	1.936	62.300	0.205	1.06	0.36	0.00497	319,000
80	26.7	1.934	62.217	0.180	0.929	0.51	0.00492	325,000
90	32.2	1.931	62.118	0.160	0.828	0.70	0.00486	329,000
100	37.8	1.927	61.998	0.143	0.741	0.95	0.00479	331,000
120	48.9	1.918	61.719	0.117	0.610	1.69	0.00466	332,000
140	60.0	1.908	61.386	0.0979	0.513	2.89		
160	71.1	1.896	61.006	0.0835	0.440	4.74		
180	82.2	1.883	60.586	0.0726	0.385	7.51		
200	93.3	1.869	60.135	0.0637	0.341	11.52		
212	100	1.847	59.843	0.0593	0.319	14.70		

[1]Surface tension of water in contact with air

Table A.8. Sediment Particles Grade Scale.							
Size				Approximate Sieve Mesh Openings Per Inch		Class
Millimeters	Millimeters	Microns	Inches	Tyler	U.S. Standard	Name
4000-2000			160-80			Very large boulders
2000-1000			80-40			Large boulders
1000-500			40-20			Medium boulders
500-250			20-10			Small boulders
250-130			10-5			Large cobbles
130-64			5-2.5			Small cobbles
64-32			2.5-1.3			Very coarse gravel
32-16			1.3-0.6			Coarse gravel
16-8			0.6-0.3	2.5		Medium gravel
8-4			0.3-0.16	5	5	Fine gravel
4-2			0.16-0.08	9	10	Very fine gravel
2-1	2.00-1.00	2000-1000		16	18	Very coarse sand
1-1/2	1.00-0.50	1000-500		32	35	Coarse sand
1/2-1/4	0.50-0.25	500-250		60	60	Medium sand
1/4-1/8	0.25-0.125	250-125		115	120	Fine sand
1/8-1/16	0.125-0.062	125-62		250	230	Very fine sand
1/16-1/32	0062-0031	62-31				Coarse silt
1/32-1/64	0.031-0.016	31-16				Medium silt
1/64-1/128	0.016-0.008	16-8				Fine silt
1/128-1/256	0.008-0.004	8-4				Very fine silt
1/256-1/512	0.004-0.0020	4-2				Coarse clay
1/512-1/1024	0.0020-0.0010	2-1				Medium clay
1/1024-1/2048	0.0010-0.0005	1-0.5				Fine clay
1/2048-1/4096	0.0005-0.0002	0.5-0.24				Very fine clay

Table A.9. Common Equivalent Hydraulic Units.

Volume

Unit	cubic inch	liter	U.S. gallon	cubic foot	cubic yard	cubic meter	acre-foot	sec-foot-day
liter	61.02	1	0.264 2	0.035 31	0.001 31	0.001	810.6 E-9	408.7 E-9
U.S. gallon	231	3.785	1	0.133 7	0.004 95	0.003 79	3.068 E-6	1.547 E-6
cubic foot	1,728	28.32	7.481	1	0.037 04	0.028 32	22.96 E-6	11.57 E-6
cubic yard	46,660	764.6	202	27	1	0.746 60	619.8 E-6	312.5 E-6
meter^3	61,020	1,000	264.2	35.31	1.308	1	810.6 E-6	408.7 E-6
acre-foot	75.27 E+6	1,233,000	325,900	43,560	1,613	1,233	1	0.5042
sec-foot-day	149.3 E+6	2,447,000	646,400	86,400	3,200	2,447	1.983	1

Discharge (Flow Rate, Volume/Time)

Unit	gallon/min	liter/sec	acre-foot/day	foot^3/sec	million gal/day	meter^3/sec
gallon/minute	1	0.063 09	0.004 419	0.002 228	0.001 440	63.09 E-06
liter/second	15.85	1	0.070 05	0.035 31	0.022 82	0.001
acre-foot/day	226.3	14.28	1	0.504 2	0.325 9	0.014 28
feet^3/second	448.8	28.32	1.983	1	0.646 3	0.028 32
million gal/day	694.4	43.81	3.068	1.547	1	0.043 82
meter^3/second	15,850	1,000	70.04	35.31	22.82	1

APPENDIX B

EXTREME EVENTS

(page intentionally left blank)

APPENDIX B

EXTREME EVENTS

B.1 INTRODUCTION

In 1994, AASHTO introduced an entirely new set of specifications based on the concept of load and resistance factor design (LRFD) methodology. The factors were developed from the theory of reliability based upon current statistical knowledge of loads and structural performance. In the evaluation of scour at bridge structures, there are two conditions, or limit states, that are of primary interest in design. The design flood for scour is used in the evaluation of these limit states:

1. Service Limit States, or limit states relating to stress, deformation and cracking

2. Strength Limit States, or limit states relating to strength and stability

Extreme-Event Limit States relate to events with return periods in excess of the design life of the bridge. There are generally three such limit states that may involve consideration of the effect of scour at bridges:

1. A flood event exceeding the scour design flood (the check flood for scour, or superflood, is used to evaluate scour for this event as described in Chapter 2).

2. An earthquake

3. A vessel collision with the bridge

Recent research (NCHRP 2003) has recommended that extreme live loads (i.e., vehicular loads) and extreme wind loads on the bridge structure also be considered in the evaluation of extreme event limit states.

In addition to the above, there are other conditions possibly relating to scour that the designer may determine are significant for a specific watershed or river system, such as the occurrence of ice loads or debris accumulations.

Scour reduces the resistance factor by changing the conditions of the bridge substructure. Loading from extreme events are related to scour with regard to the possibility that they could occur at the same time that a flood event is occurring. The loss of foundation support due to scour could then affect the stability of the foundation in resisting the additional loading. Recommendations for the consideration of the joint probability of one of these events in combination with a scour-producing flood event are discussed in the following sections.

B.2 CHANGES IN FOUNDATIONS DUE TO LIMIT STATE FOR SCOUR

In accordance with the standards set forth in the AASHTO LRFD Specifications (AASHTO 1994), the consequences of changes in foundation conditions resulting from the design flood for scour shall be considered at strength and service limit states. The consequences of changes in foundation conditions due to scour resulting from the check flood for bridge scour and from hurricanes shall be considered at the extreme event limit state.

Scour is not a force effect, but by changing the conditions of the substructure it may have a significant effect in altering the force effects acting on structures. The AASHTO LRFD Specifications, Section 3, sets forth detailed requirements for applying loads and load factors to bridge foundations. The extreme event limit states and the loads to be applied for these limit states are explained in this section based on the 1994 AASHTO Specifications.

The strength and service limit states are used in the design of a bridge foundation. Structures designed to resist damage from scour will be designed under this provision using normal design considerations and factors of safety selected by the foundation engineer. The assumption is made that all material in the scour prism has been removed and is unavailable for foundation support.

Scour shall be considered in extreme event load combinations as outlined below:

Extreme Event I - Load combination including earthquake

This extreme event limit state includes water loads and earthquakes. The probability of a major flood and an earthquake occurring at the same time is very small. Therefore, consideration of basing water loads and scour depths on mean discharges may be warranted (when considering the joint probability of an earthquake and scour). Mean discharges are considered to be normal (non-flood) flows representing the typical or daily flows in the river.

Extreme Event II - Load combination related to ice load, collision by vessels and vehicles, and certain hydraulic loads with a reduced live load other than that which is a part of the vehicular collision load

This extreme event limit state is a load combination for extreme events such as ice loads, collision by vessels and vehicles, and the check flood for scour. Its application for the check flood for scour involves a reduced live load on the structure of 50 percent. The assumption is made that all material in the scour prism has been removed and is unavailable for foundation support. The structure is to remain stable for this condition, but is not required to have any reserve capacity to resist loads.

The recurrence interval of these extreme events is expected to exceed the design life of the bridge. The joint probability of these events is extremely low, and, therefore, the events are specified to be applied separately.

The Engineer is cautioned to consider the following when applying the above noted AASHTO specifications to the evaluation of the joint probability of a flood and another extreme event. These considerations incorporate recommendations from some of the papers presented at a conference on "The Design of Bridges for Extreme Events" sponsored by the Federal Highway Administration (FHWA 1996).

- There are several current studies underway to evaluate the joint probability of extreme events. Until further and more definitive conclusions are drawn from these studies, judgment is necessary in evaluating site-specific factors on a case by case basis that could affect the safety of the traveling public.

- A differentiation must be made between long-term scour (degradation) and short-term scour (local scour and general (contraction) scour). It is reasonable to consider expected long-term degradation in evaluating the joint probability of occurrence of scour with an earthquake or vessel collision event since it is associated with a period of many years. On the other hand, live-bed local scour and contraction scour may occur only for a period of hours or days before the scour hole refills; consequently, the joint probability of this type of scour with an earthquake or vessel collision is very low. In some cases, clear-water scour holes may occur and not refill or refill very slowly. While the joint probability of the occurrence of a 100-year flood/clear-water scour hole and another extreme event is very low, the engineer may wish to consider a clear-water scour hole associated with a lesser flood event.

- The probability of the simultaneous occurrence of an extreme vessel collision load (by a ship or barge transiting the navigable channel at normal operating speeds) and short- term scour resulting from a 100-year flood is very low and can be neglected as a load combination. The probability of the simultaneous occurrence of a vessel collision load from a single (empty) hopper barge floating in the waterway at the speed of the current and both long- and short-term scour is valid and should be considered in the design where applicable.

B.3 NCHRP PROJECT 12-48: COMBINATIONS OF EXTREME EVENTS FOR HIGHWAY BRIDGES

The magnitude and consequences of extreme events such as vessel collisions, winds, earthquakes, and scour caused by flooding often govern the design of highway bridges. If these events are considered to occur simultaneously, the resulting loading condition may dominate the design. This superpositioning of extreme load values frequently increases construction costs unnecessarily because a simultaneous occurrence of two or more extreme events is unlikely. The reduced probability of simultaneous occurrence for each load combination may be determined using statistical procedures.

NCHRP Project 12-48 was initiated in 1998 to develop a design procedure for the application of extreme event loads and load combinations to highway bridges. This objective was achieved with a recommended design procedure consistent with the uniform reliability methodologies and philosophy included in the AASHTO LRFD Bridge Design Specifications. The resulting report, "Design of Highway Bridges for Extreme Events," (NCHRP Report 489) (NCHRP 2003) provides recommendations for four extreme event load combinations to maintain a consistent level of safety against failure caused by scour in combination with the following transient loads:

1. Live loads
2. Wind loads
3. Vessel collision
4. Earthquake

The extreme events of concern are transient loads with relatively low rates of occurrences and uncertain intensity levels. Once an extreme event occurs, its time duration is also a random variable with varying length, depending on the nature of the event. For example, truck loading events are normally of very short duration (on the order of a fraction of a second to 2 to 3 seconds) depending on the length of the bridge, the speed of traffic, and the number of trucks crossing the bridge simultaneously. Windstorms have varying ranges of time duration and may last for a few hours. Most earthquakes last for 10 to 15 seconds while ship collisions are instantaneous events.

On the other hand, the effects of scour may last for a few days to a few months for live bed scour, and possibly for the remainder of the life of a bridge pier under clear water conditions.

The transient nature of these loads, their low rate of occurrence, and their varying duration times imply that the probability of the simultaneous occurrence of two events is generally small. The exceptions are:

1. When one of the loads occurs frequently (e.g., truckloads)
2. When two loads are correlated (ship collision and windstorm)
3. When one of the loads lasts for long time periods (e.g., scour or, to a lesser extent, wind)

Even when two (or more) load types occur simultaneously, there is little chance that the intensities of both events will be close to their maximum lifetime values. For example, the chances are very low that the trucks crossing a bridge are very heavily loaded at the time of the occurrence of a high-velocity windstorm. On the other hand, because ship collisions are more likely to occur during a windstorm, the effect of high wind velocities may well combine with high-impact loads from ship collisions; therefore, ship collisions and high wind loads are considered to be correlated events.

In the case of scour, once a bridge's foundations have been weakened due to scour, the bridge would be exposed to a higher risk of failure given the occurrence of any other extreme event. Although scour occurs due to floods that may follow heavy windstorms, the time lag between the occurrence of a flood after the storm justifies assuming independence between extreme wind loads and scour events.

NCHRP Report 489 (TRB 2003) describes the calibration process used to provide a set of design loads associated with appropriate load factors to provide an "acceptably safe" envelope to all these possible combinations, with the goal of providing reliability index values commensurate with AASHTO LRFD methodology. The term "acceptably safe" is used because absolute safety is impossible to achieve. Using a bridge design life of 75 years, *NCHRP Report 489 recommends the following load factors for the four combinations of extreme events that include a scour component.* These event combinations are referred to as Extreme Events III through VI in NCHRP Report 489, as described below. The variables in the load factor recommendations are:

> **DC** DEAD LOAD OF STRUCTURAL COMPONENTS AND NONSTRUCTURAL ATTACHMENTS
> **SC** SCOUR
> **LL** VEHICULAR LIVE LOAD
> **WS** WIND LOAD ON STRUCTURE
> **CV** VESSEL COLLISION FORCE
> **EQ** EARTHQUAKE

1. Extreme Event III:

This event is a combination of either: a) extreme event scour plus dead loads, or b) extreme event scour plus dead loads and vehicular live loads.

> a) 1.25(DC); 2.0(SC)
> b) 1.25(DC) + 1.75(LL); 1.80(SC)

A scour factor equal to 1.80 is recommended for use in combination with a live load factor equal to 1.75. The lower scour load factor for the combination of scour and live loads as compared with the load factor proposed for scour alone reflects the lower probability of having the maximum possible 75-year live load occur when the scour erosion is also at its maximum 75-year depth.

2. Extreme Event IV:

This event consists of extreme event scour in combination with extreme wind load on the structure.

1.25(DC) + 1.40(WS); 0.70(SC)

A scour factor equal to 0.70 is recommended in combination with a wind load factor equal to 1.40. The lower scour factor observed in combination involving wind loads as compared with those involving live loads reflect the lower number of wind storms expected in the 75-year design life of the structure.

3. Extreme Event V:

This event consists of extreme event scour in combination with vessel collision.

1.25(DC) + 1.00(CV); 0.60(SC)

A scour factor equal to 0.60 is recommended in combination with vessel collision forces. The lower scour factor observed in combinations that involve collisions reflects the lower number of collisions excepted in the 75-year bridge
design life.

4. Extreme Event VI:

This event consists of extreme event scour in combination with earthquake loading.

1.25(DC) + 1.00(EQ); 0.25(SC)

A scour factor equal to 0.25 is recommended in combination with earthquakes. The lower scour factor with earthquakes reflects the fact that as long as a total washout of the foundation does not occur, bridge columns subjected to scour exhibit lower flexibilities that will help reduce the inertial forces caused by earthquakes. This reduction in inertial forces partially offsets the scour-induced reduction in soil depth and the resulting soil resisting capacity.

As of this writing, the load recommendations associated with Extreme Events III through VI as recommended in NCHRP Report 489 have not been formally adopted by AASHTO or accepted by FHWA. Therefore, the information in this appendix is provided for information only.

B.4   DESIGN FLOOD EXCEEDANCE PROBABILITY

A flood event with a recurrence interval of T years has a 1/T probability of being exceeded in any one year. The 100-year recurrence interval flood is often used as a hydraulic design value and to establish other types of flooding potential. Regardless of the flood design level, there is a chance, or probability, that it will be exceeded in any one year and the probability increases depending on the life of the structure. The probability that a flood event frequency will be exceeded in N years depends on the annual probability of exceedance as defined by:

$$P_N = 1 - (1 - P_a)^N \qquad (B.1)$$

where:

    P_N = Probability of exceedance in N years
    P_a = Annual probability of exceedance (1/T)
    N = Number of years
    T = Flood event frequency of exceedance

The number of years, N, can be assumed to equal the bridge design life or remaining life. Table B.1 shows the probability of exceedance of various flood frequencies for time periods (that may be assumed to equal the bridge design life) ranging from 1 to 100 years. For example a 100-year flood has an annual (N = 1) probability of exceedance of 1.0 percent, but has a 39.5 percent chance of exceedance in 50 years. A 200-year flood has a 22.2 percent chance of being exceeded in 50 years and a 31.3 percent chance of being exceeded in 75 years.

The probability of exceedance may be applied to an individual bridge or for a population of similar bridges. Therefore, if a 200-year design flood condition is used for a population of bridges with expected design lives of 75 years, then that flood condition will be exceeded at approximately 31.3 percent of the bridges over their lives. Because design flood conditions are exceeded at many bridges during their useful lives, factors of safety, conservative design relationships, and LRFD are used to provide adequate levels of safety and reliability in bridge design.

Table B.1. Probability of Flood Exceedance of Various Flood Levels.							
Flood Frequency	Probability of Exceedance in N Years (or Assumed Bridge Design Life)						
Years	N = 1	N = 5	N = 10	N = 25	N = 50	N = 75	N = 100
10	10.0%	41.0%	65.1%	92.8%	99.5%	100.0%	100.0%
25	4.0%	18.5%	33.5%	64.0%	87.0%	95.3%	98.3%
50	2.0%	9.6%	18.3%	39.7%	63.6%	78.0%	86.7%
100	1.0%	4.9%	9.6%	22.2%	39.5%	52.9%	63.4%
200	0.5%	2.5%	4.9%	11.8%	22.2%	31.3%	39.4%
500	0.2%	1.0%	2.0%	4.9%	9.5%	13.9%	18.1%

B.5 REFERENCES

1. American Association of State Highway and Transportation Officials, 1994, "LRFD Bridge Design Specifications and Commentary," First Edition, Washington, D.C.

2. Federal Highway Administration, 1996, "The Design of Bridges for Extreme Events," Conference Proceedings, Washington, D.C.

3. National Cooperative Highway Research Program, 2003, "Design of Highway Bridges for Extreme Events," NCHRP Report 489, Transportation Research Board, Washington, D.C. (Ghosn, M., F. Moses, and J. Wang).

APPENDIX C

CONTRACTION SCOUR AND CRITICAL VELOCITY EQUATIONS

(page intentionally left blank)

APPENDIX C

Contraction Scour and Critical Velocity Equations

C.1 CONTRACTION SCOUR

Contraction scour occurs when the flow area of a stream at flood stage is reduced, either by a natural contraction or bridge. It also occurs when overbank flow is forced back to the channel by roadway embankments at the approaches to a bridge. From continuity, a decrease in flow area results in an increase in average velocity and bed shear stress through the contraction. Hence, there is an increase in erosive forces in the contraction and more bed material is removed from the contracted reach than is transported into the reach. This increase in transport of bed material from the reach lowers the natural bed elevation. As the bed elevation is lowered, the flow area increases and, in the riverine situation, the velocity and shear stress decrease until relative equilibrium is reached; i.e., the quantity of bed material that is transported into the reach is equal to that removed from the reach, or the bed shear stress is decreased to a value such that no sediment is transported out of the reach.

In coastal waterways which are affected by tides, as the cross-sectional area increases the discharge from the ocean may increase and thus the velocity and shear stress may not decrease. Consequently, relative equilibrium may not be reached. Thus, at tidal inlets contraction scour may result in a continual lowering of the bed (long-term degradation).

Live-bed contraction scour is typically cyclic; for example, the bed scours during the rising stage of a runoff event and fills on the falling stage. The contraction of flow due to a bridge can be caused by either a natural decrease in flow area of the stream channel or by abutments projecting into the channel and/or piers blocking a portion of the flow area. Contraction can also be caused by the approaches to a bridge cutting off floodplain flow. This can cause clear-water scour on a setback portion of a bridge section or a relief bridge because the floodplain flow does not normally transport significant concentrations of bed material sediments. This clear-water picks up additional sediment from the bed in the bridge opening. In addition, local scour at abutments may well be greater due to the clear-water floodplain flow returning to the main channel at the end of the abutment.

Other factors that can cause contraction scour are (1) natural stream constrictions, (2) long highway approaches to the bridge over the floodplain, (3) ice formations or jams, (4) natural berms along the banks due to sediment deposits, (5) debris, (6) vegetative growth in the channel or floodplain, and (7) pressure flow.

Contraction Scour Equations. There are two forms of contraction scour depending upon the competence of the uncontracted approach flow to transport bed material into the contraction.

Live-bed scour occurs when there is streambed sediment being transported into the contracted section from upstream. In this case, the scour hole reaches equilibrium when the transport of bed material out of the scour hole is equal to that transported into the scour hole from upstream.

Clear-water scour occurs when the bed material sediment transport in the uncontracted approach flow is negligible or the material being transported in the upstream reach is transported through the downstream reach at less than the capacity of the flow. In this case, the scour hole reaches equilibrium when the average bed shear stress is less than that required for incipient motion of the bed material.

Contraction scour equations are based on the principle of conservation of sediment transport (continuity). As scour develops, the shear stress in the contracted section decreases as a result of a larger flow area and decreasing average velocity. For **live-bed** scour, maximum scour occurs when the shear stress reduces to the point that sediment transported in equals the bed sediment transported out and the conditions for sediment continuity are in balance. For **clear-water** scour, the transport into the contracted section is essentially zero and maximum scour occurs when the shear stress reduces to the critical shear stress of the bed material in the bridge cross-section.

The information in this appendix is provided as background on the development and derivation of the contraction scour equations. Chapter 6 provides all the information necessary to estimate contraction scour.

C.2 LIVE-BED CONTRACTION SCOUR EQUATION

Live-bed contraction scour occurs at a bridge when there is transport of bed material in the upstream reach into the bridge cross section. With live-bed contraction scour the area of the contracted section increases until, in the limit, the transport of sediment out of the contracted section equals the sediment transported in. Normally, the width of the contracted section is constrained and depth increases until the limiting conditions are reached.

Laursen derived the following live-bed contraction scour equation based on a simplified transport function, transport of sediment in uniform flow upstream and downstream of a long contraction, and other simplifying assumptions (Laursen 1960).

$$\frac{y_2}{y_1} = \left(\frac{Q_2}{Q_1}\right)^{6/7} \left(\frac{W_1}{W_2}\right)^{k_1} \left(\frac{n_2}{n_1}\right)^{k_2} \qquad (C.1)$$

$$y_s = y_2 - y_o = \text{(Average scour depth)} \qquad (C.2)$$

where:

y_1 = Average depth in the upstream main channel, m
y_2 = Average depth in the contracted section, m
y_o = Existing depth in the contracted section before scour, m
Q_1 = Flow in the upstream channel transporting sediment, m³/s
Q_2 = Flow in the contracted channel, m³/s. Often this is equal to the total discharge unless the total flood flow is reduced by relief bridges, water overtopping the approach roadway, or in the setback area
W_1 = Bottom width of the upstream main channel, m
W_2 = Bottom width of main channel in the contracted section, m
n_1 = Manning n for upstream main channel
n_2 = Manning n for contracted section
k_1 & k_2 = Exponents determined below depending on the mode of bed material transport

V_*/ω	k_1	k_2	Mode of Bed Material Transport
<0.50	0.59	0.066	Mostly contact bed material discharge
0.50 to 2.0	0.64	0.21	Some suspended bed material discharge
>2.0	0.69	0.37	Mostly suspended bed material discharge

V_* = $(gyS_1)^{1/2}$ shear velocity in the upstream section, m/s
ω = Median fall velocity of the bed material based on the D_{50}, m/s (see Figure 6.8 in Chapter 6)
g = Acceleration of gravity (9.81 m/s^2)
S_1 = Slope of energy grade line of main channel, m/m
D_{50} = Median diameter of the bed material, m

C.3 CLEAR-WATER CONTRACTION SCOUR EQUATIONS

Clear-water contraction scour occurs in a bridge opening when (1) there is no bed material transport from the upstream reach into the downstream reach or (2) the material being transported in the upstream reach is transported through the downstream reach mostly in suspension and at less than capacity of the flow. With **clear-water** contraction scour the area of the contracted section increases until, in the limit, the velocity of the flow (V) or the shear stress (τ_o) on the bed is equal to the critical velocity (V_c) or the critical shear stress (τ_c) of a certain particle size (D) in the bed material. Normally, the width (W) of the contracted section is constrained and the depth (y) increases until the limiting conditions are reached.

Following a development given by Laursen (1963) equations for determining the clear-water contraction scour in a long contraction were developed in metric units. For equilibrium in the contracted reach:

$$\tau_o = \tau_c \tag{C.3}$$

where:

τ_o = Average bed shear stress, contracted section, Pa (N/m^2)
τ_c = Critical bed shear stress at incipient motion, Pa (N/m^2)

The average bed shear stress using y for the hydraulic radius (R) and Manning equation to determine the slope (S_f) can be expressed as follows:

$$\tau_o = \gamma\, y\, S_f = \frac{\rho\, g\, n^2\, V^2}{y^{1/3}} \tag{C.4}$$

For noncohesive bed materials and fully developed clear-water contraction scour, the critical shear stress can be determined using Shields relation (Laursen 1963, FHWA 2001).

$$\tau_c = K_s\, (\rho_s - \rho)\, g\, D \tag{C.5}$$

The bed in a long contraction scours until $\tau_o = \tau_c$ resulting in

$$\frac{\rho\, g\, n^2\, V^2}{y^{1/3}} = K_s\, (\rho_s - \rho)\, g\, D \tag{C.6}$$

Solving for the depth (y) in the contracted section gives

$$y = \left[\frac{n^2 V^2}{K_s (S_s - 1) D} \right]^3 \tag{C.7}$$

In terms of discharge (Q) the depth (y) is

$$y = \left[\frac{n^2 Q^2}{K_s (S_s - 1) D W^2} \right]^{3/7} \tag{C.8}$$

where:

y	=	Average equilibrium depth in the contracted section after contraction scour, m
S_f	=	Slope of the energy grade line, m/m
V	=	Average velocity in the contracted section, m/s
D	=	Diameter of smallest nontransportable particle in the bed material, m
Q	=	Discharge, m³/s
W	=	Bottom width of contracted section, m
g	=	Acceleration of gravity (9.81 m/s²)
n	=	Manning roughness coefficient
K_s	=	Shield's coefficient
S_s	=	Specific gravity (2.65 for quartz)
γ	=	Unit weight of water (9800 N/m³)
ρ	=	Density of water (1000 kg/m³)
$ρ_s$	=	Density of sediment (quartz, 2647 kg/m³)

Equations C.7 and C.8 are the basic equations for the **clear-water** scour depth (y) in a long contraction. Laursen, in English units used a value of 4 for K_s ($ρ_s-ρ$)g in Equation C.5; D_{50} for the size (D) of the smallest nonmoving particle in the bed material and Strickler's approximation for Manning n (n = 0.034 $D_{50}^{1/6}$) (Laursen 1963). Laursen's assumption that $τ_c$ = 4 D_{50} with S_s = 2.65 is equivalent to assuming a Shields parameter K_s = 0.039.

From experiments in flumes and studies in natural rivers with bed material of sand, gravel cobbles, and boulders, Shield's coefficient (K_s) to initiate motion ranges from 0.01 to 0.25 and is a function of particle size, Froude Number, and size distribution (FHWA 2001, Parker et al. 1982, Andrews 1983, Neill 1968). Some typical values for K_s for Fr. < 0.8 and as a function of bed material size are (1) K_s = 0.047 for sand (D_{50} from 0.065 to 2.0 mm); (2) K_s = 0.03 for median coarse-bed material (2 mm > D_{50} < 40 mm) and (3) K_s = 0.02 for coarse-bed material (D_{50} > 40 mm).

In metric units, Strickler's equation for n as given by Laursen is 0.041 $D_{50}^{1/6}$, where D_{50} is in meters. Research discussed in HDS 6 (FHWA 2001) recommends the use of the effective mean bed material size (D_m) in place of the D_{50} size for the beginning of motion (D_m = 1.25 D_{50}). Changing D_{50} to D_m in the Strickler's equation gives n = 0.040 $D_m^{1/6}$. Substituting K_s = 0.039 into Equations C.7 and C.8 gives the following equations for y:

$$y = \left[\frac{V^2}{40 D_m^{2/3}} \right]^3 \tag{C.9}$$

$$y = \left[\frac{Q^2}{40 \, D_m^{2/3} \, W^2}\right]^{3/7} \tag{C.10}$$

$$y_s = y - y_o = \text{(average scour depth)} \tag{C.11}$$

where:

A	=	Discharge through contraction, m^3/s
D_m	=	Diameter of the bed material (1.25 D_{50}) in the contracted section, m
W	=	Bottom width in contraction, m
y_o	=	Average existing depth in the contracted section, m

The **clear-water** contraction scour equations assume homogeneous bed materials. However, with clear-water scour in stratified materials, using the layer with the finest D_{50} would result in the most conservative estimate of contraction scour. Alternatively, the clear-water contraction scour equations could be used sequentially for stratified bed materials.

Equations C.8 and C.10 do not give the distribution of the contraction scour in the cross section. In many cases, assuming a uniform contraction scour depth across the opening would not be in error (e.g., short bridges, relief bridges and bridges, with simple cross sections and on straight reaches). However, for wide bridges, bridges on bends, bridges with large overbank flow, or crossings with a large variation in bed material size distribution, the contraction scour depths will not be uniformly distributed across the bridge opening. In these cases, Equations C.7 or C.9 can be used if the distribution of the velocity and/or the bed material is known. Equations C.8 or C.10 are used to determine the average contraction scour depth in the section.

Both the **live-bed** and **clear-water** contraction scour equations are the best that are available and should be regarded as a first level of analysis. If a more detailed analysis is warranted, a sediment transport model could be used (FHWA 2001).

C.4 CRITICAL VELOCITY OF THE BED MATERIAL

The velocity and depth given in Equation C.7 are associated with initiation of motion of the indicated particle size (D). Rearranging Equation C.7 to give the critical velocity (V_c) for beginning of motion of bed material of size D results in

$$V_c = \left[\frac{K_s^{1/2} (S_s - 1)^{1/2} D^{1/2} y^{1/6}}{n}\right] \tag{C.11}$$

Using $K_s = 0.039$, $S_s = 2.65$, and $n = 0.041 \, D^{1/6}$

$$V_c = 6.19 \, y^{1/6} \, D^{1/3} \tag{C.12}$$

where:

V_c	=	Critical velocity above which bed material of size D and smaller will be transported, m/s
K_s	=	Shields parameter
S_s	=	Specific gravity of the bed material
D	=	Size of bed material, m
y	=	Depth of flow, m
n	=	Manning roughness coefficient

C.5 REFERENCES

Andrews, E.C., 1983, "Entrainment of Gravel from Naturally Sorted Riverbed Material," Bulletin Geological Society of America, Vol. 94, Oct.

Federal Highway Administration (FHWA), 2001, "River Engineering for Highway Encroachments, Highways in the River Environment," FHWA NHI-01-004, Federal Highway Administration, Hydraulic Design Series No. 6, U.S. Department of Transportation, Washington, D.C. (Richardson, E.V., D.B. Simons, and P.F. Lagasse).

Laursen, E.M., 1960, "Scour at Bridge Crossings," Journal Hydraulic Division, American Society of Civil Engineers, Vol. 86, No. HY 2.

Laursen, E.M., 1963, "An Analysis of Relief Bridge Scour," Journal Hydraulic Division, American Society of Civil Engineers, Vol. 89, No. HY3.

Neill, C.R., 1968, "A Re-examination of the beginning of Movement for Coarse Granular bed Materials," Report No. INT68, Hydraulics Research station, Wallingford, United Kingdom, June.

Parker, G., Klingeman, P.C., and Mclean, D.G., 1982, "Bedload and Size Distribution in Paved Gravel-bed Streams," Journal Hydraulic Division, ASCE, Vol. 108, No. HY4.

APPENDIX D

COMPREHENSIVE SCOUR PROBLEM

(page intentionally left blank)

APPENDIX D

Comprehensive Scour Problem

D.1 GENERAL DESCRIPTION OF PROBLEM

This example problem parallels the total scour workshop in National Highway Institute (NHI) training course 135046 "Stream Stability and Scour at Highway Bridges." The U.S. Army Corps of Engineers HEC-RAS computer program was used to obtain the hydraulic variables (USACE 2010a). The solution follows Steps 1-6 of the specific design approach of Chapter 2 (Section 2.4). **Data for this problem have been altered for instructional purposes.**

PROBLEM STATEMENT

Your organization has been tasked with conducting a scour evaluation of the Ordbend Bridge on the Mainstream River. You have completed a site visit and stream reconnaissance, as well as a qualitative assessment using procedures outlined in HEC-20 "Stream Stability at Highway Bridges" (FHWA 2012b). The next task is to complete a total scour analysis. You have assembled a number of historic maps and photographs and made the following notes relevant to the bridge site:

- The Mainstream River valley and location of the Ordbend Bridge reach as shown in Figure D.1.
- Flow is from top to bottom in Figure D.1.
- An old bridge, located just downstream of the Ordbend Bridge site was replaced in 1970-71 with a bridge that had a better alignment with flow conditions and the existing highway.
- Following construction of the Ordbend Bridge, the flow alignment in the upstream channel shifted east as the result of closure of a split flow channel around an upstream mid-channel bar.
- Potential channel instability and the change in flow characteristics require an evaluation of the scour critical status of the bridge. If the bridge is scour critical, a Plan of Action will need to be developed.
- This reach of the river can be classified as a meandering river.
- The sinuosity of this reach is 1.87 and is classified as a Type D channel using procedures outlined in HEC-20, Chapter 5 (FHWA 2012b). The bridge reach is a single-phase channel that is wider at bends with chutes formed across many of the point bars.

You have also assembled the following information specific to the bridge (see Figure D.2):

- The design discharge is 120,000 cfs (3400 m^3/s)
- Bridge length is 1200 ft (365.8 m)
- Spill through abutments (2H:1V)
- 10 equal spans, 9 piers total (1 in left overbank, 5 in channel, and 3 in right overbank)
- Piers are 3 ft (0.91 m) circular columns
- Left abutment set back approximately 153 ft (46.6 m) from left bank
- Right abutment set back approximately 431 ft (131.4 m) from right bank

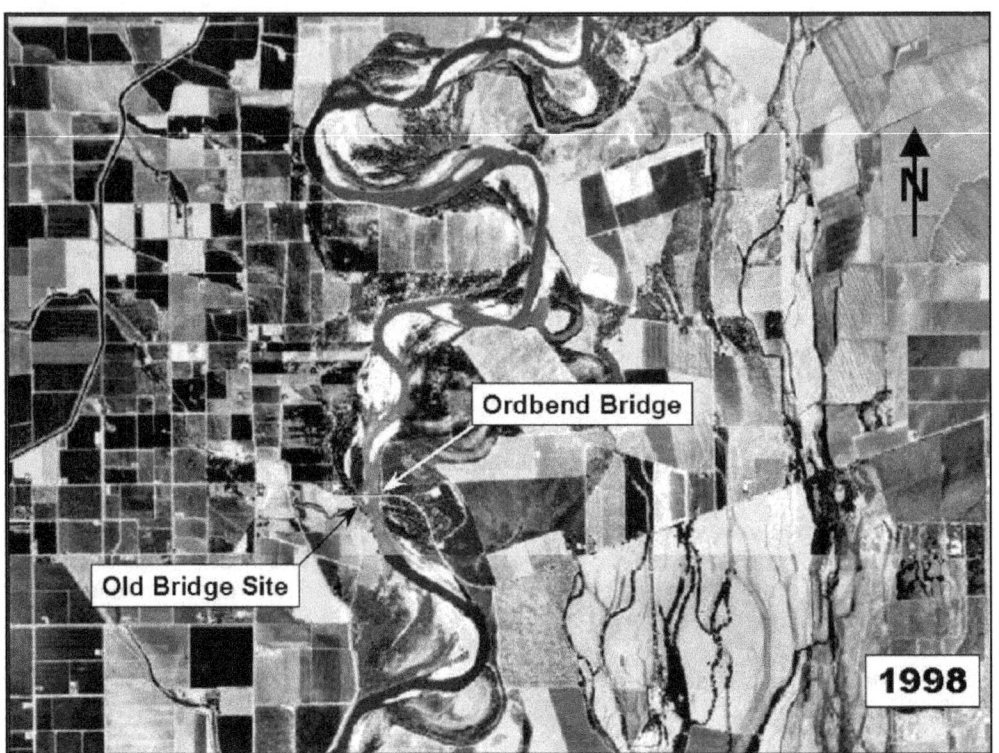

Figure D.1. Aerial photo of the bridge reach of the Mainstream River and floodplain in 1998.

Figure D.2. View of Ordbend bridge looking upstream showing bridge characteristics.

Field reconnaissance and sediment sampling have provided the following information (see Figure D.3):

- Bed Material D_{84} = 4.5 mm
- Bed Material D_{50} = 2.0 mm
- Bed Material D_{16} = 0.62 mm
- Bank Material D_{50} = 0.35 mm

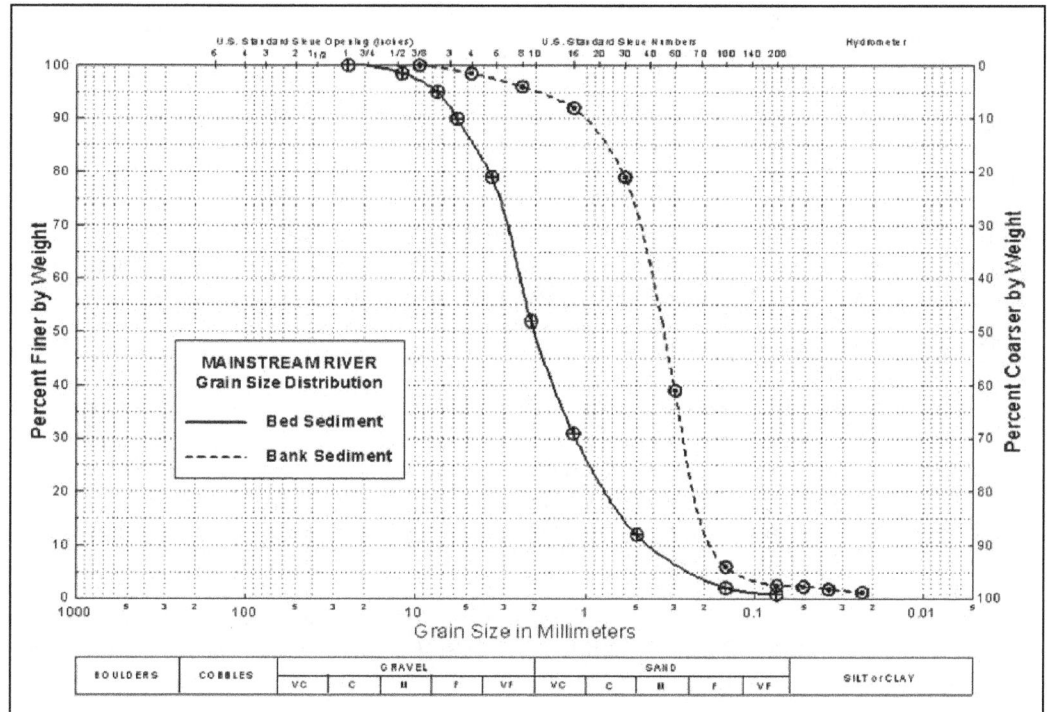

Figure D.3. Grain size distribution curves of bed and left bank samples for Mainstream River.

REQUIREMENTS

Your task is to perform the scour computations for the Mainstream River crossing near Ordbend, USA. Required computations:

- Main channel contraction scour
- Left overbank contraction scour
- Right overbank contraction scour
- Pier scour using maximum channel velocity (for zero and 15 degree angle of attack)
- Pier scour for lane expansion (bridge widening)
- Left abutment scour using HIRE equation
- Right abutment scour using HIRE equation
- Plot total scour
- Useful constants: g = 32.2 ft/s² = 9.81 m/s², ρ = 1.94 slugs/ft³ = 1000 kg/m³, γ = 62.4 lb/ft³ = 9810 N/m³.

D.2 STEP 1: DETERMINE SCOUR ANALYSIS VARIABLES

From Level 1 and Level 2 analysis - a site investigation of the crossing was conducted to identify potential stream stability problems at this crossing. Evaluation of the site indicates that the Mainstream River is a highly sinuous, actively meandering river subject to frequent chute and neck cutoffs. The presence of remnant channels indicates that there is a strong potential for lateral shifting of the channel.

Review of bridge inspection reports for bridges located upstream and downstream of the crossing indicates no long-term aggradation or degradation in this reach. At the bridge site, bedrock is approximately 150 ft (46 m) below the channel bed.

Although the bed material includes fine to medium gravel, no armoring potential is expected. At low flow, the bed for this channel consists of dunes. At higher flows, the bed will be either plane bed or antidunes.

The left and right banks are relatively well vegetated and stable; however, there are isolated portions of the bank which appear to have been undercut and are eroding. Brush and trees grow to the edge of the banks. Banks will require riprap protection if disturbed.

HYDRAULIC CHARACTERISTICS

Hydraulic characteristics at the bridge were determined using HEC-RAS (USACE 2010a). Three cross sections were used for this analysis and are denoted as "EXIT" for the section downstream of the bridge, "CROSSING" for the full-valley section at the bridge, and "APPROACH" for the approach section located approximately one bridge length upstream of the bridge. The bridge geometry was superimposed on the full-valley section and is denoted "BRIDGE" (See Figures D.4 and D.5).

Hydraulic variables for performing the various scour computations were determined from the output and from Figures D.5, D.6, and D.7. These variables, which will be used to compute contraction scour and local scour, are presented in Tables D.1 through D.3.

Contraction scour could occur both in the main channel and on the left and right overbanks of the bridge opening. For the main channel, contraction scour could be either clear-water or live-bed depending on the magnitude of the channel velocity and the critical velocity for sediment movement. A computation should be performed to determine the sediment transport characteristics of the main channel and the appropriate contraction scour equation.

In the overbank areas adjacent to the left and right abutments, contraction scour would be the result of clear-water conditions. This is because the overbank areas upstream of the bridge are vegetated, and because the velocities in these areas will be low. Thus, returning overbank flow which will pass under the bridge adjacent to the abutments will not be transporting significant amounts of material to replenish the scour on the overbank adjacent to the abutments.

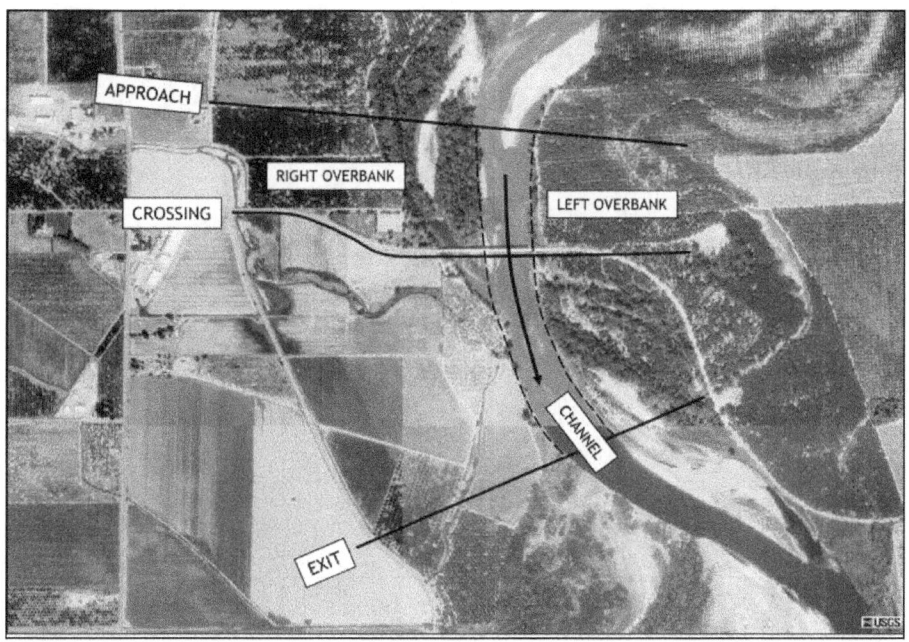

Figure D.4. Cross section layout for hydraulic model in vicinity of bridge.

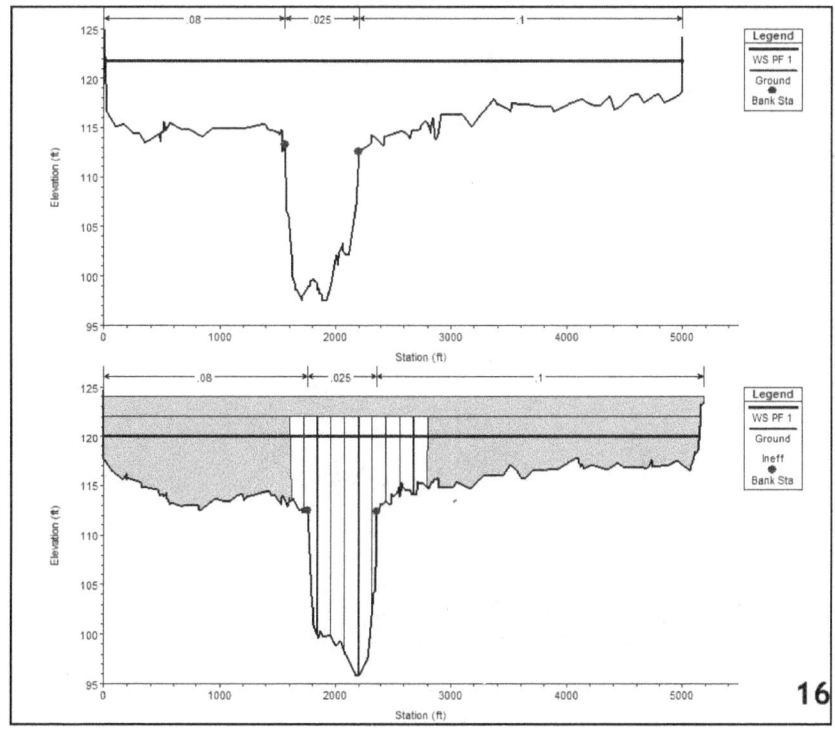

Figure D.5. Approach and Bridge cross sections.

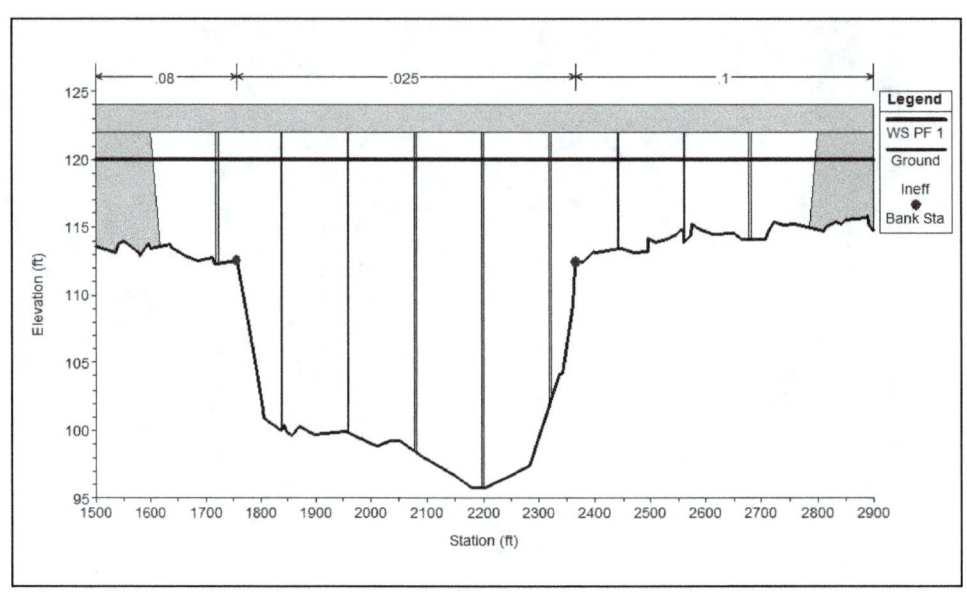

Figure D.6. Close-up of bridge cross section.

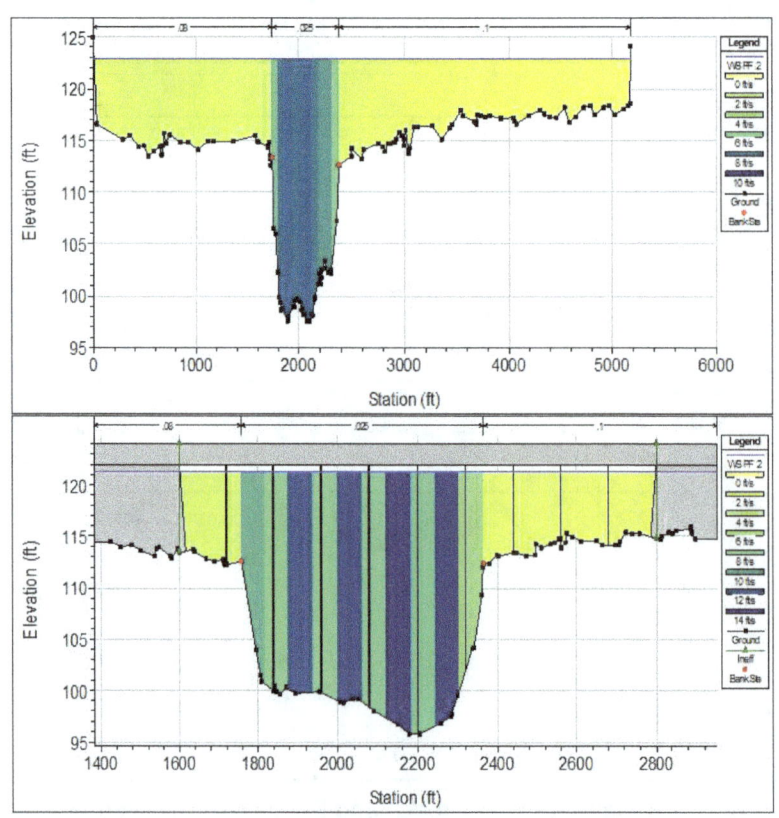

Figure D.7. Approach and bridge velocity distributions.

Table D.1. Approach and Bridge Cross Section Average Properties.

TOTAL SCOUR APPENDIX	
Approach Cross Section Average	
Q Total (cfs)	120000.00
E.G. Elev (ft)	122.0
W.S. Elev (ft)	121.3
Vel Head (ft)	0.73
E.G. Slope (ft/ft)	0.000297
Top Width (ft)	5160
Vel Total (ft/s)	3.23
Conv. Total (cfs)	6966390

(Table D.1 continued)

Bridge Cross Section Average	
Q Total (cfs)	120000
E.G. Elev (ft)	121.4
W.S. Elev (ft)	120.0
Vel Head (ft)	1.37
E.G. Slope (ft/ft)	0.000707
Top Width (ft)	1165
Vel Total (ft/s)	7.75
Conv. Total (cfs)	4513756

Table D.2. Approach and Bridge Cross Section Subarea Properties.

Approach Cross Section	Sub Area Element		
Subarea Results	Left Ovebank	Main Channel	Right Overbank
Manning n	0.08	0.025	0.10
Flow (cfs)	12032	97832	10137
Flow Area (sq ft)	10929	12919	13308
Top Width (ft)	1729	637	2795
Avg. Vel. (ft/s)	1.10	7.57	0.76
Hydr. Depth (ft)	6.3	20.3	4.8
Conv. (cfs)	698495	5679437	588458
Wetted Per. (ft)	1730	642	2796
Shear (lb/sq ft)	0.12	0.37	0.09

(Table D.2 continued)

Bridge Cross Section	Sub Area Element		
Subarea Results	Left Ovebank	Main Channel	Right Overbank
Manning n	0.08	0.025	0.10
Flow (cfs)	1694	115352	2954
Flow Area (sq ft)	1024	12028	2439
Top Width (ft)	153	608	431
Avg. Vel. (ft/s)	1.65	9.59	1.21
Hydr. Depth (ft)	6.8	20.3	5.8
Conv. (cfs)	63721	4338904	111130
Wetted Per. (ft)	167	804	460
Shear (lb/sq ft)	0.27	0.66	0.23

Table D.3. Approach and Bridge Cross Section Velocity Distribution.

	Left Sta. (ft)	Right Sta. (ft)	Inc. Width (ft)	Flow (cfs)	Area (sq/ft)	% Conv.	Hydr D. (ft)	Velocity (ft/s)
Approach Cross Section Velocity Distribution								
	0	348.2	348.2	1782	1808.51	1.5	5.4	0.99
	348.2	696.4	348.2	2806	2409	2.3	6.9	1.16
	696.4	1044.6	348.2	2501	2248	2.1	6.5	1.11
	1044.6	1392.8	348.2	2474	2233	2.1	6.4	1.11
	1392.8	1741	348.2	2469	2231	2.1	6.4	1.11
LB	1741	1804.7	63.7	6617	1050	5.5	16.5	6.3
	1804.7	1868.4	63.7	11522	1433	9.6	22.5	8.04
	1868.4	1932.1	63.7	12042	1471	10.0	23.1	8.19
	1932.1	1995.8	63.7	11060	1398	9.2	21.9	7.91
	1995.8	2059.5	63.7	11580	1437	9.7	22.6	8.06
	2059.5	2123.2	63.7	12368	1495	10.3	23.5	8.27
	2123.2	2186.9	63.7	10337	1343	8.6	21.1	7.7
	2186.9	2250.6	63.7	8566	1202	7.1	18.9	7.13
	2250.6	2314.3	63.7	8545	1198	7.1	18.8	7.13
RB	2314.3	2378	63.7	5194	892	4.3	14.0	5.82
	2378	2938	560	3856	4027	3.2	7.2	0.96
	2938	3498	560	2395	3027	2.0	5.4	0.79
	3498	4058	560	1458	2247	1.2	4.0	0.65
	4058	4618	560	1375	2169	1.2	3.9	0.63
	4618	5178	560	1052	1838	0.9	3.3	0.57

(Table D.3. Continued)

	Left Sta. (ft)	Right Sta. (ft)	Inc. Width (ft)	Flow (cfs)	Area (sq/ft)	% Conv.	Hydr D. (ft)	Velocity (ft/s)
Bridge Cross Section								
	2	1600	1598	0	9385	0	5.8	0.00
	1600	1639.25	39.25	350	256	0.3	6.5	1.37
	1639.25	1678.5	39.25	399	275	0.3	6.5	1.45
	1678.5	1717.75	39.25	443	295	0.4	7.3	1.50
	1717.75	1757	39.25	460	301	0.4	7.6	1.53
LB	1757	1817.8	60.8	6246	864	5.2	14.2	7.23
	1817.8	1878.6	60.8	11038	1149	9.2	19.9	9.61
	1878.6	1939.4	60.8	11318	1226	9.4	20.2	9.23
	1939.4	2000.2	60.8	11550	1181	9.6	20.4	9.78
	2000.2	2061	60.8	12051	1274	10.1	21.0	9.46
	2061	2121.8	60.8	12983	1267	10.8	21.9	10.25
	2121.8	2182.6	60.8	14467	1422	12.1	23.4	10.17
	2182.6	2243.4	60.8	15038	1383	12.6	23.9	10.87
	2243.4	2304.2	60.8	13402	1359	11.2	22.4	9.86
RB	2304.2	2365	60.8	7303	902	6.1	15.7	8.10
	2365	2452	87	701	613	0.6	6.9	1.14
	2452	2539	87	602	560	0.5	5.9	1.08
	2539	2626	87	550	530	0.5	5.5	1.04
	2626	2713	87	503	502	0.4	5.0	1.00
	2713	2800	87	386	428	0.3	4.9	0.90
	2800	5186	2386	0	8492	0.0	4.2	0.00

Because of this, three computations for contraction scour will be required. The first computation, which will be illustrated in Step 3A will determine the magnitude of the contraction scour in the main channel. The second computations, which are illustrated in Step 3B will utilize the clear-water equation for the left and right overbank areas. Hydraulic data for these computations are presented in Tables D.1 and D.2 for the channel and overbank contraction scour computations.

Table D.3 lists the hydraulic variables which will be used to estimate the local scour at the piers (Step 4). These hydraulic variables were determined from a plot of the velocity distribution derived from the HEC-RAS output (Figure D.7). Only one set of pier scour computations will be completed because the possibility of thalweg shifting and lateral migration will require that all of the piers be set assuming that any pier could be subjected to the maximum scour producing variables.

Estimating local scour at the left and right abutments will be illustrated in Step 5 using the HIRE equation. Scour variables derived from the HEC-RAS output for these computations are presented in Table D.3 Bridge cross section.

D.3 STEP 2: DETERMINE THE MAGNITUDE OF LONG-TERM DEGRADATION OR AGGRADATION

Evaluation of stage discharge relationships and cross sectional data obtained from other agencies do not indicate progressive aggradation or degradation. Also, long-term aggradation or degradation are not evident at neighboring bridges. Based on these observations, the channel is relatively stable vertically, at present. Furthermore, there are no plans to change the local land use in the watershed. The forested areas of the watershed are government-owned and regulated to prevent wide spread fire damage, and instream gravel mining is prohibited. These observations indicate that future aggradation or degradation of the channel, due to changes in sediment delivery from the watershed, are minimal.

Based on these observations, and due to the lack of other possible impacts to the river reach, it is determined that the channel will be relatively stable vertically at the bridge crossing and long-term aggradation or degradation potential is considered to be minimal. However, there is evidence that the channel is unstable laterally. This will need to be considered when assessing the total scour at the bridge.

D.4 STEP 3A: COMPUTE THE MAGNITUDE OF CONTRACTION SCOUR IN MAIN CHANNEL

As a precursor to the computation of contraction scour in the main channel under the bridge, it is first necessary to determine whether the flow condition in the main channel is either live-bed or clear-water. This is determined by comparing the critical velocity for sediment movement at the approach section to the average channel velocity of the flow at the approach section as computed using the HEC-RAS output. This comparison is conducted using the average velocity in the main channel of the approach section to the bridge. If the average computed channel velocity is greater than the critical velocity, the live-bed equation should be used. Conversely, if the average channel velocity is less than the critical velocity, the clear-water equation is applicable. The following computations are based on the quantities tabulated in Tables D.1 and D.2.

Compute Main Channel Contraction Scour (Check for live-bed or clear-water)

$$V_c = 11.17 y^{1/6} D_{50}^{1/3}$$

$V_c = 11.17 \times 20.3^{1/6}\, 0.0066^{1/3} = 3.46$ ft/s

$V = 7.57$ ft/s $> V_c$, therefore use live-bed contraction scour equation.

For live-bed scour, k_1 is determined from V_*/ω

$$V_* = \sqrt{\tau_o / \rho} = (g y_1 S_1)^{1/2}$$

$V_* = \sqrt{0.37 / 1.94} = (32.2 \times 20.3 \times 0.000297)^{1/2} = 0.44$ ft/s

Determine ω from HEC-18 Figure 6.8.

$\omega = 0.20$ m/s $= 0.66$ ft/s

$V_*/\omega = 0.44/0.66 = 0.67$, therefore $k_1 = 0.64$

$$y_2 = y_1 \left(\frac{Q_2}{Q_1}\right)^{6/7} \left(\frac{W_1}{W_2}\right)^{k_1}$$

$$y_2 = 20.3 \left(\frac{115352}{97832}\right)^{6/7} \left(\frac{637}{608 - 5 \times 3}\right)^{0.64} = 24.5 \text{ ft}$$

$y_s = y_2 - y_0 = 24.5 - 20.3 = 4.2$ ft

D.5 STEP 3B: COMPUTE CONTRACTION SCOUR FOR LEFT AND RIGHT OVERBANK AREAS

Clear-water contraction scour could occur in the overbank areas between the left abutment and the left bank and right abutment and right bank of bridge opening. Although the bed material in the overbank area is soil, it is protected by vegetation. Therefore, there would be no bed-material transport into the set-back bridge opening (clear-water conditions). The subsequent computations are based on the discharge and depth of flow passing under the bridge in the left and right overbanks. These hydraulic variables were determined from the HEC-RAS output and are tabulated in Table D.2 for the Bridge cross section.

Compute left overbank contraction scour.

$$y_2 = \left[\frac{0.0077 Q^2}{D_m^{2/3} W^2}\right]^{3/7}$$

$$y_2 = \left[\frac{0.0077 \times 1694^2}{(.00115 \times 1.25)^{2/3} (153-3)^2}\right]^{3/7} = 6.4 \text{ ft}$$

$y_s = y_2 - y_0 = 6.4 - 6.8 = -0.4$ ft, therefore, $y_s = 0.0$

Compute right overbank contraction scour.

$$y_2 = \left[\frac{0.0077 Q^2}{D_m^{2/3} W^2}\right]^{3/7}$$

$$y_2 = \left[\frac{0.0077 \times 2954^2}{(.00115 \times 1.25)^{2/3} (431-9)^2}\right]^{3/7} = 4.3 \text{ ft}$$

$y_s = y_2 - y_0 = 4.3 - 5.8 = -1.5$ ft, therefore, $y_s = 0.0$

Discussion of Overbank Scour Computations

For both the left and right overbank areas the clear-water contraction scour equation indicates negative contraction scour. This is an indication that there will not be any contraction scour on either overbank. **It is not interpreted as an indication of aggradation**.

D.6 STEP 4: COMPUTE THE MAGNITUDE OF LOCAL SCOUR AT PIERS

It is anticipated that any pier under the bridge could potentially be subject to the maximum flow depths and velocities derived from the HEC-RAS hydraulic model (Table D.3 Bridge cross section). Therefore, only one computation for pier scour is conducted and assumed to apply to each of the nine piers for the bridge. The pier scour computation is based on the hydraulic characteristics of flow distribution with the highest velocity and associated hydraulic depth.

Calculate pier scour for circular column pier.

$$\frac{y_s}{a} = 2.0 K_1 K_2 K_3 \left(\frac{y_1}{a}\right)^{0.35} Fr_1^{0.43}$$

$$Fr_1 = \frac{V_1}{\sqrt{gy_1}}$$

$$Fr_1 = \frac{10.9}{\sqrt{32.2 \times 23.9}} = 0.39$$

$$\frac{y_s}{a} = 2.0 \times 1.0 \times 1.0 \times 1.1 \left(\frac{23.9}{3}\right)^{0.35} 0.39^{0.43} = 3.0$$

$y_s = 3.0 \times 3 = 9.0$ ft

The DOT is considering a lane expansion at this bridge, the proposed design includes replacing the single circular column piers with circular column bents. Each bent would have 4 circular columns 3 ft (0.91 m) in diameter, spaced 12 ft (3.66 m) on center.

Calculate pier scour for 15-degree angle of attack at a 4-column bent. (Hint: Check the column spacing)

$$K_2 = \left(Cos\theta + \frac{L}{a} Sin\theta\right)^{0.65}$$

$$K_2 = \left(Cos15 + \frac{12}{3} Sin15\right)^{0.65} = 1.57$$

$$\frac{y_s}{a} = 2.0 \times 1.0 \times 1.57 \times 1.1 \left(\frac{23.9}{3}\right)^{0.35} 0.39^{0.43} = 4.8$$

$y_s = 4.8\, a = 4.8 \times 3.0 = 14.4$ ft

Discussion of Pier Scour Computation

Although the estimated local pier scour would probably not occur at each pier, the possibility of thalweg shifting, which was identified in the Level 1 analysis, precludes setting the piers at different depths even if there were a substantial savings in cost. This is because any of the piers could be subjected to the worst-case scour conditions.

It is also important to assess the possibility of lateral migration of the channel. This possibility can lead to directing the flow at an angle to the piers, thus increasing local scour if the column bent piers are installed for bridge widening. Countermeasures to minimize this problem could include riprap for the channel banks both up- and downstream of the bridge, and installation of guide banks to align flow through the bridge opening.

The possibility of lateral migration precludes setting the foundations for the overbank piers at a higher elevation. Therefore, in this example the foundations for the overbank piers should be set at the same elevation as the main channel piers.

D.7 STEP 5: DETERMINE THE FOUNDATION ELEVATION FOR THE ABUTMENTS

Calculate left abutment scour using HIRE scour equation.

L = 1598 ft

y_1 = 6.5 ft

L/y_1 = 246 > 25, therefore, HIRE equation is applicable.

$$y_s = 4.0 y_1 Fr^{0.33} \frac{K_1}{0.55} K_2$$

$$Fr = \frac{V_1}{\sqrt{gy_1}} = \frac{1.37}{\sqrt{32.2 \times 6.5}} = 0.095$$

$$y_s = 4.0 \times 6.5 \times 0.095^{0.33} \times \frac{0.55}{0.55} \times 1.0 = 12.0 \text{ ft}$$

Calculate right abutment scour using HIRE scour equation.

L = 2386 ft

y_1 = 4.9 ft

L/y_1 = 487 > 25, therefore, HIRE equation is applicable.

$$y_s = 4.0 y_1 Fr^{0.33} \frac{K_1}{0.55} K_2$$

$$Fr = \frac{V_1}{\sqrt{gy_1}} = \frac{0.9}{\sqrt{32.2 \times 4.9}} = 0.072$$

$$y_s = 4.0 \times 4.9 \times 0.072^{0.33} \times \frac{0.55}{0.55} \times 1.0 = 8.2 \text{ ft}$$

Discussion Of Abutment Scour Computations

All of the abutment scour computations (left and right abutments) assumed that the abutments were set perpendicular to the flow. If the abutments were angled to the flow (skewed), a correction utilizing K_2 would be applied to the HIRE equation. However the adjustment for skewed abutments is minor when compared to the magnitude of the computed scour depths. For example, if the abutments for this example problem were angled 30° upstream (θ = 90° + 30° = 120°), the correction for skew would increase the computed depth of abutment scour by no more than 3 to 4 percent.

Plot the total scour prism using the bridge cross section provided on the next page.

D.8 STEP 6: PLOT TOTAL SCOUR DEPTH AND EVALUATE DESIGN

As a final step, the results of the scour computations are plotted on the bridge cross section and carefully evaluated (Figure D.8). For this example, only the computations for pier scour with piers aligned with the flow were plotted and the abutment scour computations reflect the results from the HIRE equation. The topwidth of the local scour holes is suggested as 2.0 y_s.

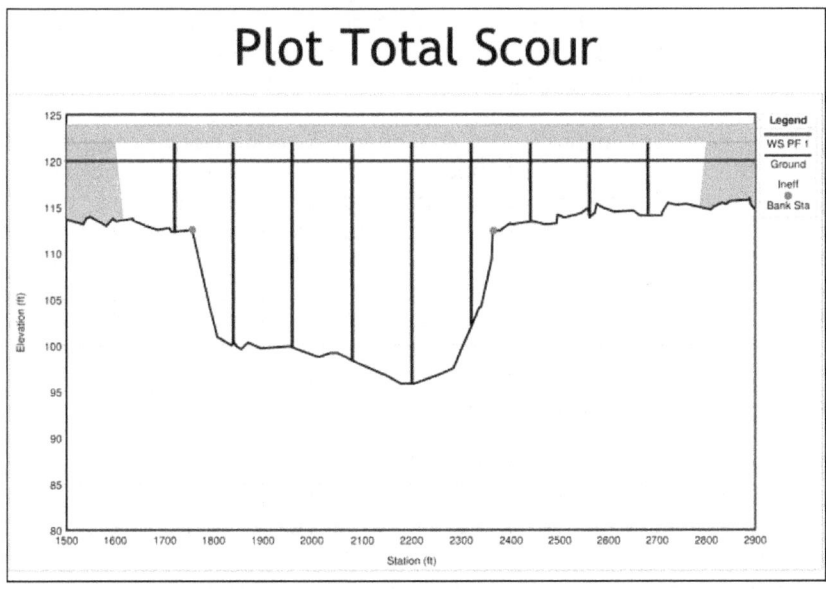

Figure D.8. Bridge cross section for total scour.

The total scour prism is made up of contraction scour in the channel, scour at the abutments, and scour at the piers. The total scour prism is shown in Figure D.9.

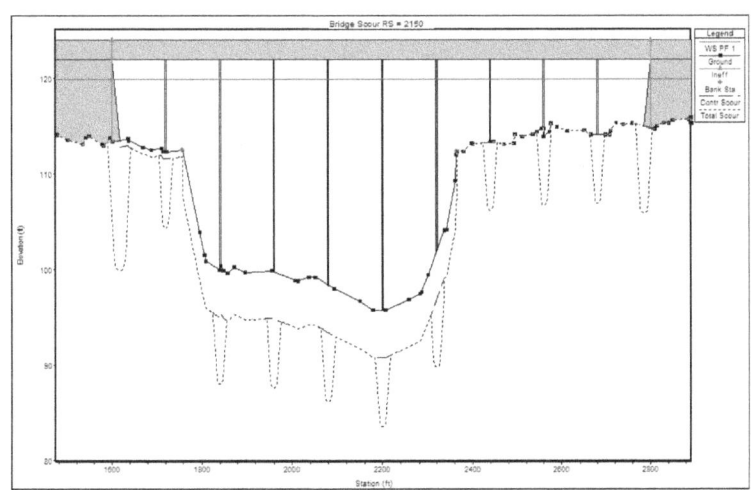

Figure D.9. Total scour prism.

It is important to evaluate the results of the scour computations carefully. For example, although the total scour plot indicates that the total scour at the overbank piers is less than for the channel piers, this does not indicate that the foundations for the overbank piers can be set at a higher elevation. Due to the possibility of channel and thalweg shifting, all of the piers should be set to account for the maximum total scour. Also, the computed contraction scour is distributed uniformly across the channel in Figure D.9. However, in reality this may not be what would happen. With the flow from the overbank area returning to the channel, the contraction scour could be deeper at both abutments. The use of guide banks would distribute the contraction scour more uniformly across the channel. This would make a strong case for guide banks in addition to the protection they would provide to the abutments.

The possibility of lateral migration of the channel could have an adverse effect on the magnitude of the pier scour. This is because lateral migration will most likely skew the flow to the piers. This problem is currently minimized by the use of circular piers at this bridge; however, if the circular column bents being considered for bridge widening are constructed, a variable angle of attack would need to be considered. One approach to this potential problem would be to install guide banks to align the flow through the bridge opening. Since the river has a history of channel migration, the bridge inspection and maintenance crews should be briefed on the nature of this problem so that any lateral migration can be identified.

D.9 COMPLETE THE GENERAL DESIGN PROCEDURE

This design problem uses Steps 1 through 6 of the specific design approach (Chapter 2) and completes Steps 1 through 6 of the general design procedure. The design must now proceed to Steps 7 and 8 of the general design procedure as outlined in Chapter 2. These steps include bridge foundation analysis and consideration of the scour design check flood (see Chapter 2, Table 2.1). This is not done for this example problem.

The total scour plot would also be compared with bridge as-built plans by a multi-disciplinary team to determine if the bridge is scour critical for the design event. If the bridge is scour critical, a Plan of Action will be required.

APPENDIX E

UNKNOWN FOUNDATIONS

(page intentionally left blank)

APPENDIX E

UNKNOWN FOUNDATIONS

E.1 INTRODUCTION

Bridges are classified as having unknown foundations when the type (spread footing, piles, columns), dimensions (length, width, thickness), reinforcing, and/or elevation are not known. They are coded as U in Item 113 of the Coding Guide (FHWA 1995). Screening conducted by bridge owners under the National Evaluation program identified approximately 60,000 bridges over waterways as having unknown foundations as of December 2010. This appendix provides a status report and guidance for protecting bridges with unknown foundations from scour.

E.2 FHWA POLICY AND TECHNICAL GUIDANCE

The National Bridge Inspection Standards (NBIS) regulation, 23 CFR 650.313.e.3, requires that bridge owners develop a plan of action (POA) for bridges identified as scour critical. A recognized concern exists that some bridges within the unknown foundation population may be scour critical, and as such, need to have a POA as required by the NBIS regulation.

Bridges with unknown foundations represent a subset of bridges over waterways that have not been evaluated for scour. FHWA has defined an approach for addressing this subset of bridges that:

1. Assists owners in developing and implementing risk-based procedures to determine enough about a bridge's foundations to conduct a scour evaluation;

2. Moves the bridge into the scour program for evaluation if the owner is comfortable with the risk-based assessment of the bridge foundations (this is the equivalent of recoding the bridge to a 6);

3. Recodes the bridge accordingly after evaluation for scour vulnerability.

The FHWA memorandum dated January 9, 2008, "Technical Guidance for Bridges over Waterways with Unknown Foundations," provides a process that should be considered by bridge owners to identify foundation characteristics such as width, depth and length for bridge foundations identified as unknown. The goal of this process is to reduce or eliminate the population of bridges over waterways identified as having unknown foundations, which in turn would allow bridge owners to evaluate these bridges for their scour vulnerability. Recommended action items described in the January 9, 2008 memorandum are briefly summarized below:

1. Screen all bridges coded U to ensure that they are correctly identified as having unknown foundations. Emphasis is placed on mining historical records that may be housed in district or local offices, and cross-referencing construction dates with known-foundation bridges constructed during the same period. Similar to current foundation practices, historical practices were very repetitive and rather simple in concept.

2. For bridges over waterways that are determined to be correctly identified as having unknown foundations:

   a. Prioritize bridges based on their functional classification, e.g., Principal Arterial – Interstate; Principal Arterial – Other Freeways or Expressways; Other Principal Arterial; Major Collector; and Minor Collector.

   b. Consider using the following criteria for determining, with a reasonable accuracy, foundation characteristics:

      - Collect and document historical knowledge of foundation and design practices for the period of original construction.

      - Consider geologic, subsurface conditions, bridge standards, and information that may be available from nearby bridges.

      - Consider applying "proven" surface and subsurface NDT tools to confirm foundation type and determine foundation length.

   c. Conduct a scour evaluation based on this determination and consider recoding the bridge for Item 113 according to the outcome of the evaluation.

3. For bridges that were previously coded as U for Item 113 and whose foundations are completely and accurately identified after completing the screening:

   a. Conduct scour evaluations following the guidance presented in FHWA publication Hydraulic Engineering Circular No. 18.

   b. Code Item 113 according to the outcome of the evaluation.

The January 9, 2008 memorandum also advised bridge owners that the FHWA was "contemplating amending the NBIS regulations so that any remaining bridge reported as having unknown foundations after November 2010 would be kept with a Code U for Item 113, considered scour critical and subject to the plan of action requirement of the NBIS regulation, 23 CFR 650.313(e)(3), until properly designed countermeasures are installed to protect the bridge foundations or until the bridge is replaced."

Subsequent to the January 9, 2008 memorandum, the FHWA assembled an interdisciplinary Unknown Foundations Team to develop a plan for needed direction, provide expert advice, and assist bridge owners with technology and process development and implementation. The interdisciplinary team features geotechnical, hydraulics and structural specialists from the Office of Bridge Technology, the Resource Center, Turner-Fairbank Highway Research Center, and the division offices.

The expectation of the Unknown Foundations Team is to provide practical and cost-effective technical guidance that can be used by bridge owners to evaluate the scour vulnerability of bridges over waterways with unknown foundations. The guidance includes risk-based considerations to help bridge owners with coding of Item 113. For example, a flowchart-based approach to categorizing unknown foundation bridges into three risk categories (A – High; B – Moderate; C- Low) is presented in the FHWA memorandum dated October 29, 2009, "Additional Guidance for Assessment of Bridges Over Waterways with Unknown Foundations."

The Unknown Foundations Team has set up a web site where all information relative to the unknown foundations initiative is housed. It can be accessed at:

www.fhwa.dot.gov/unknownfoundations

The unknown foundations web site provides information, technical guidance, and reference materials organized into four focus areas:

1. Categorization and Prioritization – Provides references pertaining to risk assessment, risk management, and risk reduction.

2. Determination of Foundation – Provides information and references on methods for determining foundation type, size, and condition, focusing primarily on:

    - Geophysical methods and nondestructive testing (NDT) techniques (discussed further in Section G.4).
    - Inferential methods (e.g., reverse-engineering approaches)

3. Determination of Scour Potential Without Determining Foundation – Provides information on hydraulic vulnerability assessment methods developed by New York State DOT, and inspection/assessment guidelines for single span bridges where abutment scour assessment and countermeasure applications are appropriate.

4. Plans of Action for Unknown Foundation Bridges – Provides recommended procedures for developing Plans of Action (POAs) for unknown foundation bridges (discussed further in Section G.3), as well as bridge management tools applicable to unknown foundation POAs such as real-time flood watch and scour forecasting software.

The FHWA organized and sponsored a 2-day conference, the Unknown Foundations Summit, in November, 2005 in Denver, Colorado. The conference was designed to address all four focus areas identified above, and consisted of presentations by select consultants and State DOT personnel on state-of-the-art technologies, methodologies, and management decision-making strategies for unknown foundation bridges.

Also featured during the Summit were focused panel discussions to: 1) identify appropriate guidance for discovering unknown foundations, and 2) define effective strategies for managing unknown foundations and the associated risks. The proceedings of the 2005 Unknown Foundations Summit are available on the FHWA Unknown Foundations web site.

E.3   PLANS OF ACTION FOR UNKNOWN FOUNDATION BRIDGES

The Coding Guide recommends development and implementation of a Plan of Action (POA) for existing bridges having an Item 113 code U. Current guidance provided by the FHWA Unknown Foundation Team for developing POAs for unknown foundation bridges includes the following:

1. A bridge coded U in Item 113 can simply be changed to a scour critical code (e.g., 3) for the NBI and subjected to a POA as described for scour critical bridges.

2. A bridge may remain coded U in Item 113, with a POA developed based on a risk assessment and owner-defined criteria considering known information about the bridge.

The POA for a bridge that remains coded U in Item 113 may be different than for a bridge determined to be scour critical. The POA developed should be based on the known information of the bridge and the owner-determined risk from scour. The POAs for bridges over waterways with unknown foundations should contain minimum requirements commensurate with the consequences of loss of service of the structure to ensure a reasonable level of safety to the traveling public. The steps below provide FHWA's most current guidance to bridge owners in developing a POA for a bridge coded U in Item 113.

Step 1: Assess bridges with unknown foundations in accordance with guidance provided on the Unknown Foundations web site (summarized in Section G.2). For bridges that remain coded U in Item 113 after a risk-based assessment, FHWA recommends that a POA be developed based on the risk categories defined by bridge owners during initial categorization and grouping (e.g. A - High Risk, B - Moderate Risk, C - Low Risk).

Step 2: Develop a POA based upon the defined risk category that considers safety to the traveling public and the consequences of loss of service of the structure. The POA may be less detailed than for a scour critical bridge based on the defined risk categories, but it should contain elements that protect users during and after a scour event, and provide a proactive plan for addressing the bridge scour concerns in the future. Examples for lowest and highest risk categories are shown below.

Lowest Risk Categories

- Assumes that the bridge has performed well and has no history of scour related problems.
- For bridges considered as low risk, plans of action may be as simple as monitoring bridges for scour during routine biennial inspections and after major events.
- If scour or a rainfall event has been observed in excess of predetermined monitoring triggers, then the bridge should be considered for an in-depth foundation investigation.
- Any information on observed or inspected conditions would be identified on the bridge inspection report so that inspectors could monitor the bridge for changes.

Highest Risk Categories

- Assumes that the bridge has performed satisfactorily, but because of owner-defined criteria, it has been identified as high risk.
- Plans of actions may be similar to those for bridges determined to be scour critical. At a minimum, the bridge should be monitored on a more frequent basis than a bridge in a moderate to low risk category.
- Also, a bridge in this category should be considered for an in-depth foundation investigation if any significant changes in the streambed occur, and scheduled for timely design and construction of a new bridge or countermeasures to make the bridge safe from scour and stream instability.

Step 3: Coordinate a global action plan for all bridges coded U in Item 113 within a state or region, whether assessed through this guidance or not. The plan should:

- Identify the scour critical and unknown foundation bridges;
- Define major events or monitoring triggers; and
- Provide information for requesting technical assistance or conducting an in-depth foundation investigation.

E.4   NONDESTRUCTIVE TESTING

NCHRP Project 21-5, initiated in 1992 and concluded in 1995 (NCHRP 1995, 1996), identified and tested the following NDT methods:

- Sonic echo/impulse response
- Bending wave method
- Ultraseismic test method
- SASW method
- Dynamic foundation response method
- Borehole parallel seismic test method
- Borehole sonic method
- Borehole radar method
- Induction field method

As a result of the above research, a second phase of this project (NCHRP 21-5(2)) was initiated to research and develop equipment, field techniques, and analysis methods for the most promising technologies. The methods selected were:

- Ultraseismic (including sonic echo/impulse response and bending wave methods)
- Borehole parallel seismic and induction field methods

In general the results of testing NTD methods were not as satisfactory as the initial research indicated. The results of NCHRP Project 21-5 indicate that of all the surface and borehole methods, the Parallel Seismic test was found to have the broadest applications for determining the bottom depth of substructures. Of the surface tests (no boring required), the Ultraseismic test has the broadest application for determining the depths of unknown bridge foundations but will provide no information on piles constructed below larger substructure elements (pile caps). The Sonic Echo/Impulse Response, Bending Wave, Spectral Analysis of Surface Wave, and Borehole Radar methods all had more specific applications. The research reports are available from the Unknown Foundations web site.

The November, 2005 FHWA Unknown Foundations Summit provided a forum designed to include information on technological advances regarding NDT methods for characterizing unknown foundation bridges. In general the research results available at that time were not conclusive and did not change previous FHWA guidance. The complete proceedings of the 2005 Unknown Foundation Summit conference are available at the FHWA Unknown Foundations web site.

For the above reasons, it is recommended that at this time a Plan of Action, as described in Section G.3, continue to be used as the primary measure to protect bridges with unknown foundations from failure due to scour.

E.5 OTHER TEST PROCEDURES

E.5.1 Core Drilling

A simple method used by one State Highway Agency (SHA) to explore unknown foundations is to use a drilling rig to core the bridge deck and to continue down through the pier or abutment footing into the supporting soil or rock under the foundation. This procedure has been used successfully to determine the foundations of some 40 structures and to reclassify the structures as known foundations to allow a scour evaluation to be performed.

E.5.2 Forensic Engineering

There may be a considerable amount of information in the files of the bridge owner that can be reviewed for information pertaining to the bridge foundations even though as-built plans are no longer available:

- Inspection records may indicate channel bed elevations taken over a period of time. In one state, a concerted effort was made to record channel bed elevations at many bridges immediately after a major flood occurred in 1973. This information now serves as a benchmark for assessing current conditions. If the channel bed is now four or five feet higher than it was in 1973, and the bridge was not damaged in the 1973 flood, this information becomes very useful in assessing the risk posed to the structure by the river.

- Inspectors may have documented exposed foundations in the aftermath of previous floods. While the foundation may no longer be visible, this knowledge of the elevation of the top or bottom of a footing will help the engineer to determine necessary information about the bridge foundation.

- In live-bed streams, the channel bed under bridge foundations is subject to scour and subsequent infilling of material back into the scour hole. The infill material is likely to be soft fine material that can be easily probed with a reinforcing rod. Careful probing can reveal the elevation of the tops of footings located below the channel bed. Inspection records will often contain basic information about the bridge foundation and whether it is a spread footing or on piles. This information can be used to estimate the footing dimensions within a reasonable degree of accuracy so that an assessment can be made as to whether worst-case scour conditions are likely to exceed the bottom of the footing.

E.6 REFERENCES

1. National Cooperative Highway Research Program, 1995, "Determination of Unknown Subsurface bridge Foundations-Final Report," NCHRP Project 21-5, Transportation Research Board, National Academy of Science, Washington, D.C. (Olson, L.D., F. Jalinoos, and M.F. Aouad).

2. National Cooperative Highway Research Program, 1996, "Nondestructive Destructive of Unknown Subsurface Bridge Foundations - Results of NCHRP Project 21-5." Research Results Digest No. 213, Transportation Research Board, National Academy of Science, Washington, D.C.

www.ingramcontent.com/pod-product-compliance
Lightning Source LLC
Chambersburg PA
CBHW080725230426
43665CB00020B/2617